2006 IBC®
STRUCTURAL/SEISMIC DESIGN MANUAL

2

BUILDING DESIGN EXAMPLES FOR LIGHT-FRAME, TILT-UP AND MASONRY

SECOND EDITION

STRUCTURAL ENGINEERS ASSOCIATIONS SEA NATIONAL COUNCIL
STRUCTURAL ENGINEERS ASSOCIATIONS SEA CALIFORNIA
ICC INTERNATIONAL CODE COUNCIL®

Copyright

Publisher

Structural Engineers Association of California (SEAOC)
1414 K Street, Suite 260
Sacramento, CA 95814
Telephone: (916) 447-1198; Fax: (916) 443-8065
E-mail: lee@seaoc.org; Web address: www.seaoc.org

The Structural Engineers Association of California (SEAOC) is a professional association of four regional member organizations (Southern California, Northern California, San Diego, and Central California). SEAOC represents the structural engineering community in California. This document is published in keeping with SEAOC's stated mission: "to advance the structural engineering profession; to provide the public with structures of dependable performance through the application of state-of-the-art structural engineering principles; to assist the public in obtaining professional structural engineering services; to promote natural hazard mitigation; to provide continuing education and encourage research; to provide structural engineers with the most current information and tools to improve their practice; and to maintain the honor and dignity of the profession."

Editor

International Code Council

Disclaimer

First Printing: September 2009

Table of Contents

Preface

This is the second volume in the three-volume 2006 *IBC Structural/Seismic Design Manual.* It has been developed and funded by the Structural Engineers Association of California (SEAOC). It is intended to provide guidance on the interpretation and use of the seismic requirements in the 2006 *International Building Code (IBC),* published by the International Code Council, Inc.

The 2000 *IBC Structural/Seismic Design Manual* was developed to fill a void that exists between the commentary of SEAOC's Blue Book, which explained the basis for the code provisions, and everyday structural engineering design practice. The 2006 *IBC Structural/Seismic Design Manual* illustrates how the provisions of the code are used. *Volume 1: Code Application Examples,* provides step-by-step examples for using individual code provisions, such as computing base shear or building period. *Volumes 2 and 3: Building Design Examples,* furnish examples of seismic design ofcommontypes of buildings. In Volumes 2 and 3, important aspects of whole buildings are designed to show, calculation-by-calculation, how the various seismic requirements of the code are implemented in a realistic design.

The examples in the 2006 *IBC Structural/Seismic Design Manual* do not necessarily illustrate the only appropriate methods of design and analysis. Proper engineering judgment should always be exercised when applying these examples to real projects. The 2006 *IBC Structural/Seismic Design Manual* is not meant to establish a minimum standard of care but, instead, presents reasonable approaches to solving problems typically encountered in structural/seismic design.

The example numbers used in the prior Seismic Design Manuals—1997 UBC and 2000 IBC Volume 2 building design example problems have been retained herein to provide easy comparison to revised code requirements.

SEAOC, NCSEA, and ICC intend to update the 2006 *IBC Structural/Seismic Design Manual* with each new edition of the building code.

Acknowledgments

Authors

The 2006 *IBC Structural/Seismic Design Manual—Volume 2* was written by a group of highly qualified structural engineers. They were selected by a steering committee set up by the SEAOC Board of Directors and were chosen for their knowledge and experience with structural engineering practice and seismic design. The consultants for Volumes 1, 2, and 3 are:

Jon P. Kiland, *Co-Project Manager*
Rafael Sabelli, *Co-Project Manager*
Douglas S. Thompson
Dan Werdowatz
Matt Eatherton
John W. Lawson
Joe Maffei
Kevin Moore
Stephen Kerr

Reviewers

A number of SEAOC members and other structural engineers helped check the examples in this volume. During its development, drafts of the examples were sent to these individuals. Their help was sought in review of code interpretations as well as detailed checking of the numerical computations.

Seismology Committee

Close collaboration with the SEAOC Seismology Committee was maintained during the development of the document. The 2004-2005 and 2005-2006 committees reviewed the document and provided many helpful comments and suggestions. Their assistance is gratefully acknowledged.

Production and Art

ICC

Suggestions for Improvement

In keeping with SEAOC's and NCSEA's Mission Statements: "to advance the structural engineering profession" and "to provide structural engineers with the most current information and tools to improve their practice," SEAOC and NCSEA plan to update this document as structural/seismic requirements change and new research and better understanding of building performance in earthquakes becomes available.

Comments and suggestions for improvements are welcome and should be sent to the following:

Structural Engineers Association of California (SEAOC)
Attention: Executive Director
1414 K Street, Suite 260
Sacramento, California 95814
Telephone: (916) 447-1198; Fax: (916) 932-2209
E-mail: lee@seaoc.org; Web address: www.seaoc.org

Errata Notification

SEAOC and NCSEA have made a substantial effort to ensure that the information in this document is accurate. In the event that corrections or clarifications are needed, these will be posted on the SEAOC web site at *http://www.seaoc.org* or on the ICC website at *http://www.iccsafe.org*. SEAOC, at its sole discretion, may or may not issue written errata.

Introduction

Volume 1 of the 2006 *IBC Structural/Seismic Design Manual: Code Application Examples* deals with interpretation and use of the structural/seismic provisions of the 2006 *International Building Code®* (IBC). The 2006 *IBC Structural/Seismic Design Manual* primarily addresses the major structural/seismic provisions of Chapter 16 of the IBC, with specific examples that explain their proper application.

Volume 2 presents seismic design of new light-frame, masonry, and tilt-up buildings for the requirements of the IBC as illustrated in this document. Seven examples are shown: (1) a two-story wood light-frame residence, (2) a large three-story wood frame building, (3) a three-story cold formed steel light-frame building, (4) a one-story masonry (concrete block) building with panelized wood roof, (5) a one-story tilt-up building with panelized wood roof, (6) the design of a tilt-up wall panel with large openings, and (7) the design of various one-story structures for wind loads.

The selected buildings are, for the most part, representative of construction types found in seismic areas in the western and mid-western states. Designs have been taken from real-world buildings, although some simplifications were necessary in order to illustrate significant points without presenting repetitive or unnecessarily complicated design aspects.

The examples are not complete building designs, or even complete seismic designs; they are, instead, examples of the significant seismic design aspects of a particular type of building.

In developing these examples, SEAOC has endeavored to illustrate correct use of the *minimum* provisions of the code. The document is intended to help the reader understand and correctly use the design provisions of IBC Chapters 16 (Structural Design), 19 (Concrete), 21 (Masonry), 22 (Steel), and 23 (Wood). Design practices of an individual structural engineer or office, which may result in a more seismic-resistant design than required by the minimum requirements of the IBC, are not given. However, these considerations are discussed as alternatives, when appropriate.

In some examples, the performance characteristics of the structural system are discussed. Several examples include a brief review of past earthquake behavior and mention of design improvements added to recent codes. SEAOC believes it is essential that structural engineers not only know how to correctly interpret and apply the provisions of the code, but that they also understand their bases.

While the 2006 *IBC Structural Seismic Design Manual* is based on the 2006 IBC, the primary referenced document is ASCE/SEI 7-05.

The 2006 *IBC Structural/Seismic Design Manual—Volume 2* is intended for use by practicing structural engineers and structural designers, building departments, other plan review agencies, and structural engineering students.

How to Use This Document

ASCE/SEI 7-05 notation is generally used throughout. Some other notation is also defined in the following pages, or in the examples.

Throughout the document, reference to specific code provisions and equations is given in the right-hand margin under the category Code Reference. For example, "ASCE/SEI 7-05 Section 12.3" is given as §12.3 with ASCE/SEI 7-05 being understood. "Equation (12-4-1)" is designated Eq 12.4-1. The phrase "T 15.2.1" is understood to be Table 15.2.1 and Figure 22-1 is designated F 22-1.

The 2006 *IBC Structural/Seismic Design Manual—Volume 2* primarily references the ASCE/SEI 7-05, unless otherwise indicated. References to IBC sections, tables, and equations are enclosed in parentheses. Occasionally, reference is made to other codes and standards (e.g., ACI 318-99 or 1997 NDS). When this is so, these documents are clearly identified.

Generally, each design example is presented in the following format. First, there is an "Overview" of the example. This is a description of the building to be designed. This is followed by an "Outline" indicating the tasks or steps to be illustrated in each example. Next, "Given Information" provides the basic design information, including plans and sketches given as the starting point for the design. This is followed by "Calculations and Discussion," which provides the solution to the example. Some examples have a subsequent section designated "Commentary" that is intended to provide a better understanding of aspects of the example and/or to offer guidance to the reader on use of the information generated in the example. Finally, references and suggested reading are given under "References." Some examples also have a "Foreword" and/or "Factors Influencing Design" section that contains remarks on salient points about the design.

Notation

The following notations are used in this document. These are generally consistent with those used in the ASCE/SEI 7-05 and other codes such as ACI and AISC. The reader is cautioned that the same notation may be used more than once and may carry entirely different meanings in different situations. For example, *E* can mean the tabulated elastic modulus under the AISC definition (steel) or it can mean the earthquake load under §12.4.2 of the ASCE/SEI 7-05. When the same notation is used in two or more definitions, each definition is prefaced with a brief description in parentheses (e.g., steel or loads) before the definition is given.

A = area of floor or roof supported by a member

A_b = area of anchor, in square inches

A_c = the combined effective area, in square feet, of the shear walls in the first story of the structure

A_{ch} = cross-sectional area of a structural member measured out-to-out of transverse reinforcement

A_{cv} = net area of concrete section bounded by web thickness and length of section in the direction of shear force considered

A_g = gross area of section

A_g = the gross area of that wall in which A_o is identified

A_i = the floor area in square feet of the diaphragm level immediately above the story under consideration

A_o = total area of opening in a wall that receives positive external wind pressure

A_s = area of non-prestressed tension reinforcement

A_{sh} = total cross-sectional area of transverse reinforcement (including crossties) within spacing s and perpendicular to dimension of b_c

$A_{s,min}$ = area having minimum amount of flexural reinforcement

A_T = tributary area

A_v = area of shear reinforcement within a distance s, or area of shear reinforcement perpendicular to flexural tension reinforcement within a distance s for deep flexural members

A_x = the torsional amplification factor at Level x – §12.8.4.3

a = (concrete) depth of equivalent rectangular stress block

a	=	(concrete spandrel) shear span, distance between concentrated load and face of supports
a_c	=	coefficient defining the relative contribution of concrete strength to wall strength
a_d	=	incremental factor relating to the *P*-delta effects as determined in §12.8.7
a_p	=	amplification factor related to the response of a system or component as affected by the type of seismic attachment determined in §13.3.1
b	=	(concrete) width of compression face of member
C_d	=	deflection amplification factor in Tables 12.2-1 or 15.4-1 or 15.4-2
C_e	=	snow exposure factor
C_s	=	the seismic response coefficient determined in §12.8.1.1 and §19.3.1
C_t	=	building period coefficient – §12.8.2.1
C_t	=	snow thermal factor
C_{vx}	=	vertical distribution factor – §12.8.3
c	=	distance from extreme compression fiber to neutral axis of a flexural member
D	=	dead load, the effect of dead load
d	=	effective depth of section (distance from extreme compression fiber to centroid of tension reinforcement)
d_b	=	(anchor bolt) anchor shank diameter
d_b	=	(concrete) bar diameter
E	=	(steel) modulus of elasticity
E	=	combined effect of horizontal and vertical earthquake induced forces (§12.4)
E_c	=	modules of elasticity of concrete, in psi
EI	=	flexural stiffness of compression member
E_m	=	seismic load effect including overstrength factors (§§12.4.3 and 12.14.2.2.2)
E_s	=	(concrete) modulus of elasticity of reinforcement

F	=	load due to fluids
F_a	=	site coefficient defined in §11.4.3
F_a	=	axial compressive stress that would be permitted if axial force alone existed
F_b	=	bending stress that would be permitted if bending moment alone existed
F_{EXX}	=	classification number of weld metal (minimum specified strength)
F_i, F_n, F_x	=	portion of the seismic base shear, V, induced at Level i, n, or x (§12.8.3)
F_i	=	the design force applied to Level i
F_p	=	seismic design force, induced by the parts being connected, centered at the component's center of gravity, and distributed relative to the component's mass distribution as in §12.8.3
F_{px}	=	the diaphragm design force
F_u	=	specified minimum tensile strength, ksi
F_{ut}	=	minimum specified tensile strength of the anchor
F_v	=	long period site coefficient (at 1.0 second period) §11.4.3
F_x	=	the design lateral force applied at Level x
F_x	=	the lateral force induced at any Level i – §12.8.3
F_y	=	specified yield strength of structural steel
F_{ye}	=	expected yield strength of steel to be used
F_{yh}	=	(steel) specified minimum yield strength of transverse reinforcement
f_1	=	live load reduction factor – IBC §1605
f_2	=	snow load reduction factor – IBC §1605
f_a	=	computed axial stress
f_b	=	bending stress in frame member
f'_c	=	specified compressive strength of concrete
f_{ct}	=	average splitting tensile strength of lightweight aggregate concrete
f'_m	=	specified compressive strength of masonry
f_r	=	modulus of rupture of concrete

f_y	=	(concrete) specified yield strength of reinforcing steel
g	=	acceleration due to gravity (gravitational acceleration constant 32.2 ft/sec^2 or 386.4 in/sec^2)
H	=	load due to lateral pressure of soil and water in soil
h	=	average roof height of structure relative to the base elevation
h	=	overall dimensions of member in direction of action considered
h_c	=	(concrete) cross-sectional dimension of column core, or shear wall boundary zone, measured center-to-center of confining reinforcement
h_i, h_n, h_x	=	height in feet above the base to Level i, n or x, respectively
h_r	=	height in feet of the roof above the base
h_{sx}	=	the story height below Level x
h_w	=	height of entire wall or of the segment of wall considered
I	=	the seismic occupancy importance factor importance factor determined in §11.5.1
I	=	moment of inertia of section resisting externally applied factored loads
I	=	importance factor determined in §11.5.1
I_{cr}	=	moment of inertia of cracked section transformed to concrete
I_g	=	(concrete, neglecting reinforcement) moment of inertia of gross concrete section about centroidal axis neglecting reinforcement
I_{se}	=	moment of inertia of reinforcement about centroidal axis of member cross section.
I_p	=	component importance factor that is either 1.00 or 1.5, as determined in §13.3.1
K	=	(steel) effective length factor for prismatic member
k	=	a distribution exponent – §12.8.3
L	=	live load, except roof live load, including any permitted live load reduction (i.e. reduced design live load). Live load related internal moments or forces. Concentrated impact loads
L_o	=	unreduced design live load
L_r	=	roof live load including any permitted live load reduction

l_u	=	unsupported length of compression member
l_w	=	length of entire wall, or of segment of wall considered, in direction of shear force
Level i	=	level of the structure referred to by the subscript i. "i = 1" designates the first level above the base
Level n	=	that level that is uppermost in the main portion of the structure
Level x	=	that level that is under design consideration. "x = 1" designates the first level above the base
M	=	(steel) maximum factored moment
M_{cr}	=	moment at which flexural cracking occurs in response to externally applied loads
M_n	=	(steel) nominal moment strength at section
M_{ot}	=	(wood) overturning moment
M_{pa}	=	nominal plastic flexural strength modified by axial load
M_t	=	torsional moment
M_{ta}	=	accidental torsional moment
M_u	=	(concrete) factored moment at section
M_u	=	(steel) required flexural strength on a member or joint
M_y	=	moment corresponding to onset of yielding at the extreme fiber from an elastic stress distribution
N	=	number of stories
P	=	(wind) design wind pressure
P_{DL}, P_{LL}, P_{seis}	=	unfactored axial load in frame member
P_b	=	nominal axial load strength at balanced strain conditions
P_c	=	(concrete) critical load
P_c	=	(concrete anchorage) design tensile strength
P_n	=	nominal axial load strength at given eccentricity, or nominal axial strength of a column
P_o	=	nominal axial load strength at zero eccentricity

P_u	=	(concrete) factored axial load, or factored axial load at given eccentricity
P_u	=	(steel) nominal axial strength of a column, or required axial strength of a column or a link
P_u	=	(concrete anchorage) required tensile strength from loads
P_y	=	nominal axial yield strength of a member, which is equal to $F_y A_g$
P_x	=	total unfactored vertical design load at and above Level x
P_E	=	axial load on member due to earthquake
P_{LL}	=	axial live load
Q_E	=	the effect of horizontal seismic forces
R	=	the response modification factor from Table 12.2-1 or 15.4-1 or 15.4-2
R_n	=	nominal strength
R_p	=	component response modification factor that varies from 1.00 to 3.50 as set forth in Table 13.5-1 or Table 13.6-1
$R, R_1 R_2$	=	live load reduction in percent – IBC §1607.9.2/1607.11.2
r	=	rate of reduction equal to 0.08 percent for floors
r	=	(steel) radius of gyration of cross section of a compression member
r_y	=	radius of gyration about y axis
S	=	snow load
S_{DS}	=	design, 5% damped, spectral response acceleration parameter at short period (0.2 second) as determined in §11.4.4
S_S, S_s	=	the mapped MCE, 5% damped, spectral response acceleration parameter at short period (0.2 second) as determined in §11.4.1
S_{D1}	=	design, 5% damped, spectral response acceleration parameter at long design (1-second period) as determined in §11.4.4
S_1	=	the mapped MCE, 5% damped, spectral response acceleration parameter at long period (1-second period) as determined in §11.4.1
S_{MS}	=	adjusted MCE, 5% damped, spectral response acceleration parameter at short period (0.2 second), adjusted for site class effects as determined in §11.4.3

S_{M1} = adjusted MCE, 5% damped, spectral response acceleration parameter at long period (1-second period), adjusted for site class effects as determined in §11.4.3

s = spacing of shear or torsion reinforcement in direction parallel to longitudinal reinforcement, or spacing of transverse reinforcement measured along the longitudinal axis

T = elastic fundamental period of vibration, in seconds, of the structure in the direction under consideration, see §1617.4.2 for limitations

T_a = approximate fundamental period as determined in accordance with §12.8.2.1

T_O = $0.2\ (S_{D1}\ /\ S_{DS})$

T_S = $(S_{D1}\ /\ S_{DS})$

U = required strength to resist factored loads or related internal moments and forces

V = the total design seismic lateral force or shear at the base of the building or structure

V_c = (concrete) nominal shear strength provided by concrete

V_c = (concrete anchorage) design shear strength

V_n = (concrete) nominal shear strength at section

V_n = (steel) nominal shear strength of a member

V_{px} = the portion of the seismic shear force at the level of the diaphragm, required to be transferred to the components of the vertical seismic-lateral-force-resisting system because of the offsets or changes in stiffness of the components above or below the diaphragm

V_s = (concrete) nominal shear strength provided by shear reinforcement

V_u = (concrete anchorage) required shear strength from factored loads

V_u = (concrete) factored shear force at section

V_u = (loads) factored horizontal shear in a story

V_u = (steel) required shear strength on a member

V_x = the seismic design story shear (force) in story x, (i.e., between Level x and x-1)

W = the total effective seismic dead load (weight) defined in §12.7.2 and §12.14.8.1

W	=	(wind) load due to wind pressure
W_p	=	component operating weight
w_c	=	weights of concrete, in pcf
w_i, w_x	=	that portion of W located at or assigned to Level i or x, respectively
w_p	=	the weight of the smaller portion of the structure
w_p	=	the weight of the diaphragm and other elements of the structure tributary to the diaphragm
w_{px}	=	the weight of the diaphragm and elements tributary thereto at Level x, including applicable portions of other loads defined in §12.7.2
w_w	=	weight of the wall tributary to the anchor
w_z	=	column panel zone width
X	=	height of upper support attachment at Level x as measured from the base
Y	=	height of lower support attachment at Level Y as measured from the base
Z	=	(steel) plastic section modulus
z	=	height in structure at point of attachment of component, §13.3.1
ϕ	=	(concrete) capacity-reduction or strength-reduction factor
ϕ_b	=	(steel) resistance factor for flexure
ϕ_c	=	(steel) resistance factor for compression
ϕ_V	=	resistance factor for shear strength of panel-zone of beam-to-column connections
ρ	=	a redundancy factor obtained in accordance with §12.3.4
ρ	=	(concrete) ratio of nonprestressed tension reinforcement (A_s/b_d)
ρ_b	=	reinforcement ratio producing balanced strain conditions
λ	=	lightweight aggregate concrete factor; 1.0 for normal-weight concrete, 0.75 for "all lightweight" concrete, and 0.85 for "sand-lightweight" concrete
λ_p	=	limiting slenderness parameter for compact element

ℓ_n	=	clear span measured face-to-face of supports
ℓ_u	=	unsupported length of compression member
ℓ_w	=	length of entire wall or of segment of wall considered in direction of shear force
μ	=	coefficient of friction
Δ	=	design story drift, shall be computed as the differences of the deflections at the center of mass at the top and bottom of the story under consideration. Note: Where ASD is used, Δ shall be computed using earthquake forces without dividing by 1.4, see §12.8.6
Δ_a	=	allowable story drift, as obtained from Table 12.12-1
Ω_o	=	system overstrength factor as given in Table 12.2-1
δ_x	=	inelastic deflections of Level x – §12.8.6
δ_{ave}	=	the average of the displacements at the extreme points of the structure at Level x
δ_{max}	=	the maximum displacement at Level x
δ_{xa}	=	deflection at Level x of structure A
δ_{xe}	=	the deflections determined by an elastic analysis of the seismic-force-resisting system
δ_{yA}	=	deflection at Level y of structure A
δ_{yB}	=	deflection at Level y of structure B
θ	=	stability coefficient – §12.8.7

Definitions

Active Fault/Active Fault Trace. A fault determined to be active by the authority having jurisdiction from properly substantiated data.

Allowable Stress Design. A method of proportioning structural members, such that elastically computed stresses produced in the members by nominal loads do not exceed specified allowable stresses (also called working stress design).

Attachments, Seismic. Means by which components and their supports are secured or connected to the seismic-force-resisting system of the structure. Such attachments include anchor bolts, welded connections, and mechanical fasteners.

Balcony, Exterior. An exterior floor projecting from and supported by a structure without additional independent supports.

Base. The level at which the horizontal seismic ground motions are considered to be imparted to the structure.

Base Shear. Total design lateral force or shear at the base.

Boundary Elements. Chords and collectors at diaphragm and shear wall edges, interior openings, discontinuities, and re-entrant corners.

Boundary Members. Portions along wall and diaphragm edges strengthened by longitudinal and transverse reinforcement and/or structural steel members.

Brittle. Systems, members, materials, and connections that do not exhibit significant energy dissipation capacity in the inelastic range.

Cantilevered Column System. A structural system relying on column elements that cantilever from a fixed base and have minimal rotational resistance capacity at the top with lateral forces applied essentially at the top and are used for lateral resistance.

Collector. A diaphragm or shear wall element parallel to the applied load that collects and transfers shear forces to the vertical-force-resisting elements or distributes forces within a diaphragm or shear wall.

Collector Elements. Members that serve to transfer forces between floor or horizontal diaphragms and vertical members of the lateral-force-resisting system.

Component. A part or element of an architectural, electrical, mechanical, or structural system.

- **Component, equipment.** A mechanical or electrical component or element that is part of a mechanical and/or electrical system within or without a building system.
- **Component, flexible.** Component, including its attachments, having a fundamental period greater than 0.06 second.
- **Component, rigid.** Component, including its attachments, having a fundamental period less than or equal to 0.06 second.

Confined Region. The portion of a reinforced concrete component in which the concrete is confined by closely spaced special transverse reinforcement restraining the concrete in directions perpendicular to the applied stress.

Coupling Beam. A beam that is used to connect adjacent concrete wall piers to make them act together as a unit to resist lateral forces.

Dead Loads. The weight of materials of construction incorporated into the building, including but not limited to walls, floors, roofs, ceilings, stairways, built-in partitions, finishes, cladding, and other similarly incorporated architectural and structural items, and fixed service equipment, including the weight of cranes.

Deck. An exterior floor supported on at least two opposing sides by an adjacent structure, and/or posts, piers, or other independent supports.

Deformability. The ratio of the ultimate deformation to the limit deformation.

> **High deformability element.** An element whose deformability is not less than 3.5 when subjected to four fully reversed cycles at the limit deformation.
> **Limited deformability element.** An element that is neither a low deformability nor a high deformability element.
> **Low deformability element.** An element whose deformability is 1.5 or less.

Deformation.

> **Limit deformation.** Two times the initial deformation that occurs at a load equal to 40 percent of the maximum strength.
> **Ultimate deformation.** The deformation at which failure occurs and which shall be deemed to occur if the sustainable load reduces to 80 percent or less of the maximum strength.

Design Earthquake. The earthquake effects that buildings and structures are specifically proportioned to resist.

Design Strength. The product of the nominal strength and a resistance factor (or strength reduction factor).

Designated Seismic System. Those architectural, electrical, and mechanical systems and their components that require design in accordance with Chapter 13 that have a component importance factor, I_p, greater than 1.0.

Diaphragm, Flexible. A diaphragm is flexible for the purpose of distribution of story shear and torsional moment when the lateral deformation of the diaphragm is more than two times the average story drift of the associated story, determined by comparing the computed maximum in-plane deflection of the diaphragm itself under lateral force with the story drift of adjoining vertical lateral-force-resisting elements under equivalent tributary lateral force.

Diaphragm, Rigid. A diaphragm that does not conform to the definition of flexible diaphragm.

Duration of Load. The period of continuous application of a given load, or the aggregate of periods of intermittent applications of the same load.

Element.

> **Ductile element.** An element capable of sustaining large cyclic deformations beyond the attainment of its strength.

Limited ductile element. An element that is capable of sustaining moderate cyclic deformations beyond the attainment of nominal strength without significant loss of strength.
Nonductile element. An element having a mode of failure that results in an abrupt loss of resistance when the element is deformed beyond the deformation corresponding to the development of its nominal strength. Nonductile elements cannot reliably sustain significant deformation beyond that attained at their nominal strength.

Equipment Support. Those structural members or assemblies of members or manufactured elements, including braces, frames, lugs, snubbers, hangers, or saddles that transmit gravity load and operating load between the equipment and the structure.

Essential Facilities. Buildings and other structures that are intended to remain operational in the event of extreme environmental loading from flood, wind, snow, or earthquakes.

Factored Load. The product of a nominal load and a load factor.

Flexible Equipment Connections. Those connections between equipment components that permit rotational and/or translational movement without degradation of performance.

Frame.

Braced frame. An essentially vertical truss, or its equivalent, of the concentric or eccentric type that is provided in a building frame system or dual frame system to resist shear.
Concentrically braced frame (CBF). A braced frame in which the members are subjected primarily to axial forces.
Eccentrically braced frame (EBF). A diagonally braced frame in which at least one end of each brace frames into a beam a short distance from a beam-column or from another diagonal brace.
Ordinary concentrically braced frame (OCBF). A steel concentrically braced frame in which members and connections are designed in accordance with the provisions of AISC Seismic without modification.
Special concentrically braced frame (SCBF). A steel or composite steel and concrete concentrically braced frame in which members and connections are designed for ductile behavior.

Frame, Moment.

Intermediate moment frame (IMF). A moment frame in which members and joints are capable of resisting forces by flexure as well as along the axis of the members.
Ordinary moment frame (OMF). A moment frame in which members and joints are capable of resisting forces by flexure as well as along the axis of the members.
Special moment frame (SMF). A moment frame in which members and joints are capable of resisting forces by flexure as well as along the axis of the members.

Frame System.

Building frame system. A structural system with an essentially complete space frame system providing support for vertical loads. Seismic force resistance is provided by shear walls or braced frames.
Dual frame system. A structural system with an essentially complete space frame system providing support for vertical loads. Seismic force resistance is provided by a moment-resisting frame and shear walls or braced frames.
Space frame system. A structural system composed of interconnected members, other than bearing walls, that is capable of supporting vertical loads and that also may provide resistance to seismic forces.

Gravity Load (*W*). The total dead load and applicable portions of other loads.

Guard. A building component or a system of building components located at or near the open sides of elevated walking sufaces that minimizes the possibility of a fall from the walking surface to a lower level.

Hazardous Contents. Material that is highly toxic or potentially explosive and in sufficient quantity to pose a significant life-safety threat to the general public if an uncontrolled release were to occur.

Impact Load. The load resulting from moving machinery, elevators, craneways, vehicles, and other similar forces and kinetic loads, pressure, and possible surcharge from fixed or moving loads.

Importance Factor. A factor assigned to each structure according to its occupancy category in accordance with Table 11.5-1.

Inverted Pendulum-type Structures. Structures that have a large portion of their mass concentrated near the top and, thus, have essentially one degree of freedom in horizontal translation. The structures are usually T-shaped with a single column supporting the beams or framing at the top.

Isolation Interface. The boundary between the upper portion of the structure, which is isolated, and the lower portion of the structure, which moves rigidly with the ground.

Isolation System. The collection of structural elements that includes individual isolator units, structural elements that transfer force between elements of the isolation system and connections to other structural elements.

Isolator Unit. A horizontally flexible and vertically stiff structural element of the isolation system that permits large lateral deformations under design seismic load. An isolator unit may be used either as part of or in addition to the weight-supporting system of the building.

Joint. A portion of a column bounded by the highest and lowest surfaces of the other members framing into it.

Limit State. A condition beyond which a structure or member becomes unfit for service and is judged to be no longer useful for its intended function (serviceability limit state) or to be unsafe (strength limit state).

Live Loads. Those loads produced by the use and occupancy of the building or other structure and do not include construction or environmental loads such as wind load, snow load, rain load, earthquake load, flood load, or dead load.

Live Loads (Roof). Those loads produced 1) during maintenance by workers, equipment, and materials; and 2) during the life of the structure by movable objects such as planters and by people.

Load and Resistance Factor Design (LRFD). A method of proportioning structural members and their connections using load and resistance factors such that no applicable limit state is reached when the structure is subjected to appropriate load combinations. The term "LRFD" is used in the design of steel and wood structures.

Load Factor. A factor that accounts for deviations of the actual load from the nominal load, for uncertainties in the analysis that transforms the load into a load effect, and for the probability that more than one extreme load will occur simultaneously.

Loads. Forces or other actions that result from the weight of building materials, occupants and their possessions, environmental effect, differential movement, and restrained dimensional changes. Permanent loads are those loads in which variations over time are rare or of small magnitude. Other loads are variable loads (see also "Nominal loads").

Loads Effects. Forces and deformations produced in structural members by the applied loads.

Maximum Considered Earthquake. The most severe earthquake effects considered by this code.

Nominal Loads. The magnitudes of the loads specified in this chapter (dead, live, soil, wind, snow, rain, flood, and earthquake.)

Nonbuilding Structure. A structure, other than a building, constructed of a type included in Chapter 15.

Other Structures. Structures, other than buildings, for which loads are specified in this chapter.

P-Delta Effect. The second order effect on shears, axial forces, and moments of frame members induced by axial loads on a laterally displaced building frame.

Panel (Part of a Structure). The section of a floor, wall, or roof located between the supporting frame of two adjacent rows of columns and girders or column bands of floor or roof construction.

Resistance Factor. A factor that accounts for deviations of the actual strength from the nominal strength and the manner and consequences of failure (also called strength reduction factor).

Seismic Design Category. A classification assigned to a structure based on the design spectral response acceleration parameters per Tables 11.6-1 and 11.6-2.

Seismic-force-resisting system. The part of the structural system that has been considered in the design to provide the required resistance to the seismic forces prescribed herein.

Seismic Forces. The assumed forces prescribed herein, related to the response of the structure to earthquake motions, to be used in the design of the structure and its components.

Seismic Response Coefficient. Coefficient C_s, as determined from §12.8.1.1.

Shallow Anchors. Shallow anchors are those with embedment length-to-diameter ratios of less than 8.

Shear Panel. A floor, roof, or wall component sheathed to act as a shear wall or diaphragm.

Shear Wall. A wall designed to resist lateral forces parallel to the plane of the wall.

Shear Wall-frame Interactive System. A structural system that uses combinations of shear walls and frames designed to resist lateral forces in proportion to their rigidities, considering interaction between shear walls and frames on all levels.

Site Class. A classification assigned to a site based on the types of soils present and their engineering properties as defined in §11.4.2.

Site Coefficients. The values of F_a and F_v indicated in Tables 11.4-1 and 11.4-2, respectively.

Special Transverse Reinforcement. Reinforcement composed of spirals, closed stirrups, or hoops and supplementary cross-ties provided to restrain the concrete and qualify the portion of the component, where used, as a confined region.

Story Drift Ratio. The story drift divided by the story height.

Strength, Nominal. The capacity of a structure or member to resist the effects of loads, as determined by computations using specified material strengths and dimensions and formulas derived from accepted principles of structural mechanics or by field tests or laboratory tests of scaled models, allowing for modeling effects and differences between laboratory and field conditions.

Strength Design. A method of proportioning structural members such that the computed forces produced in the members by factored loads do not exceed the member design strength (also called load and resistance factor design.) The term "strength design" is used in the design of concrete and masonry structural elements.

Strength Required. Strength of a member, cross section, or connection required to resist factored loads or related internal moments and forces in such combinations as stipulated by these provisions.

Torsional Force Distribution. The distribution of horizontal seismic forces through a rigid diaphragm when the center of mass of the structure at the level under consideration does not coincide with the center of rigidity (sometimes referred to as a diaphragm rotation).

Toughness. The ability of a material to absorb energy without losing significant strength.

Wall, Load-bearing. Any wall meeting either of the following classifications:

1. Any metal or wood stud wall that supports more than 100 pounds per linear foot (1459 N/m) of vertical load in addition to its own weight.
2. Any masonry or concrete wall that supports more than 200 pounds per linear foot (2919 N/m) of vertical load in addition to its own weight.

Wall, Nonload-bearing. Any wall that is not a load-bearing wall.

Wind-restraint Seismic System. The collection of structural elements that provides restraint of the seismic-isolated-structure for wind loads. The wind-restraint system may be either an integral part of isolator units or a separate device.

Wood Structural Panel. A panel manufactured from veneers; or wood strands or wafers; or a combination of veneer and wood strands or wafers; bonded together with waterproof synthetic resins or other suitable bonding systems.

References

The following codes and standards are referenced in this document. Other reference documents are indicated at the end of each design example.

ACI-318, 2005, American Concrete Institute, *Building Code Requirements for Structural Concrete,* Farmington Hills, Michigan

AISC 360, 2005, American Institute of Steel Construction, *Manual of Steel Construction,* Thirteenth Edition, Chicago, Illinois

AISI-Lateral, 2004, American Iron and Steel Institute, Standard for cold-formed steel framing—Lateral Design, Washington, D.C.

IBC, 2006, International Code Council, *International Building Code.* Falls Church, Virginia.

MSJC, 2005, *Building Code Requirements for Masonry Structures* (ACI 530-05/ASCE 5-05/TMS 402-05) Farmington Hills, Michigan

NDS, 2005, American Forest & Paper Association, *National Design Specification for Wood Construction,* Washington, D.C.

SEAOC Blue Book, 1999, *Recommended Lateral Force Requirements and Commentary.* Structural Engineers Association of California, Sacramento, California.

SDPWS, American Forest and Paper Association Supplement Special Design Provisions for Wind and Seismic, Washington, D.C.

Design Example 1A
Wood Light-frame Residence

Figure 1A-1. Wood light-frame residence

Overview

This design example illustrates the seismic design of a 2,800-square-foot single-family residence. The structure, shown in Figures 1A-1, 1A-2, 1A-3, 1A-4 and 1A-5, is of wood light-frame construction with wood structural panel shear walls, roof, and floor diaphragms. Roofing is clay tile.

IBC vs IRC

Section 101.2 of the 2006 IBC exempts detached one- and two-family dwellings not more than three stories high from the code and states that the design of these structures ". . . shall comply with the *International Residential Code.*" IBC Section 1613.1 states that detached one- and two-family dwellings in Seismic Design Categories (SDC) A, B, or C that have a mapped S_s of less than 0.4 g are exempt. Section R301.1 of the IRC states that when the structure contains structural elements that exceed the limits of §R301.1.3, "these elements shall be designed in accordance with accepted engineering practice." Section R301.2.2 of the IRC states that buildings in SDC E shall be designed in accordance with the IBC. Section R301.2.2.2.2 requires that buildings in SDCs C, D_0, D_1, and D_2 that do not meet the conventional construction requirements of the IBC shall be designed in accordance with accepted engineering practice.

Part 5 of this design example demonstrates that this residence does not meet the conventional construction provisions of either the IBC or the IRC and, hence, an engineering design per the IBC is required.

Because of the high *h*/*w* (height/width) ratios of the walls next to the garage doors, cantilevered column elements are used to provide lateral support. As shown in Figure 1-3, there is an out-of-plane offset from the cantilevered column elements on Line E to the glulam beams (GLBs) supporting the shear walls above Line D.

The residence cannot be built using conventional construction methods for reasons shown in Part 5 of this design example. The following steps illustrate a detailed analysis for some of the important seismic requirements of ASCE/SEI 7-05 that pertain to design of wood light-frame buildings. As stated in the introduction to this manual, these design examples, including this one, are not complete building designs. Many aspects of building design are not included, and only selected parts of the seismic design are illustrated. As is common for Type V construction, a complete wind design is also necessary, but is not given in this design example.

Although the code criteria only recognize two diaphragm categories, flexible and rigid, the diaphragms in this design example are judged to be semi-rigid. Consequently, the analysis in this design example will use the envelope method, which considers the worst loading condition from both the flexible and rigid diaphragm analyses for vertical-resisting elements. It should be noted that the envelope method, although not explicitly required by the code, will produce a more predictable performance than will use of only flexible or rigid diaphragm assumptions.

This design example will first determine the shear wall nailing and tiedown requirements obtained using the flexible diaphragm assumption to determine shear wall rigidities for the rigid diaphragm analysis.

The method of determining shear wall rigidities used in this design example is far more rigorous than normal practice, but is *not* the only method available to determine shear wall rigidities. The commentary at the end of this design example illustrates two other simplified approaches that would also be appropriate.

Outline

This example will illustrate the following parts of the design process

1. Design base shear and vertical distributions of seismic forces

2. Lateral forces on shear walls and shear wall nailing assuming flexible diaphragms

3. Redundancy factor ρ

4. Diaphragms

5. Whether residence meets requirements for conventional construction provisions

6. Design shear wall frame over garage on line D

7. Diaphragm shears at the low roof over garage

8. Detail the wall frame over the GLBs on line D

9. Detail the anchorage of wall frame to the GLBs on line D

10. Detail the continuous load path at the low roof above the garage doors

Given Information

Roof weights (slope 5:12)

Tile roofing	10.0 psf
½-inch sheathing	1.5
Roof framing	4.0
Insulation	1.0
Miscellaneous	0.2
Gyp ceiling	2.8
D (along slope) =	19.5 psf

Floor weights

Flooring	1.0 psf
⅝-inch sheathing	1.8
Floor framing	4.0
Miscellaneous	0.4
Gyp ceiling	2.8
	10.0 psf

D = dead load

D = (horiz. proj.) = 19.5 (13/12) = 21.1 psf (the roof and ceilings are assumed to be on a 5:12 slope, vaulted)

Weights of respective diaphragm levels, including exterior and interior walls:

For north-south direction

W_{roof} = 63,650 lb (roof and tributary walls)

W_{floor} = 38,850 lb (floor and tributary walls above and below)

W = 102,500 lb

For east-west direction

W_{roof} = 65,400 lb (roof and tributary walls)

W_{floor} = 39,900 lb (floor and tributary walls above and below)

W = 105,300 lb

Weights of diaphragms are typically determined by adding the tributary weights of the walls to the diaphragm, e.g., add one-half the height of walls at the second floor to the roof, and one-half the height of second floor walls plus one-half the height of first floor walls to second floor diaphragm. It is acceptable practice to ignore the weight of shear walls parallel to the direction of seismic forces to the upper level and add 100 percent of the parallel shear wall weight to the level below, instead of splitting the weight between floor levels. Weights of bearing partitions (not shear walls) should still be split between floors. Unlike commercial construction, the code minimum of 20 psf (vertical load) and 10 psf (lateral load) is often exceeded in residential construction.

Framing lumber is Douglas Fir-Larch grade stamped No. 1S-Dry.

DOC PS-1 or PS-2 (APA or TECO Performance-rated) wood structural panels for shear walls will be 15⁄32-inch-thick Structural-I, 32/16 span rating, five-ply with Exposure I glue, however, four-ply is also acceptable. Three-ply 15⁄32-inch sheathing has lower allowable shears in some local jurisdictions and the inner ply voids can cause nailing problems.

The roof is 15⁄32-inch-thick DOC PS-1 or PS-2 (APA or TECO Performance-rated) sheathing, 32/16 span rating with Exposure I glue.

> The floor is 19⁄32-inch-thick DOC PS-1 or PS-2 (APA or TECO Performance-rated) Sturd-I-floor 16 inches o/c rating (or APA or TECO Performance-rated sheathing, 42/20 span rating) with Exposure I glue.

Boundary members for the shear walls are 4x posts.

Common wire nails are to be used for diaphragms, shear walls, and straps.
Sinker nails are to be used for design of the shear wall sill plate nailing at the second floor. (Note: many nailing guns use the smaller diameter box and sinker nails instead of common nails. Closer nail spacing may be required for smaller diameter nails).

Seismic and site data

$S_S = 1.78$	F 22-1
$S_1 = 0.55$	F 22-2
$S_{DS} = 1.19$	Eq 11.4-3
Seismic Design Category D	T 11.6-1
$I = 1.0$	T 11.5-1

Site Class C has been determined by geotechnical investigation. Without a geotechnical investigation, Site Class D shall be used as a default value.

Figures 1A-2 through 1A-4 depict the shear walls as dark solid lines. This has been done for clarity in this example. Actual drawings commonly use other graphic depictions. Practice varies on how framing plans are actually shown and on which level the shear walls are indicated.

Actual drawings commonly do not call out shear wall lengths. However, building designers should be aware that some building departments now require shear wall lengths to be called out on plans.

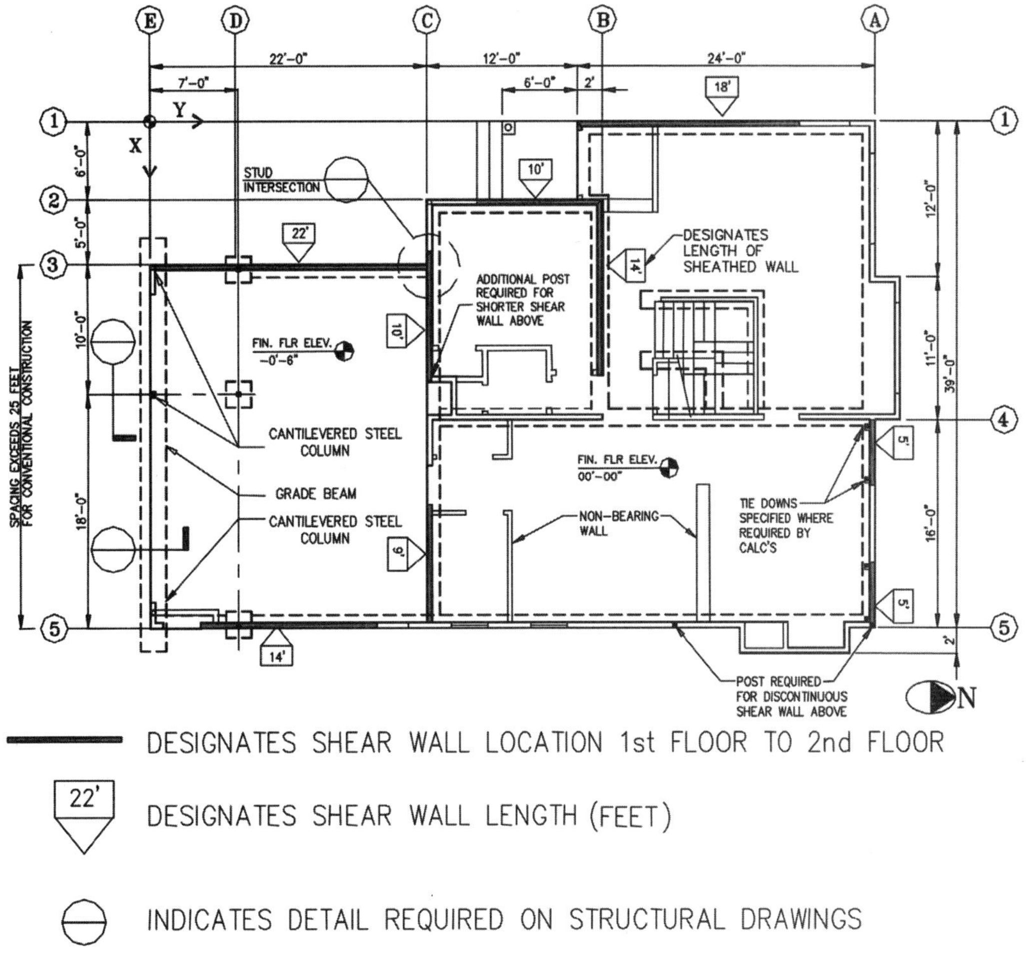

Figure 1A-2. Foundation plan (ground floor)

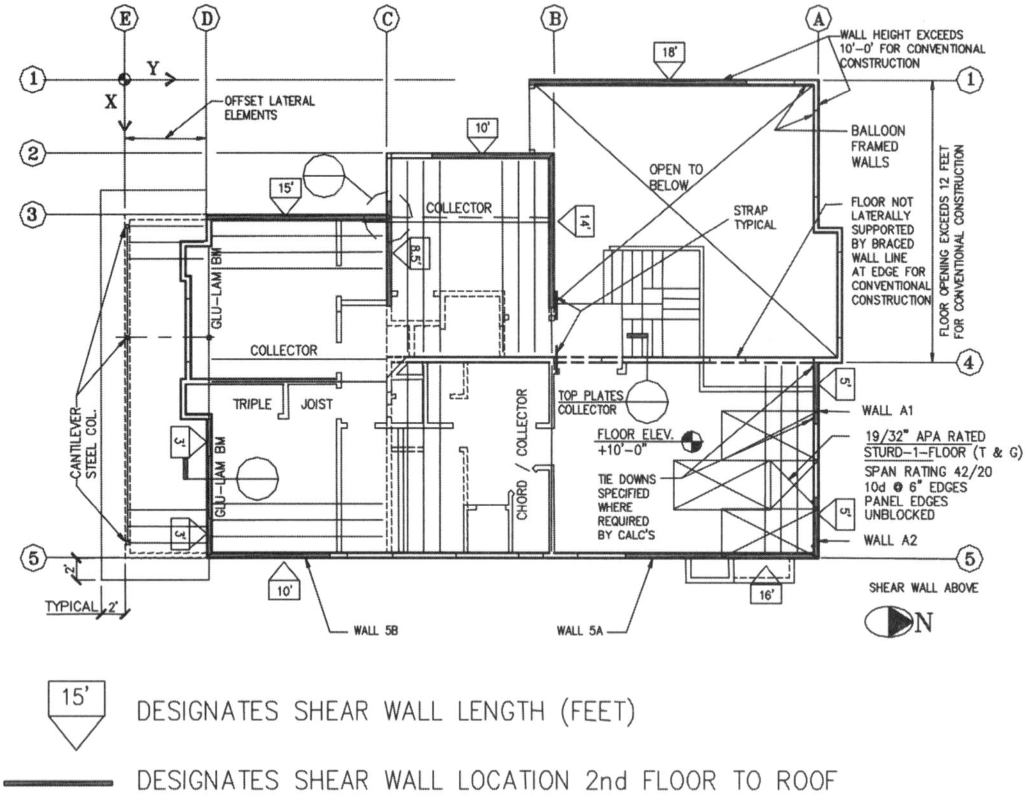

Figure 1A-3. Second floor framing plan and low roof framing plan

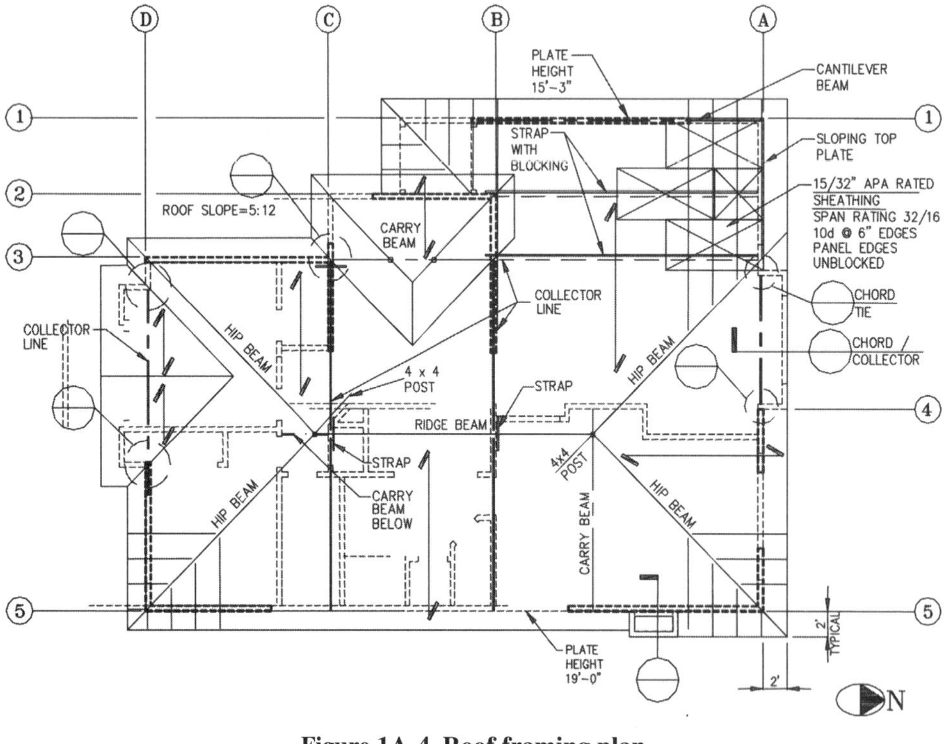

Figure 1A-4. Roof framing plan

Factors That Influence Design

Prior to starting the seismic design of the residence, three important related aspects of the design bear discussion. These are: 1) the effect of moisture content on lumber, 2) the level of engineering design required to meet code requirements in present-day California practice, and 3) the effects of box nails on wood structural panel shear walls.

Moisture content in lumber connections NDS-05 T 10.3.3

This design example is based on dry lumber. Project specifications typically call for lumber to be grade stamped S-Dry (surfaced dry). Dry lumber has a moisture content (MC) less than or equal to 19 percent. Partially seasoned or green lumber grade-stamped S-GRN (surfaced green) has an MC between 19 and 30 percent. Wet lumber has MC greater than 30 percent. Construction of structures using lumber with MC greater than 19 percent can produce shrinkage problems in the structures. Also, many engineers and building officials are not aware of the reduction requirements, or wet service factors, related to installation of nails, screws, and bolts (fasteners) into lumber with MC greater than 19 percent at time of installation. For fasteners in lumber with MC greater than 19 percent at the time of installation, the wet service factor, $C_M = 0.7$ for nails, bolts, lags, and screws (dowel-type fasteners) [National Design Specification (NDS-05) Table 10.3.3]. In other words, in lumber whose MC exceeds 19 percent, there is a 30-percent reduction in the strength of connections, diaphragms, and shear walls that is permanent. Drying the lumber after installing the connectors does not improve the connector capacity.

The effect of green lumber (mc > 19 percent) on diaphragms and shear walls is different. Recently, the APA conducted tests (APA Report T2002-53) with green lumber and dry lumber, and the tests showed that stiffness is greatly affected but the strength is, for the most part, not affected. Strength is not affected because the diaphragm and shear wall allowable shears/nominal capacities are based on their ultimate capacity and not by limit state as are connections. Therefore, it is recommended that diaphragm and shear wall deflections be modified when green lumber is used, but the shear wall and diaphragm capacities need no modification.

The engineer should exercise good judgment in determining whether it is prudent to base the structural design on dry or green lumber. Other areas of concern are geographical area and time of year the structure will be built. It is possible for green lumber (or dry lumber that has been exposed to rain) to dry out to an MC below 19 percent. For 2x framing, this generally takes 2 to 3 weeks of exposure to dry air. Thicker lumber takes even longer. Moisture content can be easily verified by a hand-held moisture meter.

Level and type of engineering design required for residences in high seismic zones

The residence in this design example was chosen because it contains many of the structural problem areas that are commonly present in residential construction. These include:

1. The discontinuous shear wall at the north end of line 5. (Although this is not a code violation per se, selection of a shear wall location that is continuous to the foundation would improve performance).

2. Lack of a lateral resisting element along line 4. (Although this is not a code violation per se, the addition of a shear wall at this location would improve performance).

3. The reduced scope of many structural engineering service contracts, such as calculation-and-sketch projects where the structural engineer provides a set of calculations and sketches of important structural details and the architect produces the actual plans and specifications. This often leads to poorly coordinated drawings and missing structural information. This method also makes structural observation requirements of the building code less effective when the engineer responsible for the design is not performing the site observation. Refer to the commentary at the end of this design example for further discussion on this subject.

An important factor in the design of residences in high seismic zones is the level of sophistication and rigor required by the designer. In this design example, a complete, rigorous analysis has been performed. In some jurisdictions, this may not be required by the building official or may not be warranted given the specifics of the design and the overall strength of the lateral-force-resisting system. The designer must choose between using the more rigorous approach of considering a rigid diaphragm with torsional resistance characteristics or the more common approach of considering flexible diaphragms with tributary mass. The former may not be necessary in some situations, but the designer needs to recognize that the laws of physics must be obeyed. In all cases, the completed structure must have a continuous lateral load path to resist lateral forces. Complete detailing is necessary, even for simple structures.

Effects of box nails on wood structural panel shear walls

This design example uses common nails for fastening wood structural panels. Based on cyclic testing of shear walls and performance in past earthquakes, the use of common nails is preferred. Special Design Provisions for Wind and Seismic (SDPWS) Table 4.3A lists nominal unit shear capacities for wood structural panel shear walls for common or galvanized box nails. IBC Table 2306.4.1 lists allowable shears for wood structural panel shear walls for common or galvanized box nails. Footnote j of Table 2306.4.1 states that the galvanized nails shall be hot-dipped or tumbled (these nails are not gun nails). Most contractors use gun nails for diaphragm and shear wall installations. Neither the SDPWS nor the IBC has a table giving allowable shears for wood structural panel shear walls or diaphragms using box nails.

Box nails have a smaller diameter shank and a smaller head size than common nails. Using 10d box nails would result in a 19-percent reduction in allowable load for diaphragms and shear walls as compared to 10d common nails. Using 8d box nails would result in a 22-percent reduction in allowable load for diaphragms and shear walls as compared to 8d common nails. This is based on comparing allowable shear values listed in Table 11Q in the NDS-05 for $^{15}/_{32}$-inch side member thickness t_s and Douglas Fir-Larch framing. In addition to the reduction of the shear wall and diaphragm capacities, when box nails are used, the walls will also drift more than when common nails are used.

A contributor to the problem is that when contractors buy large quantities of nails (for nail guns), the words *box* or *common* do not appear on the carton label. Nail length and diameters are the most common listings on the labels. Thus, it is **extremely important** to list the required nail lengths and diameters on the structural drawings for all diaphragms and shear walls. Another problem is that contractors prefer box nails because their use reduces splitting, eases driving, and costs less.

To illustrate: if an engineer designs for dry lumber (as discussed above) and common nails, and subsequently green lumber and box nails are used in the construction, the result is a compounding of the reductions. For example, for 10d nails installed into green lumber, the reduction would be 0.81 times 0.7, or a 43-percent reduction in capacity.

Calculations and Discussion	*Code Reference*

1. Design base shear and vertical distributions of seismic forces §12.8.1

This example uses the total building weight W applied to each respective direction. The results shown will be slightly conservative since W includes the wall weights for the direction of load, which can be subtracted out. This approach is simpler than using a separated building weight W for each axis under consideration.

1a. Design base shear

Period using approximate fundamental period (see Figure 1A-5 for section through structure)

$$T_a = C_t(h_n)^{\chi} = 0.020(23)^{3/4} = 0.21 \text{ sec}$$ Eq 12.8-7

where h_n is defined as the highest level of the structure. Since the highest level is not a level surface, the center of gravity (average height) of diaphragm above the first floor will be used.

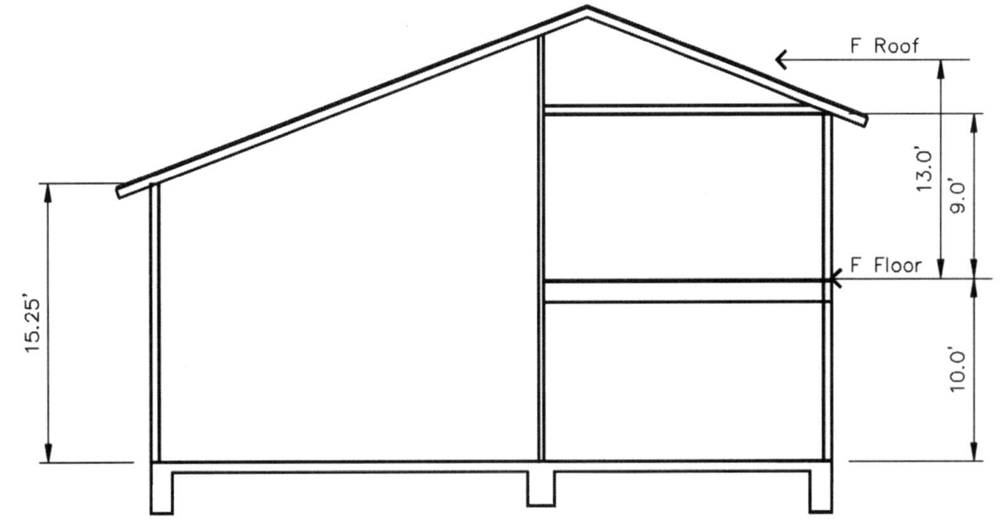

Figure 1A-5. Cross-section through residence

North-south direction

For light-framed walls with wood structural panels that are both shear walls and bearing walls

$$R = 6.5$$ T 12.2-1

Design base shear is

$$V = C_sW$$ Eq 12.8-1

where

$$C_s = \frac{S_{DS}}{\left(\frac{R}{I}\right)} \qquad \text{Eq 12.8-2}$$

(Note that design base shear in the ASCE/SEI 7-05 is on a strength design basis.)

All tables in the IBC for wood diaphragms and shear walls are based on allowable loads. All tables in the NDS-05 supplement Special Design Provisions for Wind and Seismic (SDPWS) are nominal values and must be adjusted for both strength and ASD loads. It is not known how much longer the IBC will publish ASD values since the trend is toward having all values in strength or nominal format. Since ASD is still predominantly practiced, it has been decided to have this design example in ASD format. In addition, all the manufacturers of metal hardware connectors only publish ASD values.

$$I = 1.0$$

$$R = 6.5$$

$$S_{DS} = \tfrac{2}{3}\, S_{MS} = \tfrac{2}{3}\,(1.78) = 1.19$$

$$S_{MS} = F_a S_S = 1.0 \times 1.78 = 1.78$$

where:

$$F_a = 1.0$$

$$S_S = 1.78$$

Site Class C

$$C_s = \frac{1.19}{\left(\frac{6.5}{1.0}\right)} = 0.183$$

but need not exceed

$$C_s = \frac{S_{D1}}{T\left(\frac{R}{I}\right)} \qquad \text{Eq 12.8-3}$$

$$S_1 = 0.55$$

$$S_{M1} = F_v S_1 = 1.3 \times 0.55 = 0.72$$

$$S_{D1} = \tfrac{2}{3}\, S_{M1} = \tfrac{2}{3}\,(0.72) = 0.48$$

$$F_v = 1.3$$

$$C_s = \frac{0.48}{\left(\frac{6.5}{1.0}\right)0.21} = 0.35 > 0.183 \qquad \therefore \text{ does not control}$$

$$C_s = 0.183$$

but shall not be less than

$$C_s = 0.01$$

$$\therefore V_{n\text{-}s} = 0.183W$$

In addition, equation (12.8-6) requires an additional check for C_s minimum for structures that are located where S_1 is equal to or greater than 0.6 g:

$$C_s = \frac{0.5S_1}{\left(\frac{R}{I}\right)} = \frac{0.5(1.0)}{\left(\frac{6.5}{1.0}\right)} = 0.07$$

However, this condition does not control.

Comparison of the above result with the simplified static method permitted under §12.14 shows that it is more advantageous to use the simplified static method of determining the design base shear.

$$V = \frac{FS_{DS}}{R} W = \frac{1.1(1.0)}{6.5} W = 0.169W < 0.183W \qquad \text{Eq 12.14-11}$$

where

F = 1.1 for two-story buildings

$S_{DS} = (2/3)1.0 \times 1.5 = 1.0$ §12.14.8.1

S_S = 1.5 max

$F_a = 1.0$ for $S_S \geq 1.25$ Table 11.4.1

The engineer can choose which method to use. This design example will illustrate the Equivalent Lateral Force Procedure.

Check for maximum S_S value: §12.8.1.3

Since the structure has an Out-of-Plane Offset Irregularity (ASCE 7-05 Table 12.3-1, Type 4), the structure is classified as "Irregular" and hence the S_S limitation of 1.5 does not apply.

It is desirable to keep the strength level forces throughout the design of the structure for two reasons:

1. Errors in calculations can occur and confusion on which load is being used—strength or allowable stress design. This design example will use the following format

   $V_{base\ shear}$ = strength

   F_{px} = strength

   F_x = force to wall (strength)

   v = wall shear at element level (ASD)

   $$v = \frac{F_x(0.7)}{b} \text{ (ASD)}$$

2. This design example will not be applicable in the future, when the code will be all strength design.

Seismic load effect E:

Where the effects of gravity and the seismic ground motion are additive, the seismic load E is defined as

$$E = \rho Q_E + 0.2S_{DS}\,D \qquad \text{§12.4-1}$$

Where the effects of the gravity and seismic ground motion counteract, the seismic load E is defined as

$$E = \rho Q_E - 0.2S_{DS}\,D \qquad \text{§12.4-2}$$

The redundancy ρ is assumed to be 1.0. This is the case for most Type V residential structures. The value for ρ will have to be verified. This is done later in Part 3.

The basic load combinations for allowable stress design are

$$(1.0 + 0.14S_{DS})D + 0.7\,\rho Q_E$$

and §12.4.2.3

$$(1.0 + 0.105S_{DS})D + 0.525\,\rho Q_E + 0.75L + 0.75L_r$$

$$V_{n\text{-}s} = 0.183W$$

$$\therefore\ V_{n\text{-}s} = 0.183(102{,}500\text{ lb}) = 18{,}750\text{ lb} \qquad \text{Eq 12.8-1}$$

East-west direction

Since there are different types of lateral-resisting elements in this direction, determine the controlling R value.

For light-framed walls with wood structural panels that are both shear walls and bearing walls:

$$R = 6.5$$

For cantilevered column elements with ordinary steel moment frames

$$R = 1.25 \qquad \text{T12.2-1}$$

For combinations along the same axis, the ASCE/SEI 7-05 requires the use of least value for any of the systems utilized in that same direction, therefore the value for the cantilevered column elements must be used for the entire east-west direction. This provision for combinations along the same axis first appeared in the 1994 UBC. However, ASCE/SEI 7-05 has added an exception (§12.2.3.2) when three conditions are met: 1) An Occupancy Category I or II building; 2) two stories or less in height; and 3) use of light-frame construction or flexible diaphragms. For Design Example 1, all three conditions have been met, hence the value of R may be used for each line of resistance.

The seismic base shear using $R = 6.5$ for the light-framed walls with wood structural panels will be used for the building as a whole and then the seismic load will be factored-up for the cantilever columns with a factor of (6.5/1.25).

$$\therefore\ V_{e\text{-}w} = 0.183W \qquad \text{Eq 12.8-1}$$

$$V_{e\text{-}w} = 0.183\,(105{,}300\text{ lb}) = 19{,}275\text{ lb}$$

Discussion of *R* factors

The 1999 SEAOC Blue Book added an exception for light-frame buildings in Occupancy Groups 4 and 5 and of two stories or fewer in height. This was the first time this exception appeared and actually was instigated by the first publication of this design manual where the *R* factor for the cantilever columns was used for the entire east-west direction. A higher force level could be counterproductive in terms of splitting caused by added close nailing.

An ordinary moment-resisting frame could be used with an *R* value equal to 3.5. This would produce design base shear values just under two times higher than in the north-south direction. Additionally, the architecture could be modified to provide shear wall lengths that meet the *h*/*w* ratio limit of 2:1. With the plate height at 9 feet, the minimum wall length needed would be 4 feet 6 inches. Another solution would be to increase the concrete curb height at the base of the wall such that the height-to-width (*h*/*w*) ratio limit of 2:1 is not exceeded. Note that in SDCs A through C, the *h*/*w* ratio may be 3½:1. For illustrative purposes, this design example uses the cantilevered column elements with the higher design base shear for the entire east-west direction. This conforms to the ASCE/SEI 7-05. Pre-manufactured proprietary trussed wall systems, factory-built wood shear wall systems, and steel shear wall systems are also available. Special design consideration should be given when using these systems as outlined below.

1. Building system *R* values can be based on model code agency evaluation reports.
2. Pre-manufactured systems should not be used in the same line as field-built shear walls because of deformation compatibility uncertainties.
3. Pre-manufactured systems should be limited to the first floor level only (of multistory wood frame buildings) unless testing has been done for these systems that sit on wood framing and are not rigidly attached to a concrete foundation. Some of the proprietary systems are approved for multistory wood framed buildings.
4. Many of the these systems not only exceed the aspect ratio limit of 2:1 for SDCs D, E, and F, but also exceed the aspect-ratio limit of 3½:1 for SDCs A, B, and C. Some are as narrow as 16 inches in width, leaving unanswered the question of whether this is a shear wall or a cantilever column (by comparison, if the system were a steel channel with the same width, it would be considered a cantilever column).
5. Many building officials are requesting that the same aspect-ratio limit for wood structural panel shear walls be adhered to for the pre-manufactured systems.

1b. Vertical distribution of seismic forces

The vertical distribution of seismic forces is determined from §12.8.3 as

$$F_x = C_{vx} V \qquad \text{Eq 12.8-11}$$

$$C_{vx} = \frac{w_x h_x^k}{\sum_{i=1}^{n} w_i h_i^k} \qquad \text{Eq 12.8-12}$$

where h_x is the height from the base to level *x*, in feet
k is a distribution exponent related to the building period

Since $T = 0.21$ seconds < 0.5 seconds, $k = 1$

Determination of F_x is shown in Table 1A-1a.

Table 1A-1a. Vertical distribution of seismic forces for north-south direction

Level	w_x (lb)	h_x (ft)	$w_x h_x$ (lb-ft)	$\frac{w_x h_x}{\Sigma w_i h_i}$ (%)	$F_{x_{N-S}}$ (lb)	$\frac{F_{x_{N-S}}}{w_x}$
Roof	63,650	23.0	1,463,950	79	14,800	0.233
Floor	38,850	10.0	388,500	21	3,950	0.102
Σ	102,500	—	1,852,450	100	18,750	0.183

Table 1A-1b. Vertical distribution of seismic forces for east-west direction

Level	w_x (lb)	h_x (ft)	$w_x h_x$ (lb-ft)	$\frac{w_x h_x}{\Sigma w_i h_i}$ (%)	$F_{x_{E-W}}$ (lb)	$\frac{F_{x_{E-W}}}{w_x}$
Roof	65,400	23.0	1,504,200	79	15,230	0.233
Floor	39,900	10.0	399,000	21	4,045	0.101
Σ	105,300	—	1,903,200	100	19,275	0.183

2. Lateral forces on shear walls and shear wall nailing, assuming flexible diaphragms

Determine the forces on shear walls. Because of the exemption in ASCE/SEI 7-05 (§12.3.1.1) for diaphragms of wood structural panels in one- and two-family dwellings, the diaphragm may be considered flexible. The ASCE/SEI 7-05 does not require torsional effects to be considered for flexible diaphragms. Design Example 1A will use flexible diaphragm assumptions.

The selected method of determining loads to shear walls is based on tributary areas with simple spans between supports. Another method of determining loads to shear walls can assume a continuous beam, although a continuous beam approach may not be accurate because of shear deformations in the diaphragm. The tributary area approach works with reasonable accuracy for a continuous beam with 100-percent shear deflection and zero bending deflection. This design example uses the exact tributary area to the shear walls, an approach that is fairly comprehensive. An easier and more common method would be to use a uniform load equal to the widest portion of the diaphragm, which results in conservative loads to the shear walls.

2a. Forces on east-west shear walls

Roof diaphragm

Roof area = 2164 sq ft (sf)

$$f_{p\ roof} = \frac{15{,}230\text{ lb}}{2164\text{ sf}} = 7.04\text{ psf}$$

$$w_1 = (7.04\text{ psf})(43.0\text{ ft}) = 303\text{ plf}$$

$$w_2 = (7.04\text{ psf})(37.0\text{ ft}) = 261\text{ plf}$$

$$w_3 = (7.04\text{ psf})(32.0\text{ ft}) = 226\text{ plf}$$

Check sum of forces

$$452 + 1700 + 1760 + 1977 + 2097 + 3333 + 3333 + 606 = 15{,}258 \text{ lb}$$

$$V_{Roof} = 15{,}258 \text{ lb} \approx 15{,}230 \text{ lb} \ldots o.k.$$

Note that Figures 1A-6, 1A-7, 1A-8, and 1A-9 are depicted as a continuous beam. From a technical standpoint, nodes should be shown at the interior supports. In actuality, with the tributary area approach, these are considered separate simple span beams between the shear wall supports (Figure 1A-6 has three separate single span beams).

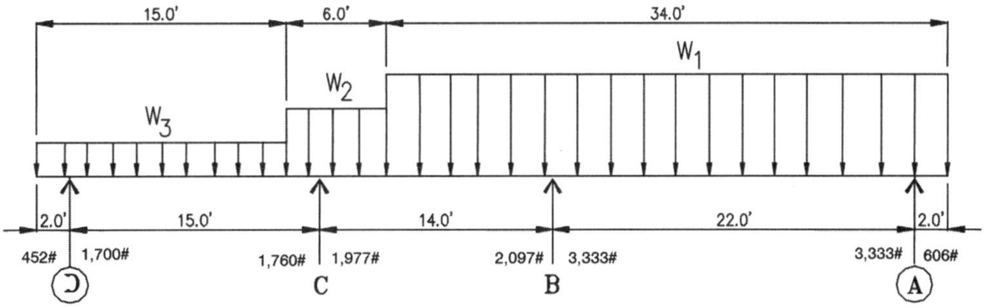

Figure 1A-6. Roof diaphragm loading for east-west forces

Floor diaphragm

Second floor area = 1542 sf

$$f_{p\,floor} = \frac{4045 \text{ lb}}{1542 \text{ sf}} = 2.62 \text{ psf}$$

$$w_4 = (2.62 \text{ psf})(16.0 \text{ ft}) = 42 \text{ plf}$$

$$w_5 = (2.62 \text{ psf})(20.0 \text{ ft}) = 53 \text{ plf}$$

$$w_6 = (2.62 \text{ psf})(33.0 \text{ ft}) = 87 \text{ plf}$$

$$w_7 = (2.62 \text{ psf})(28.0 \text{ ft}) = 74 \text{ plf}$$

$$w_8 = (2.62 \text{ psf})(32.0 \text{ ft}) = 84 \text{ plf}$$

$$P_D = (452 \text{ lb} + 1{,}700 \text{ lb}) = 2152 \text{ plf}$$

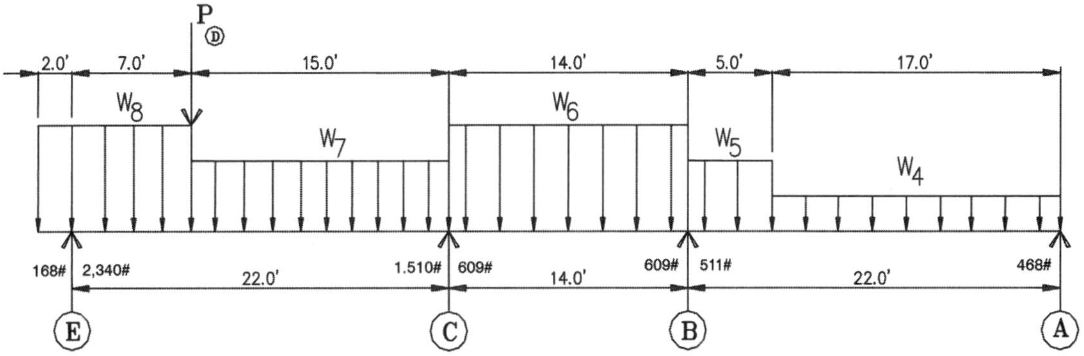

Figure 1A-7. Second floor diaphragm loading for east-west forces

Check sum of forces

$$168 + 2340 + 1510 + 609 + 609 + 511 + 468 = 6215 \text{ lb}$$

Subtract P_D from the sum of forces

$$6215 - 2152 = 4063 \text{ lb}$$

$$V_{floor} = 4063 \text{ lb} \approx 4045 \text{ lb} \ldots o.k.$$

2b. Required edge nailing for east-west shear walls using 10d common nails (T 2306.4.1)

Table 1A-2a. East-west shear walls at roof level (second floor to roof)[1, 2, 3, 4, 5, 6, 7]

Wall (grid line)	ΣF_{above} (lb)	ΣF_x (lb)	F_{tot} (lb)	b (ft)	$v = \frac{F_{tot}(0.7)^{(6)}}{b}$ (plf)	Sheathing[5] 1 or 2 sides	Allowable Shear (plf)	Edge Nail Spacing (in.)
A	0	3,939	3,939	10.0	276[6]	One	340	6[2, 4]
B	0	5,430	5,430	14.0	272[6]	One	340	6[4]
C	0	3,737	3,737	8.5	308[6]	One	340	6[4]
D	0	2,152	2,152	6.0	251[6]	One	340	6[2, 4]
Σ	0	15,258	15,258	38.5				

Notes:

1. In SDC D, E, or F, the 2006 IBC (Table 2306.4.1 footnotes and §2305.3.11) requires 3x nominal thickness stud framing at abutting panel edges and at foundation sill plates when the allowable stress design shear values exceed 350 pounds per foot.
2. Sill bolt washers: For SDC D, E, or F, §2305.3.11 requires a minimum of 3-inch-square by 0.229-inch-thick plate washers to be used for each foundation sill bolt (regardless of allowable shear values in the wall). These changes were a result of the splitting of framing studs and sill plates observed in the Northridge earthquake and in cyclic testing of shear walls. The plate washers are intended to help resist uplift forces on shear walls. Because of vertical displacements of hold-downs, these plate washers are required even if the wall has hold-downs designed to take uplift forces at the wall boundaries. The washer edges shall be parallel/perpendicular to the sill plate. Sill bolt plate washers are not required in SDCs A, B, and C.
3. Section 2305.3.11 includes an exception to the 3x foundation sill plates that allows 2x foundation sill plates when the allowable shear values are less than 600 pounds per foot, provided that sill bolts are designed for 50 percent of allowable values.
4. Refer to Design Example 2 for discussions about fasteners for pressure-preservative-treated wood and the gap at bottom of sheathing.
5. DOC PS-1 or PS-2 (APA or TECO performance) 15⁄32-in. Structural-I rated wood structural panels may be either plywood or oriented strand board (OSB) using 10d common nails with a minimum 1½-inch penetration.
6. Note that forces are strength level, and shear in wall is multiplied by 0.7 to convert to allowable stress design.
7. The 1999 SEAOC Blue Book recommends special inspection when the nail spacing is closer than 4 inches on center.

Table 1A-2b. East-west shear walls at floor level (first floor to second floor)

Wall (grid line)	ΣF_{above} (lb)	ΣF_x (lb)	F_{tot} (lb)	b (ft)	$v = \frac{F_{tot}(0.7)}{b}$ (plf)	Sheathing[5] 1 or 2 sides	Allowable Shear (plf)	Edge Nail Spacing (in.)
A	3,939	468	4,407	10.0	308[2]	One	510	4
B	5,430	1,120	6,550	14.0	328[2]	One	510	4
C	3,737	2,119	5,856	19.0	215[2]	One	510	4
D	2,152	0	0	0	0			
E	0	2,508	2,508	Frame	Frame			
Σ	15,258	6,215	19,321	43.0				

See notes for Table 1-2a.

2c. Forces on north-south shear walls

Roof diaphragm

$$f_{p\ roof} = \frac{14{,}800\text{ lb}}{2164\text{ sf}} = 6.84\text{ psf}$$

$$w_1 = (6.84\text{ psf})(55.0\text{ ft}) = 376\text{ plf}$$

$$w_2 = (6.84\text{ psf})(40.0\text{ ft}) = 274\text{ plf}$$

$$w_3 = (6.84\text{ psf})(34.0\text{ ft}) = 233\text{ plf}$$

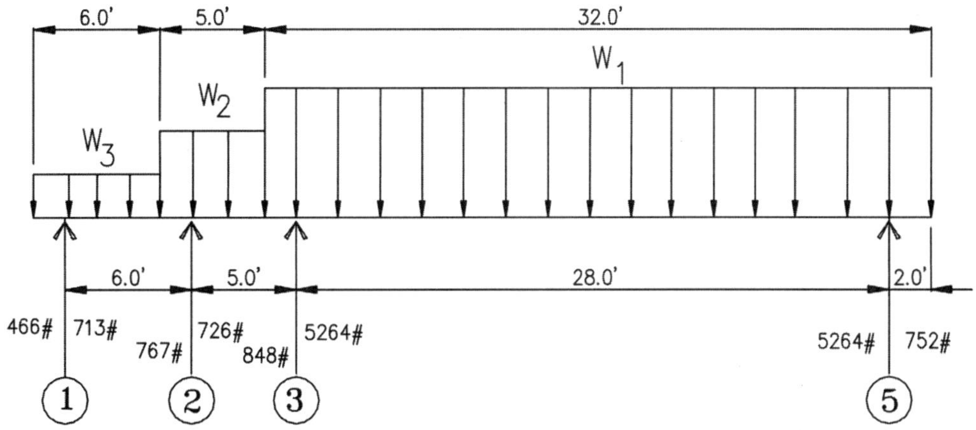

Figure 1A-8. Roof diaphragm loading for north-south forces

Check sum of forces

$$466 + 713 + 767 + 726 + 848 + 5264 + 5264 + 752 = 14{,}800\text{ lb}$$

$$V_{roof} = 14{,}800\text{ lb} \approx 14{,}800\text{ lb} \ldots o.k.$$

Floor diaphragm

$$f_{p\,floor} = \frac{3950 \text{ lb}}{1542 \text{ sf}} = 2.56 \text{ psf}$$

$$w_4 = (2.56 \text{ psf})(9.0 \text{ ft}) = 23 \text{ plf}$$

$$w_5 = (2.56 \text{ psf})(60.0 \text{ ft}) = 154 \text{ plf}$$

$$w_6 = (2.56 \text{ psf})(43.0 \text{ ft}) = 110 \text{ plf}$$

$$w_7 = (2.56 \text{ psf})(38.0 \text{ ft}) = 97.2 \text{ plf}$$

$$w_8 = (2.56 \text{ psf})(23.0 \text{ ft}) = 58.9 \text{ plf}$$

$$w_9 = (2.56 \text{ psf})(14.0 \text{ ft}) = 35.8 \text{ plf}$$

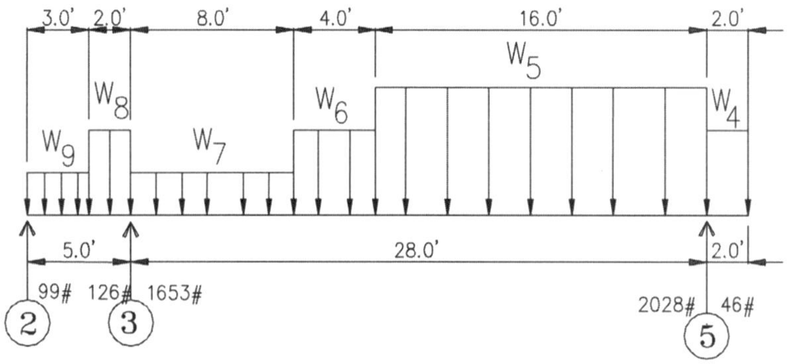

Figure 1A-9. Second floor diaphragm loading for north-south forces

Check sum of forces

$$99 + 126 + 1653 + 2028 + 46 = 3952 \text{ lb}$$

$$V_{floor} = 3952 \text{ lb} \approx 3950 \text{ lb} \ldots o.k.$$

2d. Required edge nailing for north-south shear walls using 10d common nails (T 2306.4.1)

Table 1A-3a. North-south shear walls at roof level (second floor to roof)

Wall	ΣF_{above} (lb)	ΣF_x (lb)	F_{tot} (lb)	b (ft)	$v = \frac{F_{tot}(0.7)}{b}$ (plf)	Sheathing 1 or 2 sides	Allowable Shear (plf)	Edge Nail Spacing (in.)
1	0	1,179	1,179	18.0	46	One	510	4
2	0	1,493	1,493	10.0	105	One	510	4
3	0	6,112	6,112	15.0	285	One	510	4
5	0	6,016	6,016	26.0	162	One	510	4
Σ	0	14,800	14,800	69.0				

Table 1A-3b. North-south shear walls at floor level (first floor to second floor)

Wall	ΣF_{above} (lb)	ΣF_x (lb)	F_{tot} (lb)	b (ft)	$v = \frac{F_{tot}(0.7)}{b}$ (plf)	Sheathing 1 or 2 sides	Allowable Shear (plf)	Edge Nail Spacing (in.)
2	1,493	99	1,592	10.0	111	One	510	4
3	6,112	1,779	7,891	22.0	251	One	510	4
5	6,016	2,074	8,090	14.0	405	One	510	4
Σ	13,621	3,952	17,573	46.0				

3. Redundancy coefficient ρ §12.3.4.2

The reliability/redundancy factor penalizes lateral-force-resisting systems without adequate redundancy. In Part 1 of this example, the reliability/redundancy factor was assumed to be $\rho = 1.0$. This will now be checked.

The method for determining the Redundancy Factor ρ is different in the ASCE/SEI 7-05. The code now requires structures in Seismic Design Categories D, E or F to use a $\rho = 1.3$, unless one of two exceptions is met, in which case $\rho = 1.0$.

The exceptions are:

a) Each story resisting more than 35 percent of the base shear in the direction under consideration complies with Table 12.3-3. For lateral force-resisting elements consisting of shear walls, Table 12.3-3 states that: For shear walls with a height-to-length ratio of greater than 1.0, the removal of that wall would not result in more than a 33 percent reduction in the overall story strength. From Tables 1A-1a and 1A-1b, both levels resist more than 35 percent of the base shear; therefore, determine percentage of reduction of story strength for walls with a height-to-length ratio of more than 1.0.

b) The structure is regular in plan with at least two bays of lateral-load-resisting perimeter framing on each side. For light-framed construction, the number of bays for a shear wall is defined as twice the length of the shear wall divided by the story height.

For east-west direction

Check for condition "a"

Check walls from second floor to roof (see Table 1A-2a)

Story height = 9 ft

The total shear wall length is 10.0 + 14.0 + 8.5 + 6.0 = 38.5 ft

Walls less than 9 feet long are walls C and D.

If wall C is removed, the reduction of story strength is 8.5/38.5 = 0.22 < 0.33 . . . *o.k.*

If wall D is removed, the reduction of story stength is 6.0/38.5 = 0.15 < 0.33 . . . *o.k.*

By inspection of the building plan, the removal of either wall C or D would not produce an extreme torsional irregularity.

Check walls from first floor to second floor (see Table 1A-2b)

Story height = 9 ft

There are no walls less than 9 ft long

Check cantilever columns:

Lateral load to line E is 2508 lb

Lateral load to each cantilever column is 2508/3 = 836 lb

If a single cantilever column is removed, the reduction of story strength is $^{836}/_{19,275} = 0.04$ < 0.33 . . . *o.k.*

By inspection of the building plan, the removal of a single cantilever column would not produce an extreme torsional irregularity.

For the east-west direction, condition "a" of §12.3.4.2 has been met.

Check for condition "b"

Check walls from second floor to roof

Determine number of bays of seismic-force-resisting systems:

Wall	ℓ (ft)	h (ft)	$2\ell/h$
A	10.0	9.0	2.2
B	14.0	9.0	3.1
C	8.5	9.0	1.9
D	6.0	9.0	1.3
Σ			8.5

The code requires at least 2 bays on each side or a total of 4 bays. We have 8.5 bays; therefore we have redundancy.

Check walls from first floor to second floor:

Determine number of bays of seismic-force-resisting systems:

Wall	ℓ	h	$2\ell/h$
A	10.0	9.0	2.2
B	14.0	9.0	3.1
C	19.0	9.0	4.2
Σ			9.5

The code requires at least 2 bays on each side or a total of 4 bays. We have 9.5 bays; therefore we have redundancy.

For the east-west direction, both conditions "a" and "b" have been met. Only one of the two conditions needs to be met; therefore the structure is redundant ($\rho = 1.0$).

For north-south direction

Check for condition "a"

Check walls from second floor to roof (see Table 1A-3a)

Story height = 9 ft

All of the walls are longer than 9 feet; therefore we have redundancy.

Check walls from first floor to second floor (see Table 1A-3b)

Story height = 9 ft

All the walls are longer than 9 feet; therefore we have redundancy.

For the north-south direction, condition "a" of §12.3.4.2 has been met.

Check for condition "b"

Check walls from second floor to roof:

Determine number of bays of seismic-force-resisting systems:

Wall	ℓ (ft)	h (ft)	$2\ell/h$
1	18.0	15.25	2.4
2	10.0	9.0	2.2
3	15.0	9.0	3.3
5	26.0	9.0	5.8
Σ			13.6

The code requires at least 2 bays on each side or a total of 4 bays. We have 13.6 bays; therefore we have redundancy.

Check walls from first floor to second floor:

Determine number of bays of seismic-force-resisting systems:

Wall	ℓ (ft)	h (ft)	$2\ell/h$
2	10.0	9.0	2.2
3	22.0	9.0	4.9
5	14.0	9.0	3.1
Σ			10.1

The code requires at least 2 bays on each side or a total of 4 bays. We have 10.1 bays; therefore we have redundancy.

For the north-south direction, both conditions "a" and "b" have been met. Only one of the two conditions needs to be met; therefore the structure is redundant ($\rho = 1.0$).

4. Diaphragms §12.10

Diaphragm shears will now be checked.

The basic equation for determining seismic forces on a diaphragm is shown below. The following will compute the seismic forces in the east-west direction

$$F_{px} = \frac{\sum_{i=x}^{n} F_i}{\sum_{i=x}^{n} w_i} w_{px} \quad \text{Eq 12.10-1}$$

Note that the forces in the east-west direction are higher.

$$F_{p\,roof} = \frac{(15{,}230 \times 65{,}400)}{65{,}400} = 15{,}230 \text{ lb}$$

$$F_{p\,roof} = \frac{15{,}230 \text{ lb}}{2164 \text{ sf}} = 7.04 \text{ psf}$$

For the uppermost level, the above calculation will always produce the same force as that computed in Equation 12.8-11.

$$F_{p\,floor} = \frac{(15{,}230 + 4045) \times 39{,}900}{(39{,}900 + 65{,}400)} = 7300 \text{ lb}$$

$$F_{p\,min} = 0.2S_{DS}Iw_{px} = 0.2(1.19)(1.0)w_{px} = 0.238(39{,}900) = 9500 \text{ (governs)} \quad \S 12.10.1.1$$

$$F_{p\,max} = 0.4S_{DS}Iw_{px} = 0.4(1.19)(1.0)w_{px} = 0.476(39{,}900) = 19{,}000 \text{ lb}$$

$$F_{p\,floor} = \frac{9500 \text{ lb}}{1542 \text{ sf}} = 6.16 \text{ psf} \quad \S 12.10.1.1$$

The designer is required to check diaphragm forces in each direction. This design example performs the check in one direction only.

4a. Roof diaphragm

Check diaphragm shear.

Based on the $F_{p\,roof} = 7.04$ psf as computed above, find roof shear to line A for the east-west direction:

1. Area of roof including overhangs: 22x43 ft
2. Wall length: 39 ft
3. Diaphragm shears are converted to allowable stress design by multiplying by 0.7

$$v = \frac{(7.04)43.0(22.0)(0.7)}{(39.0)2} = 60 \text{ plf} < 190 \text{ plf allowable}$$

From IBC Table 2306.3.1, the allowable shear of 190 plf is based on 15⁄32-inch APA or TECO Performance-rated wood structural panels (DOC PS1 or PS2) with unblocked edges and 10d nails with a minimum 1½-inch penetration, spaced at 6 inches o/c at boundaries and panel edges. APA or TECO performance-rated wood structural panels may be either plywood or oriented strand board (OSB).

The designer is required to check diaphragm forces in each direction. This design example performs the check in one direction only.

4b. Floor diaphragm

Check diaphragm shear.

Based on the $F_{p\,floor} = 6.16$ psf as computed in Part 4 above, find floor shear to line A for the east-west direction (floor area is 22 × 16).

Diaphragm shears are converted to allowable stress design by multiplying by 0.7

where

$$v = \frac{(6.16\text{ psf})16.0\text{ ft}(22.0\text{ ft})(0.7)}{2(16.0)} = 47\text{ plf} < 190\text{ plf} \qquad \text{(T 2306.3.1)}$$

Allowable shear of 190 plf is based on 15⁄32-inch APA-rated sheathing with unblocked edges and 10d nails with a minimum 1½-inch penetration spaced at 6 inches o/c at boundaries and panel edges supported on framing. APA or TECO performance-rated wood structural panels may be either plywood or OSB.

5. Does residence meet requirements of conventional construction provisions (§2308)

The IBC and the IRC have prescriptive provisions for Type V (light-frame) construction. It used to be quite common for building officials to allow developers, architects, building designers, and homeowners to build structures under similar provisions with other codes *without* any engineering design. The size and style of current single-family residences now being constructed—with vaulted ceilings and large floor openings, tile roofs, and larger window sizes—require that an engineering design be made. Because of misuse of the conventional construction requirements, more stringent limitations on the use of these provisions were placed in the 1994 UBC. These more stringent limitations became the basis of the requirements of the IBC and IRC. Following is an analysis of the construction of the residence proposed in this design example compared with conventional construction requirements and an explanation of why an engineering design is required for both vertical and lateral loads. As engineered design code changes continue to get more restrictive, the gap between the double standard (i.e., conventional construction vs. engineered design) continues to widen.

5a. IBC Provisions

The structure must be checked against the individual requirements of §2308. Additionally, because this structure is in SDC D, it must also be checked against §2308.2.2. Results of these checks are shown below.

Roof total loads

Dead load of roof exceeds the 15-psf limit (§2308.2, Item 3)

Number of stories

Conventional light-frame construction shall not exceed one story in height in Seismic Design Category D or E (§2308.12.1)

Unusually shaped buildings

Exterior braced wall panels at line D over the garage are horizontally offset from the bracing systems at the floor below by more than 4 feet and, therefore, not in one vertical plane. (§2308.9.3)

Floor opening exceeds 50 percent of the least floor dimension at line A. (§2308.12.6, Item 6)

Floor is not laterally supported by braced wall lines on all edges. (§2308.12.6, Item 2)

Cantilever column bracing at the garage door does not conform to prescribed methods. (§2308.9.3)

Stud height exceeds 10 feet without lateral support at line 1. (§2308.2, Item 2)

Braced wall lines on the second floor extend more than 1 foot over an opening below (F 2308.9.3)

Braced wall lines

Spacing between braced wall lines 3 and 5 exceeds 25-foot maximum for both levels. (F 2308.9.3)

∴ The residence cannot be designed using the conventional construction provisions of the IBC.

5b. IRC Provisions

The structure must be checked against the individual requirements of IRC §R301.2.2. Additionally, because this structure is in Seismic Category D_2, it must also be checked against IRC §R301.2.2.2. Results of these checks are shown below.

Roof total loads

Dead load of roof exceeds the 15-psf limit; therefore, wall bracing adjustment factors need to be applied to the minimum lengths. Adjustment factor (from interpolation) is 1.1 IRC T R301.2.2.2.1

Irregular buildings

Exterior braced wall panels at line D over the garage are horizontally offset from the bracing systems at the floor below and are not in one vertical plane; therefore, it is classified as an irregular building. IRC §R301.2.2.2.2.1

Floor openings exceed 12 feet and 50 percent of the least floor dimension at line A. IRC §R301.2.2.2.2.4

Floor is not laterally supported by braced wall lines on all edges. IRC §R301.2.2.2.2.2

Cantilever column bracing at the garage door does not conform to prescribed methods. IRC §R301.2.2.2.2

Braced wall lines on the second floor extend more than 1 foot over an opening below. IRC §R301.2.2.2.2.3

Braced wall lines

Minimum individual panel length is less than 4 feet on second floor at line D. IRC §R602.10.4

Braced wall not located within 12.5 feet from each end of walls at lines A and D for top story. IRC §R602.10.1

Braced wall not located at each end of wall at lines A and 5, and is less than 55 percent of the wall length for the first story IRC §R602.10.1

Braced wall lines 3 and 5 exceed 25 feet o/c for both levels IRC §R602.10.11

Since the residence is not in conformance with conventional light-frame construction provisions of IBC §2308, the exemption from the seismic provisions of §1613 does not apply

∴ The residence cannot be designed using the conventional construction provisions of the IRC.

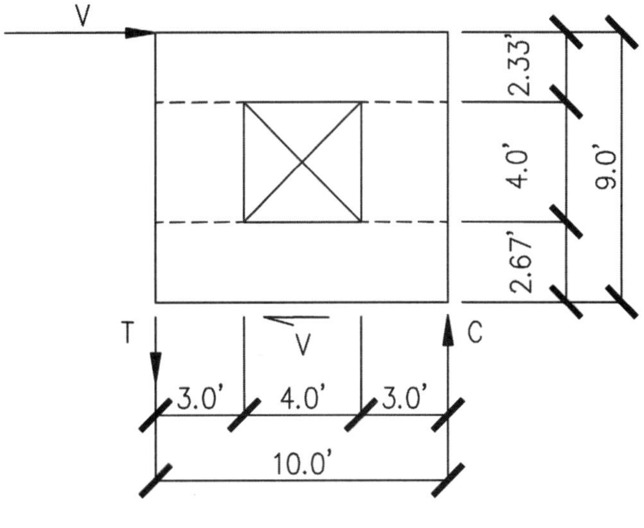

Figure 1A-10. Elevation of wall frame on line D

6. Design shear wall over garage on line D

$V = 2152$ lb (from Table 1A-2a)

Converting to allowable stress design for the wall frame

$V = 2152(0.7) = 1506$ lb (refer to Figure 1A-10)

The following sections illustrate three different methods for designing the shear wall at Line D. Any one of the three methods will meet the intent of the code.

6a. Design of shear walls

This method is the conventional approach that designs the individual wall piers as independent shear walls. (§2305.3.4)

Determine h/w aspect ratios for the shear walls

$$h/w = 9.0/3.0 = 3.0$$

Maximum $h/w = 2.0$ for seismic forces unless allowable shear values in IBC Table 2306.4.1 are multiplied by $2w/h$. (§2305.3.4)

Adjusted allowable shear (2(3.0)/9.0) (340 plf) = 225 plf. Since the actual shear in the wall is 251 plf (Table 1A-2a), the nailing needs to be increased from 6 inches o/c to 4 inches o/c. The new adjusted allowable shear (2(3.0)/9.0) (510) = 340 plf.

The SDPWS-05 has similar language (SDPWS-05 Table 4.3.4) when the h/w ratio exceeds 2.0. Instead of multiplying the allowable shears by $2w/h$, the nominal unit shear capacities are multiplied by $2b_s/h$. Note that these shear wall segments are discontinuous; therefore, design anchorage per §12.3.3.3. Determine anchorage at shear wall boundaries.

The overturning moment for each full-height shear wall segment is

$$M_{ot} = \frac{V}{2}(h) = \Omega_O \frac{2152}{2}(9.0) = 29{,}050 \text{ ft-lb} \quad \text{(strength level)}$$

where $\Omega_O = 3.0$

Because walls are narrow and only have normal tributary roof loads, the dead load of the wall will only be used for determining the shear wall segment resisting moment. For Allowable Stress Design, the critical loading conditions are:

for downward

$$(1.0 + 0.105S_{DS})D + 0.525\Omega_O Q_E + 0.75L_r$$

for upward

$$(0.6 - 0.14S_{DS})D + 0.7\Omega_O Q_E$$

where $(0.6 - 0.14S_{DS})D = (0.6 - 0.14 \times 1.19)D = 0.43D$

$$M_R = (15 \text{ psf} \times 9.0 \text{ ft})\ 3.0^2/2 = 610 \text{ ft-lb}$$

With a 4x6 post at each end of the shear wall segments, the length of the wall $L = 3.0 - \frac{3.5}{12} = 2.7$ ft

$$\text{Uplift at ends} = \frac{(29{,}050)0.7 - (610)0.43}{2.7 \text{ ft}} = 7435 \text{ lb}$$

Consult ICC Evaluation Reports for the allowable load capacity of premanufactured straps.

Allowable load per 10d common nail with 14-gage metal side plate = 119 lb NDS-05 T 11P

From part 6B with 3-inch nails, penetration factor $C_d = 1.0$

For allowable stress design, the allowable stress increase factor is 1.2 §12.4.3.3

$$\text{Number of 10d common nails required} = \frac{7435}{119\text{ lb / nail}(1.2)(1.6)} = 33\text{ nails}$$

where $C_D = 1.6$ for duration of load.

6b. Design of wall frame (with force transfer around opening)

It is possible to get the mistaken impression from IBC Figure 2305.3.5 that all a designer needs to do to reduce the *h*/*w* ratio is add some blocking and straps. This design example has a structure with 9-foot plate heights, which makes using a wall frame feasible. However, when the plate height is 8 feet, which is a more common occurrence, there are chord development and panel nailing capacity problems. Most often, the wall shears above and below the opening will be higher than in the wall piers. This design example analyzes the wall frame and neglects gravity loads, although from a technically correct standpoint, some engineers will argue that vertical loads need to be considered when determining wall shears. The standard practice of neglecting gravity loads when considering wall shears is deemed appropriate.

Using statics, determine the shears and forces in each free body panel. This is a two-step procedure as follows:

First: Find forces acting on upper left corner of wall frame (Figure 1A-11).

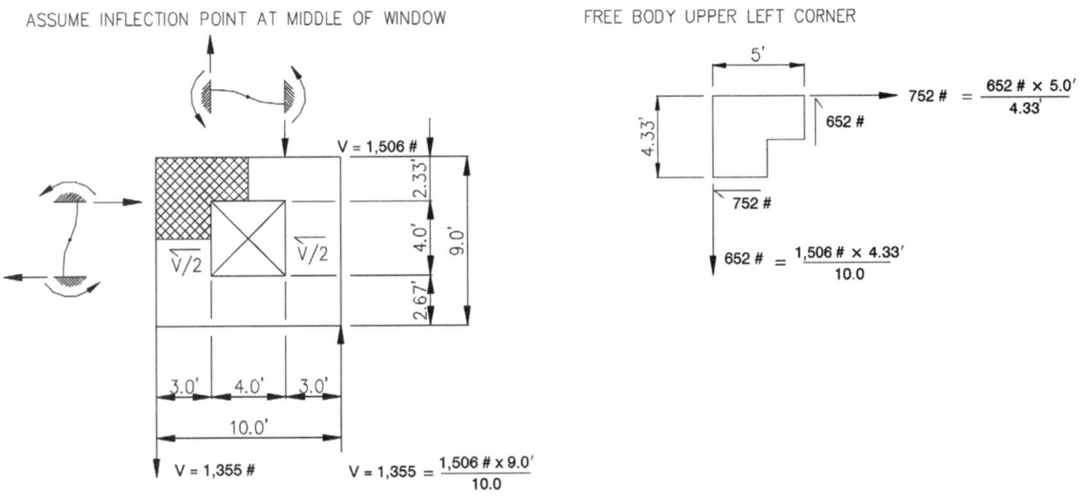

Figure 1A-11. Wall frame elevation at line D

Second: Break up wall frame into free-body panel sections and balance forces for each panel starting with upper left corner forces already determined (Figure 1A-12).

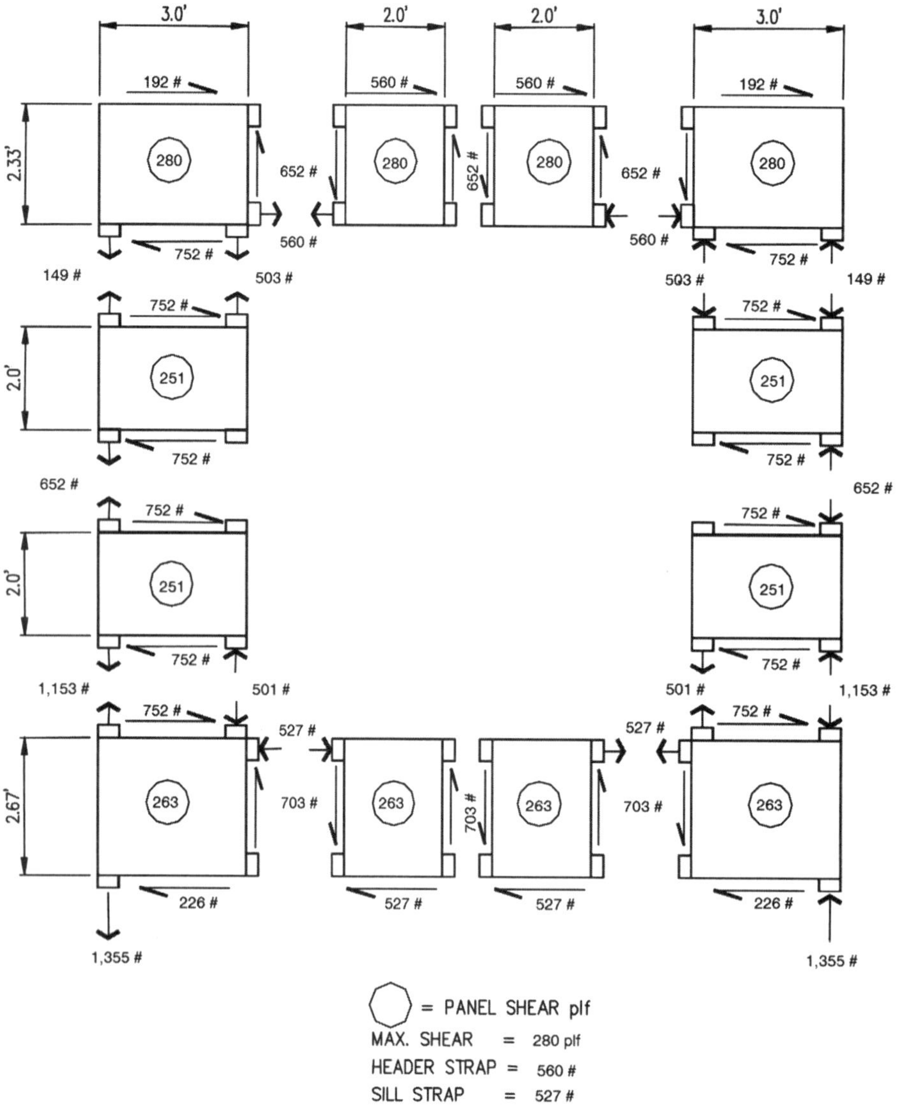

Figure 1A-12. Free-body individual panels of wall on line D

Many engineers will arbitrarily add tiedowns at the window jamb members (Figure 1A-13). However, with this type of design, the tiedowns at these locations are not necessary, but shear stresses above and below the window may become higher. Adding tiedowns at the window jambs would increase the wall frame performance and help prevent sill plate uplift at the window jambs, which occurs (to some degree) when they are not provided.

Design horizontal tie straps above and below windows (see Figure 1A-13)

Determine the tie force for the horizontal strap (from Figure 1A-12). Tie force is maximum at the header beam.

$F_{tie} = 560$ lb

Consult ICC Evaluation Reports for the allowable load capacity of premanufactured straps.

Check penetration depth factor

C_d : for 10d nail through 16-gage strap and ½-inch sheathing

penetration = 3.0 – 0.060 – 0.5 = 2.4 in

Required penetration for full value = $10D = 10 \times 0.148 = 1.5 < 2.4$ in . . . *o.k.*

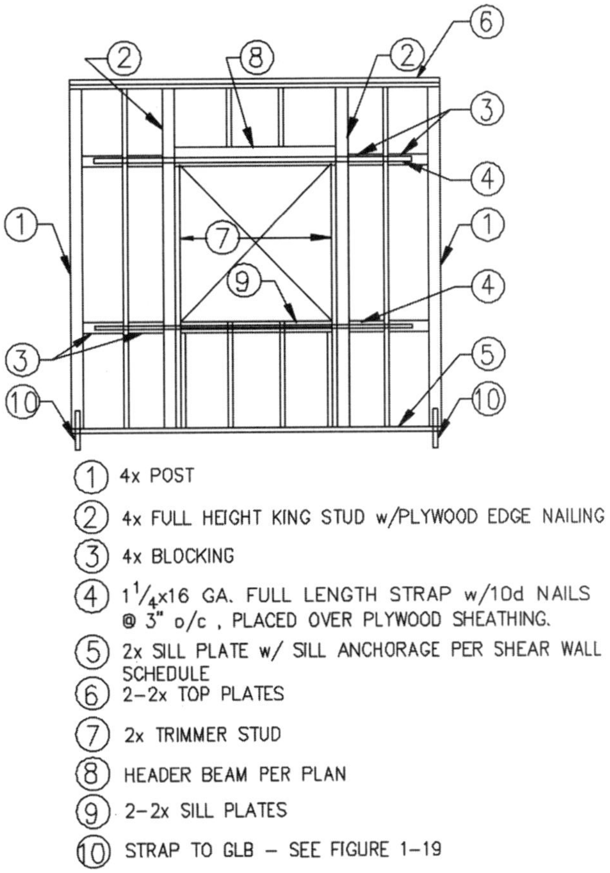

Figure 1A-13. Details of wall frame on line D at second floor

Allowable load per 10d common nail with 16-gage metal side plate = 115 lb — NDS-05 T 11P

$$\text{Number of 10d nails required each end} = \frac{560 \text{ lb}}{115 \text{ lb / nail} \times 1.6} = 3 \text{ nails}$$

(nailing does not control)

Use a continuous 16-gage by 1¼-inch strap across the opening head and sill to blocking.

Allowable strap load is (1.25)0.06 (0.6 × 33,000) = 1485 lb > 560 lb . . . *o.k.*
Note that an allowable stress increase has not been used for the metal strap.

Load combinations using allowable stress design §2.4.1

The basic load combinations of §2.4.1 do not permit stress increases unless it can be justified by structural behavior caused by rate or duration of load. Wood is the only material that has justification for duration of load.

Check shear panel nailing in wall frame

From Figure 1A-12:

Maximum panel shear = 280 plf

6-inch edge nailing with sheathing on one side . . . *o.k.*

v allowable = 340 plf UBC T 2306.4.1

Determine anchorage of wall to the supporting GLB

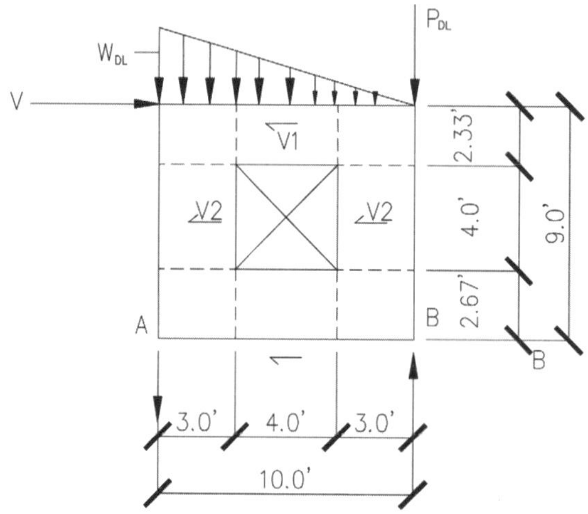

Figure 1A-14. Wall frame elevation at D at second floor

From Figure 1A-14

w_{DL} = 100 plf (triangle loading from hip roof)

P_{DL} = 700 lb

$Wall_{DL}$ = 1100 lb

Q_E = V = 2152lb

M_{ot} = 2152 lb (9 ft) = 19,368 ft-lb (strength level)

Elements supporting discontinuous systems §12.3.3.3

Because location A does not continue to the foundation, check special seismic load combination for elements supporting discontinuous systems.

For Strength Design

$$(1.2 + 0.2S_{DS})D + \Omega_O Q_E + L$$

$$(0.9 - 0.2S_{DS})\ D + \Omega_O Q_E$$

where

Q_E = The effect of horizontal seismic forces

Determine the seismic force overstrength factor Ω_O

$\Omega_O = 3.0$ for wood structural panel wall T 12.2-1

Determine anchorage at A

$$M_R = 100 \text{ plf } (10.0 \text{ ft}/2)(10.0 \times 2/3) + 1100(10.0 \text{ ft}/2) + 700 \text{ lb } (0 \text{ ft}) = 8833 \text{ ft-lb}$$

With a 4x6 post at each end wall, $L = 10.0 - \dfrac{3.5 \text{ in}}{12} = 9.7 \text{ ft}$

The critical loading condition is $(0.9 - 0.2S_{DS})D + \Omega_O E$

where

$$(0.9 - 0.2 \times 1.19)D = 0.66D$$

$$\text{Uplift at A} = \frac{(19{,}368) - (8833 \times 0.66)}{9.7 \text{ ft}} = 1400 \text{ lb (strength level)}$$

Determine anchorage at B

$$M_R = 100 \text{ plf } (10.0 \text{ ft}/2)(10.0/3) + 1100(10.0 \text{ ft}/2) + 700 \text{ lb } (10.0 \text{ ft}) = 14{,}167 \text{ ft-lb}$$

$$\text{Uplift at B} = \frac{(19{,}368) - (14{,}167 \times 0.66)}{9.7 \text{ ft}} = 1035 \text{ lb (strength level)}$$

For east-west axis of structure, $R = 1.25$ for cantilevered column systems with ordinary steel moment frames. T 12.2-1

Therefore, $\Omega_O = 1.25$

As discussed earlier in this design example, the three exception conditions of §12.2.3.2 have been met and the R of 6.5 for the building as a whole will be used.

For Allowable Stress Design

$$(1.0 + 0.10S_{DS})D + 0.525\ \Omega_O Q_E + 0.75\ (L_r)$$

$$(0.6 - 0.14S_{DS})D + 0.7\ \Omega_O Q_E$$

$$M_{ot} = 2{,}152 \text{ lb} \times 9.0 \text{ ft} \times \Omega_O = 58{,}100 \text{ ft lb}$$

Determine anchorage at A

The critical loading condition is $(0.6 - 0.14S_{DS})D + 0.7\ \Omega_O Q_E$

where $(0.6 - 0.14S_{DS})D = (0.6 - 0.14 \times 1.19)\ D = 0.43\ D$

$M_R = 8833$ ft-lb

$$\text{Uplift at A} = \frac{(58{,}100)0.7 - (8833)0.43}{9.7 \text{ ft}} = 3800 \text{ lb}$$

Determine anchorage at B

$M_R = 14{,}167$ ft-lb

$$\text{Uplift at B} = \frac{(58{,}100)0.7 - (14{,}167)0.43}{9.7 \text{ ft}} = 3600 \text{ lb}$$

Consult ICC Evaluation Reports for the allowable load capacity of premanufactured straps.

Allowable load per 10d common nail with 14-gage metal side plate = 119 lb NDS-05 T 11P

From Part 6b, with 3-inch nails penetration factor $C_d = 1.0$

For allowable stress design, the allowable stress increase factor is 1.2 §12.4.3.3

$$\text{Number of 10d common nails required} = \frac{3800 \text{ lb}}{119 \text{ lb / nail}(1.2)(1.6)} = 17 \text{ nails}$$

where $C_D = 1.6$ for duration of load

Use a continuous 14-gage by 3-inch strap bent around GLB on one side.

It is not clear to which elements the code requires the E_m force applied. The SEAOC Seismology Committee is proposing a code change to include only compression elements.

Note that §12.4.3.3 allows the combination of allowable stress increase of 1.2 with the duration of load increase in Chapter 23.

Note too that the adequacy of the GLB to resist the overturning of the wall must be checked using the special seismic load combinations. As permitted in §12.4.3.3, an allowable stress increase of 1.2 can be used in addition to the duration of load increase of 1.6 for C_D.

Also, the boundary posts at the ends and the connections of the GLB supporting the wall must be checked for the special seismic load combination. §12.3.3.3

6c. Design perforated shear wall

Strength $V = 2152$ lb (from Table 1A-2a)

Wall height $(h) = 9$ ft

Maximum opening height ratio and height

Ratio: (6.67/9.0) = 0.74

Height = 6 ft 8 inches

Percent full-height sheathing $\frac{(3.0+3.0)}{10.0} \times 100 = 60\%$

Determine adjustment factor C_o:
From IBC Table 2305.3.8.2 for 60 percent of full-height sheathing and maximum opening height ratio of 0.74 (by interpolation) the adjustment factor $C_o = 0.67$.

Anchorage for in-plane shear (§2305.3.8.2.5)

$$v = \frac{V}{C_o \Sigma L_i} \quad \text{(Eq 23-4)}$$

where ΣL_i = sum of width of perforated shear wall segments

For allowable stress design

$$v = \frac{V \times 0.7}{C_o \Sigma L_i} = \frac{2152 \times 0.7}{0.67(3.0+3.0)} = 375 \text{ plf}$$

Determine height-to-width ratios of perforated shear wall segments

$$\frac{h}{w} = \frac{9.0}{3.0} = 3.0 > 2.0$$

therefore the unadjusted shear resistance values must be multiplied by $2\frac{w}{h} = \frac{2 \times 3.0}{9.0} = 0.67$
using 15⁄32-inch structural I sheathing, the allowable shear for 10d common nails spaced at 3 inches o/c is 665 plf. The adjusted allowable is $665 \times 0.67 = 447$ plf > 375 plf . . . *o.k.*

The increased nailing requirement is to account for unequal load distribution in the perforated shear wall. Note the only 6-inch nail spacing was required for the force transfer around openings methods.

Uplift anchorage at ends of perforated shear wall (§2305.3.8.2.4)

$$T = \frac{Vh}{C_o \Sigma L_i} \quad \text{(Eq 23-3)}$$

where C_o = shear resistance adjustment factor from IBC Table 2305.3.8.2

L_i = sum of widths of perforated shear wall segments (feet)

Because this perforated shear wall is an element of an out-of-plane offset, the connections for the uplift anchorage need to have the strength to resist load combinations with the overstrength factor in §12.4.3.2 of the ASCE Standard. The above equation is empirical and is not related to any vertical load combinations on the wall. Increase horizontal load by Ω_O factor of 3.0.

For allowable stress design:

$$T = \frac{\Omega_o (V \times 0.7)h}{C_o \Sigma L_i} = \frac{3.0(2152 \times 0.7)9.0}{0.67(3.0+3.0)} = 10{,}100 \text{ lb}$$

Note that this uplift force is significantly higher than the other two methods of designing the shear wall.

Uplift anchorage between perforated shear wall ends (§2305.3.8.2.6)

In addition to the uplift anchorage at the ends of the wall, the full-height studs shall be anchored for a uniform anchorage force, *t*, equal to the horizontal anchorage force for in-plane shear previously calculated.

$$t = v = 375 \text{ plf}$$

$$t/\text{stud} = 375\ (16/12) = 500 \text{ lb}$$

7. Diaphragm shears at the low roof over garage

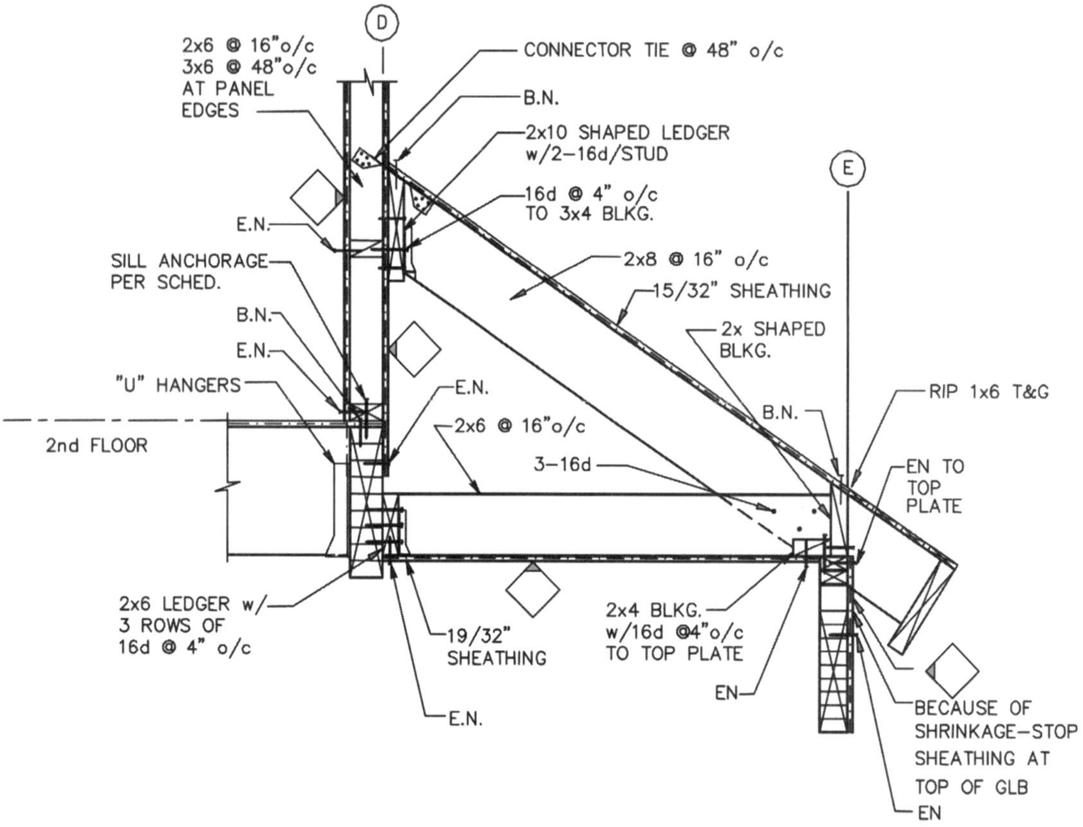

Figure 1A-15. Detail of load path for low roof over garage

Note that Figure 1A-15 depicts sheathing on both sides of the shear wall. In this design example, sheathing is only required on one side of the wall.

From Table 12.3-1, this has horizontal irregularity type 4 and from Table 12.3-2 this has vertical irregularity type 4.

The diaphragm between lateral resisting elements C and E is required to transfer the design seismic force from shear wall D because of the offset between D and E. Section 12.3.3.4 requires the diaphragm connections used in Equation 12.8-1 to be increased 25 percent unless the special load combination including overstrength factor is used.

From Part 4b in this design example

$$f_{p\,floor} = 6.16 \text{ psf}$$

From Table 12.2-1, Ω_O for cantilever column type structures is 1.25 and Ω_O is 3.0 for light framed walls sheathed with wood structural panels. The critical Ω_O factor for the diaphragm is 3.0.

$$f_p\Omega_O = 6.16 \times 3.0 = 18.5 \text{ psf}$$

For simplification of analysis, assume the diaphragm over the garage is a simple span between lateral resisting elements at lines C and E.

Load from wall D above = 2152 lb (strength level)

$$V_E = 18.5(28.0 \text{ ft})(22.0/2) + 2152 \text{ lb } (15.0 \text{ ft}/22.0 \text{ ft}) = 7165 \text{ lb}$$

$$v_E = 7165 \text{ lb}/(28.0) = 260 \text{ plf} < 430 \times 0.8$$ SDPWS T 4.2B

Increasing shears 25 percent

$$v_E = 260 \times 1.25 = 325 \text{ plf}$$

where $\phi = 0.8$

Therefore, panel edges need not be blocked.

8. Detail the wall frame over the GLB

Wall frame details must be shown on the drawings. Depending on the variations, when multiple wall frames are on a project, it is occasionally necessary to have individual details for each condition. While the detail shown in Figure 1A-16 is somewhat generic, it should be noted that a separate anchorage detail may be necessary where the end of the GLB is connected to the supporting post.

9. Detail the anchorage of wall frame to the GLB

Cross-grain shrinkage of the GLB may be a problem when using a connection of the type shown in Figure 1A-16. Also, nails above the neutral axis of the GLB should be left out of the design to avoid cross-grain tension. In other words, only the nails below the neutral axis are considered effective for uplift forces. To avoid confusion in the field, all nail holes are to be filled. It should be noted that a separate anchorage detail may be necessary where the end of the GLB is connected to the supporting post (intersection of grids D and 5).

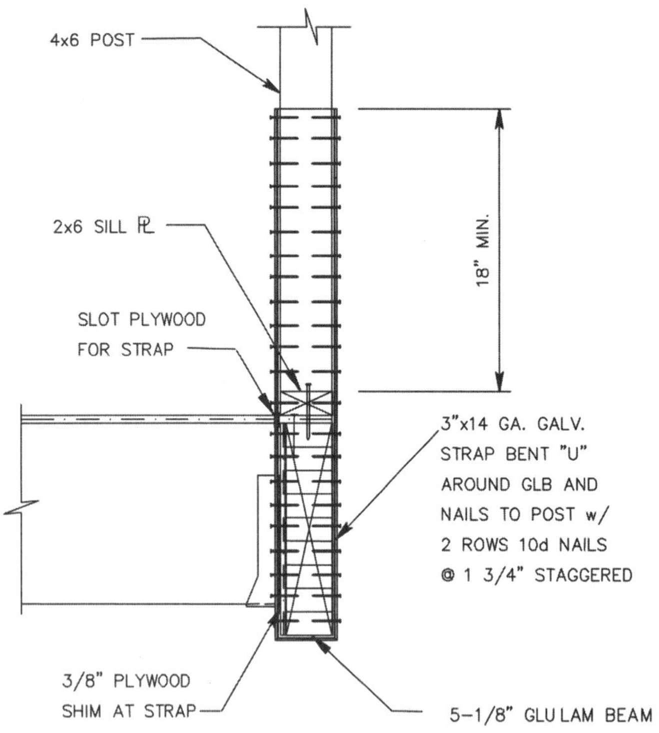

Figure 1A-16. Detail of anchorage at point A (see also Figure 1A-14)

10. Detail the continuous load path at the low roof above the garage doors

The low roof above the garage is an important part of the continuous load path. Historically, this type of detail has been mis-detailed and mis-constructed. This detail has two load paths: the loads from the roof can go either through the pitched roof or down the wall to the GLB and across the horizontal diaphragm to the exterior wall.

Figure 1A-15 shows one way that the shear transfer can be made. Also note that the chord/drag tie of the top plates will be interrupted by the GLB-to-post connection and will require detailing at grids D3 and D5.

Commentary

Following are some issues and topics related to the seismic design of wood frame residences that can be used to improve design practices and/or understanding of important aspects of design.

"Calc and sketch" philosophy.

In wood frame construction, particularly for single-family residences, it has been a common design practice to have an engineer provide only calculations and sketches for the architect to include on the architectural drawings. This is done to provide a cost savings to the owner. This approach has some significant problems based on reviews of how residential framing is actually being constructed; the "calc and sketch only" service is a practice that should be discontinued, with a few exceptions.

Architects and building officials need to be encouraged to adopt the following standards:

1. Any new building (or remodel requiring the existing building to be brought into conformance with the current building code) that cannot be clearly shown to conform with building code conventional construction framing requirements should require submittal of structural drawings and calculations signed for by a licensed professional, civil, or structural engineer.

2. Structural framing plans and details should be separate from the architectural drawings.

   Most new wood residential building designs are complex and beyond the scope and intent of the prescriptive conventional construction requirements of the IBC and IRC. Misuse of these conventional requirements has led to structures with incomplete lateral force systems, resulting in poor performance during earthquakes. Since the engineer generally is not asked to review the architect's final drawings, the use of calculations and sketches lends itself to poorly coordinated drawings and missing structural information. The common practice of referring to details on architectural drawings as "similar" leads to further confusion as to the design intent. The structural observation requirements of the code, when enforced (many jurisdictions do not require structural observation for single-family residences), are even less effective, since the architect did not design the structural system and often can not identify what is missing or incorrect.

Tiedown location.

When designing shear walls, the engineer needs to consider where the tiedown posts will actually be located. The tiedown posts occur where shear walls stack from floor to floor.

The lower level wall requires tiedown devices on each side of the tiedown post. However, the upper shear wall only requires a tiedown device on one side of the tiedown post. Since the posts must align between story levels, the upper level tiedown post will need to be offset inward so as to line up with the post below.

Based on actual tiedown post locations, the upper level shear wall design may have to be rechecked once the lower level shear wall design is complete. The use of tiedown devices on each side of the post will improve the shear wall performance, since eccentricity in the connection, as occurs when there is only a single-sided tiedown, is avoided. Double-sided tiedowns are generally preferred over single-sided.

Design comments.

This design example illustrates a detailed analysis for some of the important seismic requirements of the ASCE/SEI 7-05 and 2006 IBC. To complete this design, the engineer will have to check all the major structural elements along the various lateral load paths of the residence, including the foundations. The seismic calculations and details for this example residence are approximately 50-percent complete. Normal engineering design of this type of structure may omit many of the calculations shown in this example and rely on good engineering judgment. This design example illustrates a comprehensive approach to the engineering calculations, and fills a void in the existing engineering literature—many engineers have stated that there simply are not sufficient reference documents available on this particular subject.

In the so called "big one," it is expected that actual peak earthquake forces may be 2 to 3 times greater than the equivalent static forces required by the ASCE/SEI 7-05 and used in this example. The use of good detailing practices with ductile elements to absorb energy, clear construction documents with adequate detailing, structural site observation, and special inspection are considered every bit as important as a comprehensive set of structural calculations.

References

American Forest and Paper Association, 2005, AF&PA Supplement Special Design Provisions for Wind and Seismic (SDPWS-05) American Forest and Paper Association, Washington, D.C.

American Forest and Paper Association, 2005, *National Design Specification for Wood Construction* including Supplements, NDS-05. American Forest and Paper Association, Washington D.C.

American Forest and Paper Association, 1996, *Wood Construction Manual.* American Forest and Paper Association, Washington D.C.

American Plywood Association, 1997, *Design/ Construction Guide—Diaphragms and Shear Walls.* Form L350, Engineered Wood Association, Tacoma, Washington.

American Plywood Association, 1997, *Diaphragms and Shear Walls.* Engineered Wood Association, Tacoma, Washington.

American Plywood Association, 1993, revised, *Wood Structural Panel Shear Walls.* Report 154, Engineered Wood Association, Tacoma, Washington.

American Plywood Association, 1994, *Northridge, California Earthquake.* Report T-94-5. Engineered Wood Association, Tacoma, Washington.

American Plywood Association, *Performance Standards and Policies for Structural-Use Panels [Sheathing Standard, Sec. 2.3.31].* Standard PRP-108. Engineered Wood Association, Tacoma, Washington.

American Plywood Association, 1997, *Plywood Design Specifications,* Form Y510, Engineered Wood Association, Tacoma, Washington.

American Plywood Association, 1990, *Plywood Diaphragms,* Research Report 138. American Plywood Association, Tacoma, Washington.

American Plywood Association, 2005, Using Narrow Pieces of Wood Structural Panel Sheathing in Wood Shear Walls, APA T2005-08, The Engineered Wood Association, Tacoma, Washington.

Applied Technology Council, 1995, *Cyclic Testing of Narrow Plywood Shear Walls ATC R-1.* Applied Technology Council, Redwood City, California.

Applied Technology Council, 1981, *Guidelines for Design of Horizontal Wood Diaphragms,* ATC-7. Applied Technology Council, Redwood City, California.

Applied Technology Council, 1980, *Proceedings of a Workshop on Design of Horizontal Wood Diaphragms,* ATC-7-1. Applied Technology Council, Redwood City, California.

Building Seismic Safety Council, 1997, *National Earthquake Hazard Reduction Program, Recommended Provisions for Seismic Regulations for New Buildings.* Building Seismic Safety Council, Washington D.C.

Breyer, Donald E., Kenneth J. Fridley, David G. Pollock, Jr. and Kelly E. Cobeen, 2007, *Design of Wood Structures ASD.* McGraw-Hill Book Co., New York, New York

Bugni, David A., 1999, "A Linear Elastic Dynamic Analysis of a Timber Framed Structure." *Building Standards,* International Conference of Building Officials, Whittier, California

Cobeen, K. E., 1996, "Performance Based Design of Wood Structures." *Proceeding: Annual SEAOC Convention.* Structural Engineers Association of California, Sacramento, California.

Coil, J., 1999, "Seismic Retrofit of an Existing Multi-Story Wood Frame Structure," *Proceedings: Annual SEAOC Convention.* Structural Engineers Association of California, Sacramento, California.

Commins, A. and Gregg, R., 1996, *Effect of Hold Downs and Stud-Frame Systems on the Cyclic Behavior of Wood Shear Walls,* Simpson Strong-Tie Co., Pleasanton, California.

Countryman, D., and Col Benson, 1954, *1954 Horizontal Plywood Diaphragm Tests.* Laboratory Report 63, Douglas Fir Plywood Association, Tacoma Washington.

CUREe, 1999, *Proceedings of the Workshop on Seismic Testing, Analysis, and Design of Wood Frame Construction.* California Universities for Research in Earthquake Engineering.

Dolan, J. D., 1996, *Experimental Results from Cyclic Racking Tests of Wood Shear Walls with Openings.* Timber Engineering Report No. TE-1996-001. Virginia Polytechnic Institute and State University, Blacksburg, Virginia.

Dolan, J. D. and Heine, C. P., 1997a, *Monotonic Tests of Wood Frame Shear Walls with Various Openings and Base Restraint Configurations.* Timber Engineering Report No. TE-1997-001, Virginia Polytechnic Institute and State University, Blacksburg, Virginia.

Dolan, J. D. and Heine, C. P., 1997b, *Sequential Phased Displacement Cyclic Tests of Wood Frame Shear Walls with Various Openings and Base Restrain Configurations.* Timber Engineering Report No. TE-1997-002, Virginia Polytechnic Institute and State University, Blacksburg, Virginia.

Dolan, J. D., and Heine, C. P., 1997c, *Sequential Phased Displacement Test of Wood Frame Shear Walls with Corners.* Timber Engineering Report No. TE-1997-003, Virginia Polytechnic Institute and State University, Blacksburg, Virginia.

Earthquake Engineering Research Institute, 1996, "Reconnaissance Report: Northridge Earthquake of January 17, 1994," *Earthquake Spectra.* Vol. 11, Supplement C. Earthquake Engineering Research Institute, Oakland, California.

Faherty, Keith F., and Williamson, Thomas G., 1995, *Wood Engineering Construction Handbook.* McGraw Hill, Washington D.C.

Federal Emergency Management Agency, 2003, *National Earthquake Hazard Reduction Program, Recommended Provisions for Seismic Regulations for New Buildings and Other Structures and Commentary.* Federal Emergency Management Agency, Washington D.C.

Ficcadenti, S. K., T. A. Castle, D. A. Sandercock, and R. K. Kazanjy, 1996, "Laboratory Testing to Investigate Pneumatically Driven Box Nails for the Edge Nailing of 3/8″ Plywood Shear Walls," *Proceedings: Annual SEAOC Convention.* Structural Engineers Association of California, Sacramento, California.

Foliente, Greg C., 1994,*Analysis, Design and Testing of Timber Structures Under Seismic Loads.* University of California Forest Products Laboratory, Richmond, California.

Foliente, Greg C., 1997, *Earthquake Performance and Safety of Timber Structures.* Forest Products Society, Madison Wisconsin.

Forest Products Laboratory, 1999, *Wood HandbookPublication FPL-GTR-113.* Madison, Wisconsin.

Goers R. and Associates, 1976, *A Methodology for Seismic Design and Construction of Single-Family Dwellings.* Applied Technology Council, Redwood City, California.

Ju, S. and Lin, M., 1999, "Comparison of Building Analysis Assuming Rigid or Flexible Floors," *Journal of Structural Engineering.* American Society of Civil Engineers, Washington, D.C.

Mendes, S., 1987, "Rigid versus Flexible: Inappropriate Assumptions Can Cause Shear Wall Failures!" *Proceedings: Annual SEAOC Convention.* Structural Engineers Association of California, Sacramento, California.

Mendes, S., 1995, "Lessons Learned From Four Earthquake Damaged Multi-Story Type V Structures," *Proceedings: Annual SEAOC Convention.* Structural Engineers Association of California, Sacramento, California.

Rose, J. D., 1998, *Preliminary Testing of Wood Structural Panel Shear Walls Under Cyclic (Reversed) Loading.* Research Report 158, APA-Engineered Wood Association, Tacoma, Washington.

Rose, J. D., and E. L. Keith, 1996, *Wood Structural Panel Shear Walls with Gypsum Wallboard and Window [Sheathing Standard, Sec. 2.3.3].* Research Report 158. APA-The Engineered Wood Association, Tacoma Washington.

SEAOC, 1999, *Acceptable Diaphragm-Rigidity Assumptions for Distribution of Horizontal Forces in Light-Frame Construction.* Structural Engineers Association of California, Sacramento, California.

SEAOC, 1999, *Cantilever Column Opinion.* Structural Engineers Association of California, Sacramento, California.

SEAOC, 1997, *Seismic Detailing Examples for Engineered Light Frame Timber Construction.* Structural Engineers Association of California, Sacramento, California.

SEAOC, 1999, *Guidelines for Diaphragms and Shear Walls.* Structural Engineers Association of California, Sacramento, California.

SEAOC, 1999, *Plan Review-Codes and Practice.* Structural Engineers Association of California, Sacramento, California.

Shipp, J., and D. Thompson, 2001, *Timber Design I, II, and III.* Volumes VIII, IX, and X, Professional Engineering Development Publications, Inc., Irvine, California.

Steinbrugge, J., 1994, "Standard of Care in Structural Engineering Wood Frame Multiple Housing," *Proceedings: Annual SEAOC Convention.* Structural Engineers Association of California, Sacramento, California.

Design Example 1B
Wood Light-frame Residence

Figure 1B-1. Wood light-frame residence

Foreword

Wood light-frame residences, such as that in Figure 1B-1, have been designed using simplified design assumptions and procedures based largely on judgment and precedent.

Earlier editions of this Structural Seismic Design Manual included a rigid diaphragm analysis for Example 1, because an exemption for one- and two-family dwellings did not exist in the UBC or IBC codes. Although a rigid or semi-rigid diaphragm analysis for this type of building structure is not required by the code, it is the desire of SEAOC to include Design Example 1B showing the "envelope" technique because there may be conditions where the design professional or the building official feels that this type of analysis is warranted. In addition, design procedures in this design example (solid wall stiffness wall frame stiffness, cantilever column stiffness) can occur in larger light-framed structures.

Rigid versus flexible diaphragm assumptions

Small, light-frame detached one- and two-family dwellings have traditionally been designed using flexible diaphragm assumptions, or by a "hybrid" approach of treating closely spaced walls as a unit (i.e., as rigidly connected) and treating the remaining diaphragm as flexible. Also, light-frame detached one- and two-family dwellings have been built with the conventional construction provisions of the code, or by using the one- and two-family dwelling code, without an engineering design. These light-frame structures have historically performed satisfactorily from a life-safety standpoint when subjected to strong seismic shaking.

The Commentary of the 1999 SEAOC Blue Book (§C805.3.1) recognizes that lateral forces for many structures with wood diaphragms, mostly large buildings, may be better represented as rigid, as opposed to flexible, diaphragms. Relative to the small structure used in this example, the use of the rigid diaphragm assumptions generally will not significantly improve the seismic behavior.

While the building response remains elastic, the rigid diaphragm assumptions will better reflect the initial stiffness of the building system. However, it is not possible to accurately calculate the stiffness of all the various elements, including the stiffness contributed by finishes and nonstructural elements, and taking into account the fact that stiffness of these elements will degrade as the ground shaking intensifies. As a result, the use of the rigid diaphragm assumptions may not be significantly better than the traditional flexible diaphragm assumption for structures of this type.

This design example (1B) will be based on a strict interpretation of the code.

Overview

This design example illustrates the seismic design of a 2800-square-foot single-family residence. The structure, shown in Figures 1B-1, 1B-2, 1B-3, 1B-4 and 1B-5, is of wood light-frame construction with wood structural panel shear walls, roof, and floor diaphragms. Roofing is clay tile.

Outline

This example will illustrate the following parts of the design process

1. **Design base shear and vertical distributions of seismic forces**
2. **Lateral forces on shear walls and shear wall nailing assuming flexible diaphragms**
3. **Rigidities of shear walls and cantilever columns at garage**
4. **Centers of mass and rigidity of diaphragms**
5. **Distribution of lateral forces to the shear walls with rigid diaphragms**
6. **Diaphragm deflections and whether diaphragms are flexible or rigid**

Given Information

Roof weights (slope 5:12)

Tile roofing	10.0 psf
½-inch sheathing	1.5
Roof framing	4.0
Insulation	1.0
Miscellaneous	0.2
Gyp ceiling	2.8
D (along slope) =	19.5 psf

Floor weights

Flooring	1.0 psf
⅝-inch sheathing	1.8
Floor framing	4.0
Miscellaneous	0.4
Gyp ceiling	2.8
	10.0 psf

D = dead load

D = (horiz. proj.) = 19.5 (13/12) = 21.1 psf (the roof and ceilings are assumed to be on a 5:12 slope, vaulted)

Weights of respective diaphragm levels, including exterior and interior walls:

For north-south direction

W_{roof} = 63,650 lb (roof and tributary walls)

W_{floor} = 38,850 lb (floor and tributary walls above and below)

W = 102,500 lb

For east-west direction

W_{roof} = 65,400 lb (roof and tributary walls)

W_{floor} = 39,900 lb (floor and tributary walls above and below)

W = 105,300 lb

Weights of diaphragms are typically determined by adding the tributary weights of the walls to the diaphragm, e.g., add one-half the height of walls at the second floor to the roof, and one-half the height of second floor walls plus one-half the height of first floor walls to second floor diaphragm. It is acceptable practice to ignore the weight of shear walls parallel to the direction of seismic forces to the upper level and add 100 percent of the parallel shear wall weight to the level below, instead of splitting the weight between floor levels. Weights of bearing partitions (not shear walls) should still be split between floors. Unlike commercial construction, the code minimum of 20 psf (vertical load) and 10 psf (lateral load) is often exceeded in residential construction.

Framing lumber is Douglas Fir-Larch grade stamped No. 1S-Dry.

DOC PS-1 or PS-2 (APA or TECO performance-rated) wood structural panels for shear walls will be 15⁄32-inch-thick Structural-I, 32/16 span rating, five-ply with Exposure I glue, however, four-ply is also acceptable. Three-ply 15⁄32-inch sheathing has lower allowable shears in some local jurisdictions and the inner ply voids can cause nailing problems.

The roof is 15⁄32-inch-thick DOC PS-1 or PS-2 (APA or TECO performance-rated) sheathing, 32/16 span rating with Exposure I glue.

The floor is 19⁄32-inch-thick DOC PS-1 or PS-2 (APA or TECO performance-rated) Sturd-I-floor 16 inches o/c rating (or APA or TECO performance-rated sheathing, 42/20 span rating) with Exposure I glue.

Boundary members for the shear walls are 4x posts.

Common wire nails are to be used for diaphragms, shear walls, and straps.

Sinker nails are to be used for design of the shear wall sill plate nailing at the second floor. (Note: many nailing guns use the smaller diameter box and sinker nails instead of common nails. Closer nail spacing may be required for smaller diameter nails).

Seismic and site data

$S_S = 1.78$	F 22-1
$S_1 = 0.55$	F 22-2
$S_{DS} = 1.19$	Eq 11.4-3
Seismic Design Category D	T 11.6-1
$I = 1.0$	T 11.5-1

Site Class C has been determined by geotechnical investigation. Without a geotechnical investigation, Site Class D shall be used as a default value.

Figures 1B-2 through 1B-4 depict the shear walls as dark solid lines. This has been done for clarity in this example. Actual drawings commonly use other graphic depictions. Practice varies on how framing plans are actually shown and on which level the shear walls are indicated.

Actual drawings commonly do not call out shear wall lengths. However, building designers should be aware that some building departments now require shear wall lengths to be called out on plans.

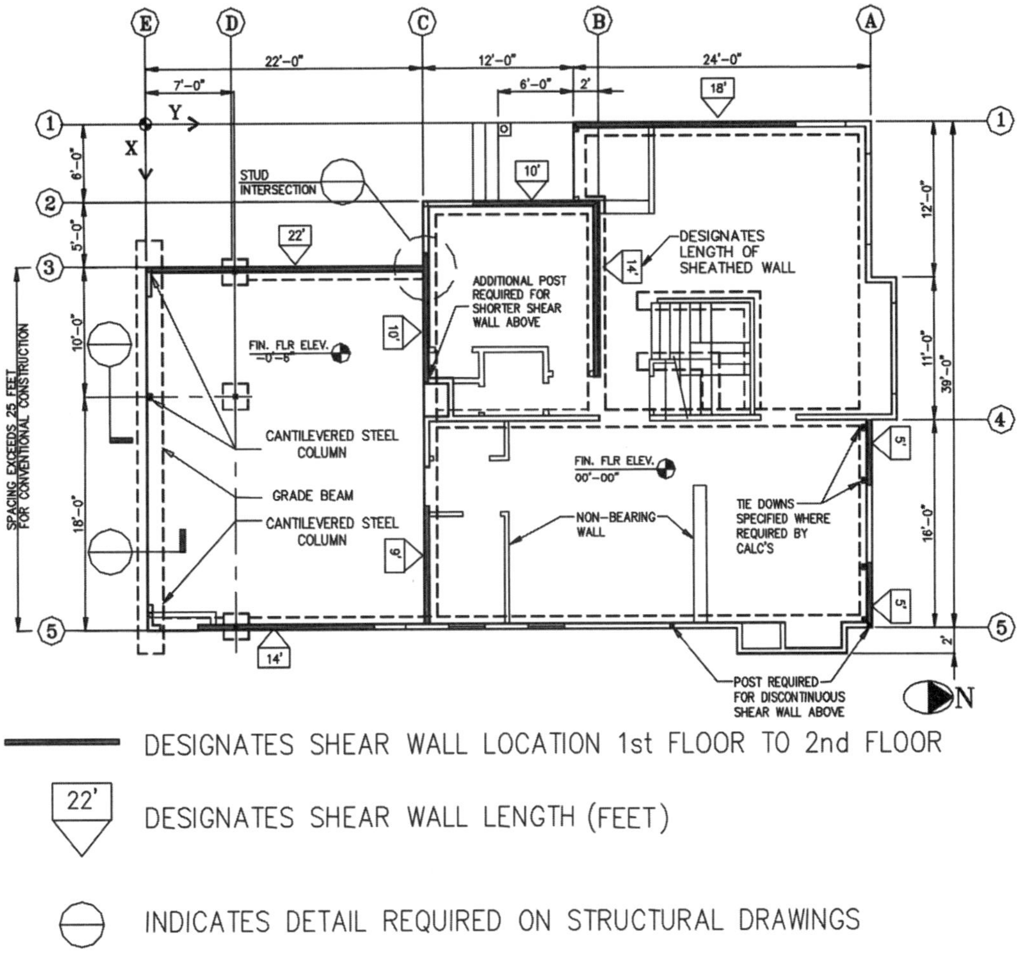

Figure 1B-2. Foundation plan (ground floor)

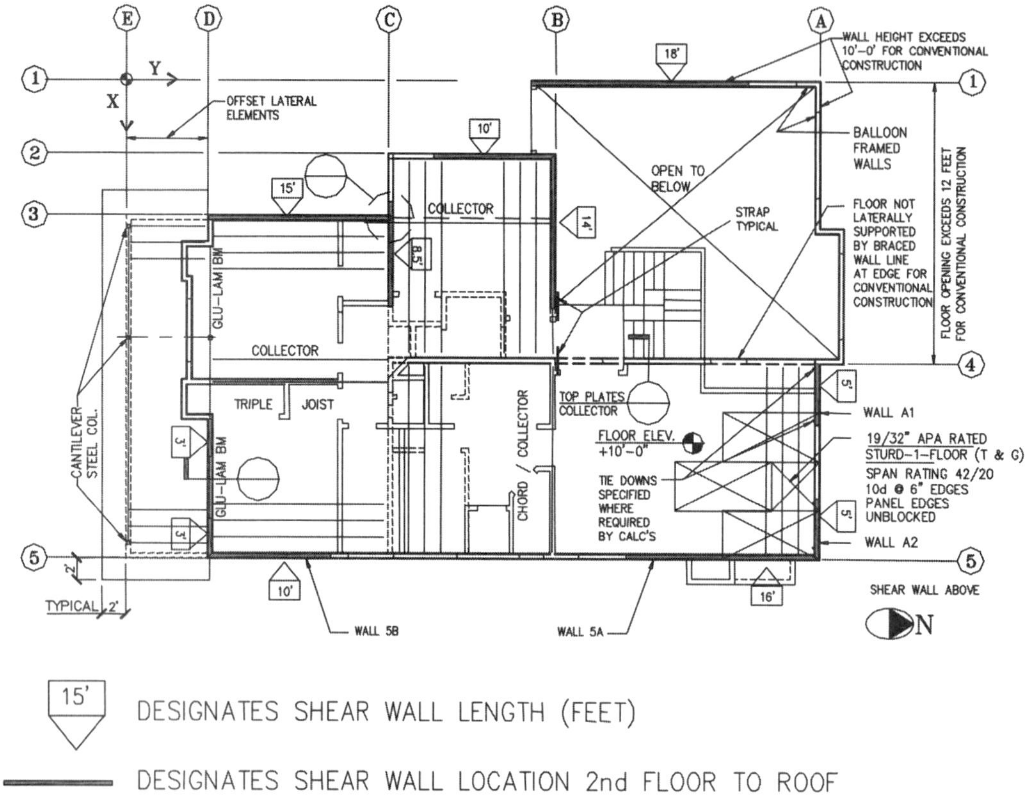

Figure 1B-3. Second floor framing plan and low roof framing plan

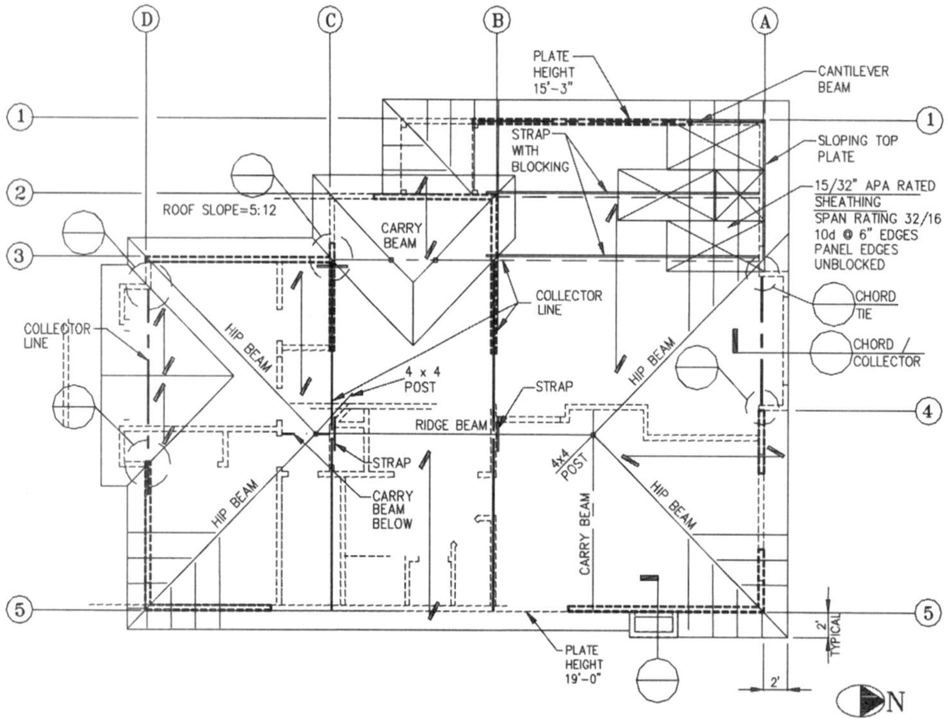

Figure 1B-4. Roof framing plan

Calculations and Discussion | **Code Reference**

1. Design base shear and vertical distributions of seismic forces §12.8.1

This example uses the total building weight W applied to each respective direction. The results shown will be slightly conservative since W includes the wall weights for the direction of load, which can be subtracted out. This approach is simpler than using a separated building weight W for each axis under consideration.

1a. Design base shear

Period using approximate fundamental period (see Figure 1B-5 for section through structure)

$$T_a = C_t(h_n)^{\chi} = 0.020(23)^{3/4} = 0.21 \text{ sec}$$ Eq 12.8-7

where h_n is defined as the highest level of the structure. Since the highest level is not a level surface, the center of gravity (average height) of diaphragm above the first floor will be used.

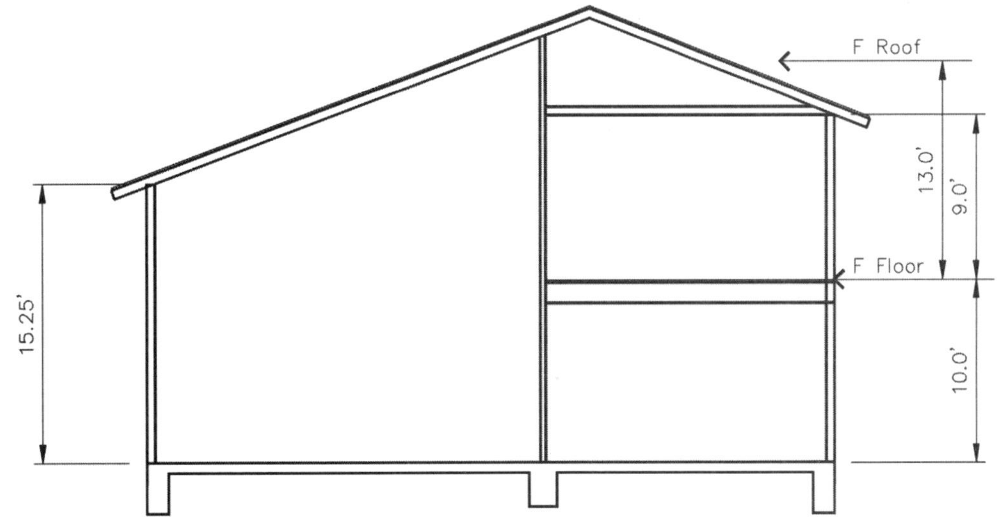

Figure 1B-5. Cross-section through residence

North-south direction

For light-framed walls with wood structural panels that are both shear walls and bearing walls

$$R = 6.5$$ T 12.2-1

Design base shear is

$$V = C_s W$$ Eq 12.8-1

where

$$C_s = \frac{S_{DS}}{\left(\frac{R}{I}\right)} \qquad \text{Eq 12.8-2}$$

(Note that design base shear in the ASCE/SEI 7-05 is on a strength design basis.)

All tables in the IBC for wood diaphragms and shear walls are based on allowable loads. All tables in the NDS-05 supplement Special Design Provisions for Wind and Seismic (SDPWS) are nominal values and must be adjusted for both strength and Allowable Stress Design (ASD) loads.

It is not known how much longer the IBC will publish ASD values since the trend is toward having all values in strength format or nominal format. Since ASD is still predominantly practiced, it has been decided to have this design example in ASD format. In addition, all the manufacturers of metal hardware connectors publish only ASD values.

$I = 1.0$

$R = 6.5$

$S_{DS} = ⅔\, S_{MS} = ⅔\,(1.78) = 1.19$

$S_{MS} = F_a S_S = 1.0 \times 1.78 = 1.78$

where

$F_a = 1.0$

$S_S = 1.78$

Site Class C

$$C_s = \frac{1.19}{\left(\frac{6.5}{1.0}\right)} = 0.183$$

but need not exceed

$$C_s = \frac{S_{D1}}{T\left(\frac{R}{I}\right)} \qquad \text{Eq 12.8-3}$$

$S_1 = 0.55$

$S_{M1} = F_v S_1 = 0.55 \times 1.3 = 0.72$

$S_{D1} = ⅔\, S_{M1} = ⅔\,(0.72) = 0.48$

$F_v = 1.3$

$$C_s = \frac{0.48}{\left(\frac{6.5}{1.0}\right)0.21} = 0.35 > 0.183 \quad \therefore \text{ does not control}$$

$C_s = 0.183$

but shall not be less than

$C_s = 0.01$

$\therefore V_{n\text{-}s} = 0.183W$

In addition, equation (12.8-6) requires an additional check for C_s minimum for structures that are located where S_1 is equal to or greater than 0.6*g*:

$$C_s = \frac{0.5S_1}{\left(\frac{R}{I}\right)} = \frac{0.5(1.0)}{\left(\frac{6.5}{1.0}\right)} = 0.07$$

However, this condition does not control.

Comparison of the above result with the simplified static method permitted under §12.14 shows that it is more advantageous to use the simplified static method of determining the design base shear.

$$V = \frac{FS_{DS}}{R}W = \frac{F(1.0)}{6.5}W = 0.169W < 0.183W \qquad \text{§12.14.8.1}$$

where

F = 1.1 for two-story buildings

$S_{DS} = (\tfrac{2}{3})1.0 \times 1.5 = 1.0$

S_S = 1.5 max

$F_a = 1.0$ for $S_S \geq 1.25$

The engineer can choose which method to use. This design example will illustrate the Equivalent Lateral Force Procedure.

Check for maximum S_S value: §12.8.1.3

Since the structure has an Out-of-Plane Offset Irregularity (ASCE7-05 Table 12.3-1, Type 4), the structure is classified as "Irregular," and hence the S_S limitation of 1.5 does not apply.

It is desirable to keep the strength level forces throughout the design of the structure for two reasons:

1. Errors in calculations can occur and confusion on which load is being used—strength or allowable stress design. This design example will use the following format

   $V_{base\ shear}$ = strength

   F_{px} = strength

   F_x = force to wall (strength)

   v = wall shear at element level (ASD)

   $$v = \frac{F_x(0.7)}{b} \text{ (ASD)}$$

2. This design example will not be applicable in the future, when the code will be all strength design.

Seismic load effect E:

Where the effects of gravity and the seismic ground motion are additive, the seismic load E is defined as

$$E = \rho Q_E + 0.2S_{DS} D \qquad \text{§12.4-1}$$

Where the effects of the gravity and seismic ground motion counteract, the seismic load E is defined as

$$E = \rho Q_E - 0.2S_{DS} D \qquad \text{§12.4-2}$$

The redundancy ρ is assumed to be 1.0. This is the case for most Type V residential structures. The value for ρ will have to be verified.

The basic load combinations for allowable stress design are

$$(1.0 + 0.14S_{DS})D + 0.7\,\rho Q_E \qquad \text{§12.4.2.3}$$

$$(1.0 + 0.105S_{DS})D + 0.525\,\rho Q_E + 0.75L + 0.75L_r$$

$$V_{n\text{-}s} = 0.183W$$

$$\therefore\ V_{n\text{-}s} = 0.183(102{,}500 \text{ lb}) = 18{,}750 \text{ lb} \qquad \text{Eq 12.8-1}$$

East-west direction

Since there are different types of lateral-resisting elements in this direction, determine the controlling R value.

For light-framed walls with wood structural panels that are both shear walls and bearing walls:

$$R = 6.5$$

For cantilevered column elements

$$R = 1.25 \qquad \text{T12.2-1}$$

For combinations along the same axis, the ASCE/SEI 7-05 requires the use of least value for any of the systems utilized in that same direction, therefore the value for the cantilevered column elements must be used for the entire east-west direction. This provision for combinations along the same axis first appeared in the 1994 UBC. However, ASCE/SEI 7-05 has added an exception (§12.2.3.2) when three conditions are met: 1) An occupancy Category I or II building; 2) two stories or less in height; and 3) use of light-frame construction or flexible diaphragms. For Design Example 1, all three conditions have been met, hence the value of R may be used for each line of resistance.

The seismic base shear using $R = 6.5$ for the light-framed walls with wood structural panels will be used for the building as a whole and then the seismic load will be factored-up for the cantilever columns with a factor of (6.5/1.25).

$$\therefore\ V_{e\text{-}w} = 0.183W \qquad \text{Eq 12.8-1}$$

$$V_{e\text{-}w} = 0.183\,(105{,}300 \text{ lb}) = 19{,}275 \text{ lb}$$

For a discussion of R factor see Design Example 1A.

1b. Vertical distribution of seismic forces

The vertical distribution of seismic forces is determined from §12.8.3 as

$$F_x = C_{vx} V \qquad \text{Eq 12.8-11}$$

$$C_{vx} = \frac{w_x h_x^k}{\sum_{i=1}^{n} w_i h_i^k} \qquad \text{Eq 12.8-12}$$

where h_x is the height from the base to level x, in feet
k is a distribution exponent related to the building period

Since $T = 0.21$ seconds < 0.5 seconds, $k = 1$

Determination of F_x is shown in Table 1B-1a.

Table 1B-1a. Vertical distribution of seismic forces for north-south direction

Level	w_x (lb)	h_x (ft)	$w_x h_x$ (lb-ft)	$\frac{w_x h_x}{\Sigma w_i h_i}$ (%)	$F_{x_{N-S}}$ (lb)	$\frac{F_{x_{N-S}}}{w_x}$
Roof	63,650	23.0	1,463,950	79	14,800	0.233
Floor	38,850	10.0	388,500	21	3,950	0.102
Σ	102,500	—	1,852,450	100	18,750	0.183

Table 1B-1b. Vertical distribution of seismic forces for east-west direction

Level	w_x (lb)	h_x (ft)	$w_x h_x$ (lb-ft)	$\frac{w_x h_x}{\Sigma w_i h_i}$ (%)	$F_{x_{E-W}}$ (lb)	$\frac{F_{x_{E-W}}}{w_x}$
Roof	65,400	23.0	1,504,200	79	15,230	0.233
Floor	39,900	10.0	399,000	21	4,045	0.101
Σ	105,300	—	1,903,200	100	19,275	0.183

2. Lateral forces on shear walls and shear wall nailing, assuming flexible diaphragms

Determine the forces on shear walls. The ASCE/SEI 7-05 does not require torsional effects to be considered for flexible diaphragms. The effects of torsion and wall rigidities will be considered in Part 5 of this design example.

The selected method of determining loads to shear walls is based on tributary areas with simple spans between supports. Another method of determining loads to shear walls can assume a continuous beam, although a continuous beam approach may not be accurate because of shear deformations in the diaphragm. The tributary area approach works with reasonable accuracy for a continuous beam with 100-percent shear deflection and zero bending deflection. This design example uses the exact tributary area to the shear walls, an approach that is fairly comprehensive. An easier and more common method would be to use a uniform load equal to the widest portion of the diaphragm, which results in conservative loads to the shear walls.

2a. Forces on east-west shear walls

Roof diaphragm

Roof area = 2164 sq ft (sf)

$$f_{p\ roof} = \frac{15,230 \text{ lb}}{2164 \text{ sf}} = 7.04 \text{ psf}$$

$$w_1 = (7.04 \text{ psf})(43.0 \text{ ft}) = 303 \text{ plf}$$

$$w_2 = (7.04 \text{ psf})(37.0 \text{ ft}) = 261 \text{ plf}$$

$$w_3 = (7.04 \text{ psf})(32.0 \text{ ft}) = 226 \text{ plf}$$

Check sum of forces

$$452 + 1700 + 1760 + 1977 + 2097 + 3333 + 3333 + 606 = 15,258 \text{ lb}$$

$$V_{Roof} = 15,258 \text{ lb} \geq 15,230 \text{ lb} \ldots o.k.$$

Note that Figures 1B-6, 1B-7, 1B-8, and 1B-9 are depicted as a continuous beam. From a technical standpoint, nodes should be shown at the interior supports. In actuality, with the tributary area approach, these are considered separate single span beams between the shear wall supports (Figure 1B-6 has three separate single-span beams).

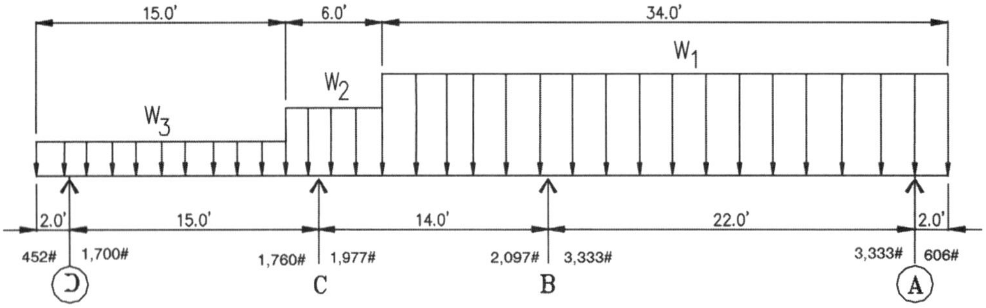

Figure 1B-6. Roof diaphragm loading for east-west forces

Floor diaphragm

Second floor area = 1542 sf

$$f_{p\ floor} = \frac{4045 \text{ lb}}{1542 \text{ sf}} = 2.62 \text{ psf}$$

$$w_4 = (2.62 \text{ psf})(16.0 \text{ ft}) = 42 \text{ plf}$$

$$w_5 = (2.62 \text{ psf})(20.0 \text{ ft}) = 53 \text{ plf}$$

$$w_6 = (2.62 \text{ psf})(33.0 \text{ ft}) = 87 \text{ plf}$$

$$w_7 = (2.62 \text{ psf})(28.0 \text{ ft}) = 74 \text{ plf}$$

$$w_8 = (2.62 \text{ psf})(32.0 \text{ ft}) = 84 \text{ plf}$$

$$P_D = (452 \text{ lb} + 1,700 \text{ lb}) = 2,152 \text{ plf}$$

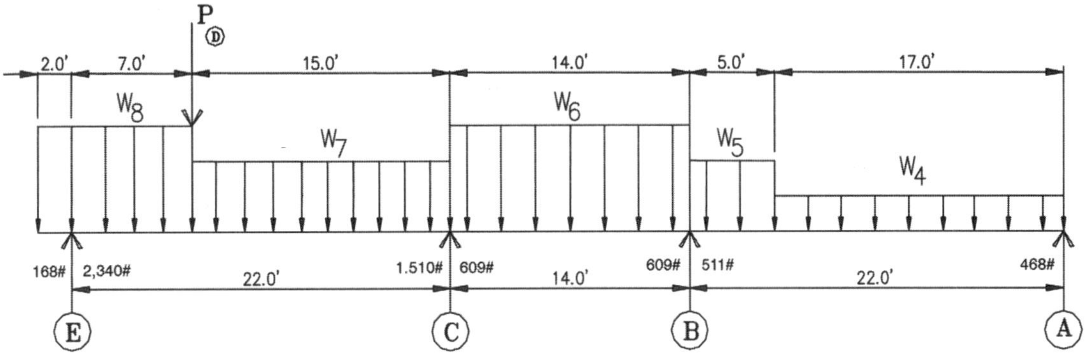

Figure 1B-7. Second floor diaphragm loading for east-west forces

Check sum of forces

$$168 + 2340 + 1510 + 609 + 609 + 511 + 468 = 6215 \text{ lb}$$

Subtract P_D from the sum of forces

$$6215 - 2152 = 4063 \text{ lb}$$

$$V_{floor} = 4063 \text{ lb} \geq 4045 \text{ lb} \ldots o.k.$$

2b. Required edge nailing for east-west shear walls using 10d common nails (T 2306.4.1)

Table 1B-2a. East-west shear walls at roof level (second floor to roof)[1, 2, 3, 4, 5, 6, 7]

Wall (grid line)	ΣF_{above} (lb)	ΣF_x (lb)	F_{tot} (lb)	b (ft)	$v = \frac{F_{tot}(0.7)^{(6)}}{b}$ (plf)	Sheathing[(5)] 1 or 2 sides	Allowable Shear (plf)	Edge Nail Spacing (in.)
A	0	3,939	3,939	10.0	276[(6)]	One	340	4[(2)(4)]
B	0	5,430	5,430	14.0	272[(6)]	One	340	4[(4)]
C	0	3,737	3,737	8.5	308[(6)]	One	340	4[(4)]
D	0	2,152	2,152	6.0	251[(6)]	One	340	4[(2)(4)]
Σ	0	15,258	15,258	38.5				

Notes:

1. In SDC D, E, or F, the 2006 IBC (Table 2306.4.1 footnotes and §2305.3.11) requires 3x nominal thickness stud framing at abutting panel edges and at foundation sill plates when the allowable stress design shear values exceed 350 pounds per foot.
2. Sill bolt washers: For SDC D, E, or F, §2305.3.11 requires a minimum of 3-inch-square by 0.229-inch-thick plate washers to be used for each foundation sill bolt (regardless of allowable shear values in the wall). These changes were a result of the splitting of framing studs and sill plates observed in the Northridge earthquake and in cyclic testing of shear walls. The plate washers are intended to help resist uplift forces on shear walls. Because of vertical displacements of hold-downs, these plate washers are required even if the wall has hold-downs designed to take uplift forces at the wall boundaries. The washer edges shall be parallel/perpendicular to the sill plate. Sill bolt plate washers are not required in SDCs A, B, and C. *(Notes continued on next page)*

3. IBC Section 2305.3.11 includes an exception to the 3x foundation sill plates that allows 2x foundation sill plates when the allowable shear values are less than 600 pounds per foot, provided that sill bolts are designed for 50 percent of allowable values.
4. Refer to Design Example 2 for discussions about fasteners for pressure-preservative-treated wood and the gap at bottom of sheathing.
5. DOC PS-1 or PS-2 (APA or TECO performance) 15⁄32-in. Structural-I rated wood structural panels may be either plywood or oriented strand board (OSB) using 10d common nails with a minimum 1½-inch penetration.
6. Note that forces are strength level, and shear in wall is multiplied by 0.7 to convert to allowable stress design.
7. The 1999 SEAOC Blue Book recommends special inspection when the nail spacing is closer than 4 inches on center.

Table 1B-2b. East-west shear walls at floor level (first floor to second floor)

Wall (grid line)	ΣF_{above} (lb)	ΣF_x (lb)	F_{tot} (lb)	b (ft)	$v = \frac{F_{tot}(0.7)}{b}$ (plf)	Sheathing[5] 1 or 2 sides	Allowable Shear (plf)	Edge Nail Spacing (in.)
A	3,939	468	4,407	10.0	308[2]	One	510	4
B	5,430	1,120	6,550	14.0	328[2]	One	510	4
C	3,737	2,119	5,856	19.0	215[2]	One	510	4
D	2,152	0	0	0	0			
E	0	2,508	2,508	Frame	Frame			
Σ	15,258	6,215	19,321	43.0				

See notes for Table 1B-2a.

2c. Forces on north-south shear walls

Roof diaphragm

$$f_{p\ roof} = \frac{14{,}800\ \text{lb}}{2164\ \text{sf}} = 6.84\ \text{psf}$$

$$w_1 = (6.84\ \text{psf})(55.0\ \text{ft}) = 376\ \text{plf}$$

$$w_2 = (6.84\ \text{psf})(40.0\ \text{ft}) = 274\ \text{plf}$$

$$w_3 = (6.84\ \text{psf})(34.0\ \text{ft}) = 233\ \text{plf}$$

Check sum of forces

$$466 + 713 + 767 + 726 + 848 + 5264 + 5264 + 752 = 14{,}800\ \text{lb}$$

$$V_{roof} = 14{,}800\ \text{lb} \approx 14{,}800\ \text{lb} \ldots \textit{o.k.}$$

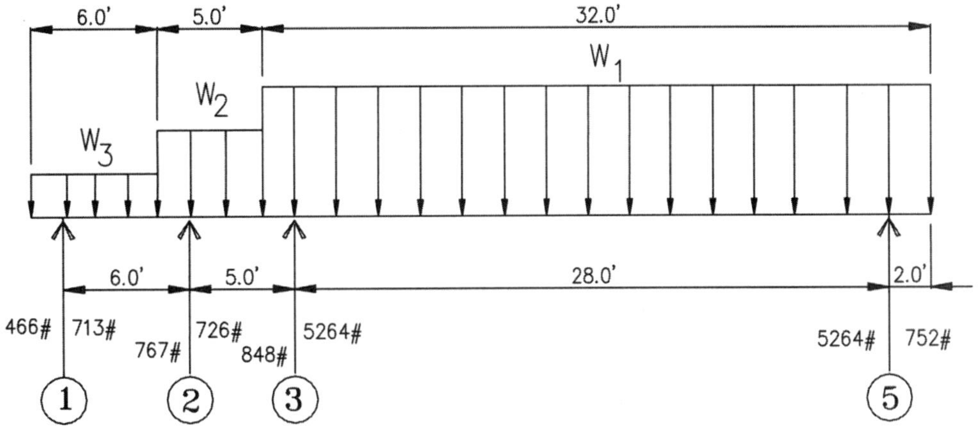

Figure 1B-8. Roof diaphragm loading for north-south forces

Floor diaphragm

$$f_{p\,floor} = \frac{3950 \text{ lb}}{1542 \text{ sf}} = 2.56 \text{ psf}$$

$$w_4 = (2.56 \text{ psf})(9.0 \text{ ft}) = 23 \text{ plf}$$

$$w_5 = (2.56 \text{ psf})(60.0 \text{ ft}) = 154 \text{ plf}$$

$$w_6 = (2.56 \text{ psf})(43.0 \text{ ft}) = 110 \text{ plf}$$

$$w_7 = (2.56 \text{ psf})(38.0 \text{ ft}) = 97.2 \text{ plf}$$

$$w_8 = (2.56 \text{ psf})(23.0 \text{ ft}) = 58.9 \text{ plf}$$

$$w_9 = (2.56 \text{ psf})(14.0 \text{ ft}) = 35.8 \text{ plf}$$

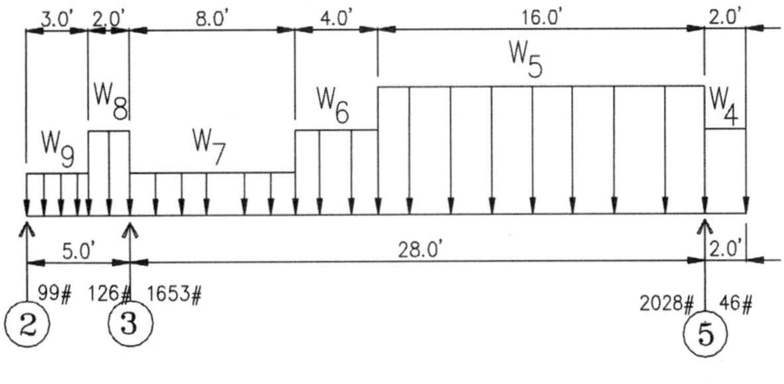

Figure 1B-9. Second floor diaphragm loading for north-south forces

Check sum of forces

$$99 + 126 + 1653 + 2028 + 46 = 3952 \text{ lb}$$

$$V_{floor} = 3952 \text{ lb} \approx 3950 \text{ lb} \ldots o.k.$$

2d. Required edge nailing for north-south shear walls using 10d common nails (T 2306.4.1)

Table 1B-3a. North-south shear walls at roof level (second floor to roof)

Wall	ΣF_{above} (lb)	ΣF_x (lb)	F_{tot} (lb)	b (ft)	$v = \frac{F_{tot}(0.7)}{b}$ (plf)	Sheathing 1 or 2 sides	Allowable Shear (plf)	Edge Nail Spacing (in.)
1	0	1,179	1,179	18.0	46	One	510	4
2	0	1,493	1,493	10.0	105	One	510	4
3	0	6,112	6,112	15.0	285	One	510	4
5	0	6,016	6,016	26.0	162	One	510	4
Σ	0	14,800	14,800	69.0				

Table 1B-3b. North-south shear walls at floor level (first floor to second floor)

Wall	ΣF_{above} (lb)	ΣF_x (lb)	F_{tot} (lb)	b (ft)	$v = \frac{F_{tot}(0.7)}{b}$ (plf)	Sheathing 1 or 2 sides	Allowable Shear (plf)	Edge Nail Spacing (in.)
2	1,493	99	1,592	10.0	111	One	510	4
3	6,112	1,779	7,891	22.0	251	One	510	4
5	6,016	2,074	8,090	14.0	405	One	510	4
Σ	13,621	3,952	17,573	46.0				

3. Rigidities of shear walls and cantilever columns at garage

3a. Estimation of wood shear wall rigidities

Determination of the rigidities of wood shear walls is often difficult and inexact, even for design loads. In addition, when walls are loaded substantially beyond their design limits, as happens under strong earthquake motions, rigidity determination becomes even more difficult. It is complicated by a number of factors that make any exact determination virtually impossible short of full-scale testing.

A well-known 4-term expression for shear wall deflection is found in IBC §2305.3.2. This expression, shown below, is used to estimate deflections of shear walls with fixed bases and free tops for design level forces.

$$\Delta = \frac{8vh^3}{EAb} + \frac{vh}{Gt} + 0.75he_n + d_a\frac{h}{b}$$ (Eq 23-2) (SDPWS Eq C4.3.2-1)

The expression above was developed from static tests of solid wood shear walls; many typically measuring 8 feet by 8 feet. Until recently, there was very little cyclic testing of wood shear walls (to simulate actual earthquake behavior) or testing of walls with narrow aspect ratios.

The American Forest and Paper Association (AF&PA) has recently published a new supplement to the National Design Specification, the Special Design Provisions for Wind and Seismic (SDPWS). In this publication, the SDPWS has changed the traditional 4-term expression to a new simplified 3-term expression for shear wall deflection. This new equation is shown below.

$$\Delta = \frac{8vh^3}{EAb} + \frac{vh}{1000G_a} + d_a\frac{h}{b}$$ SDPWS Eq 4.3-1

The new simplified 3-term equation combines the second and third terms (of the 4-term equation) into one term. Computed deflections by using either the 4-term equation or the 3-term equation produces nearly identical results at the critical strength level (1.4 times the allowable shear values for seismic).

In modern wood frame building construction, shear walls come in many forms and sizes, and these are often penetrated by ducts, windows, and door openings. Also, many walls in residences are not designed as shear walls, yet have stiffness from their finish materials (gypsum board, stucco, brick and stone veneers, etc.). In multi-story structures, walls are stacked on the walls of lower floors, producing indeterminate structural systems. In general, it is difficult to calculate wall rigidities with the IBC equation or the SDPWS equation alone. As will be shown in subsequent paragraphs, things like shrinkage can significantly effect deflection and subsequent stiffness calculations. Further, in strong earthquake motions, shear walls may see forces and displacements several times larger than those used in design, and cyclic degradation effects can occur that significantly change the relative stiffness of shear walls at the same level.

The small differences in computed shear wall deflections from using the 4-term equation versus the 3-term equation and the method of determining the tiedown displacement will impact shear wall rigidities and load distribution. It is recommended that the designer use consistent equations and tiedown displacement when computing a particular project.

It can be argued that wall rotation of the supporting wall below should be taken into account when considering shear wall rigidities. However, considering rotation of the supporting wall below would be like measuring the shear wall as the cumulative height, rather than the accepted floor-to-floor clear height. Not considering rotation of the supporting wall below is appropriate for determining relative wall rigidities.

At present, there are a number of ways to estimate shear wall rigidities, particularly when only relative rigidities are desired (see Blue Book §C805.3). These include

1. Rigidity based on estimated nail slip
2. Rigidity calculated from IBC Equation 23-2 (the four-term equation given above)
3. Rigidity incorporating both IBC Equation 23-2 and shrinkage
4. Rigidity calculated from SDPWS Equation 4.3-1 (the three-term equation given above)
5. Rigidity incorporating both SDPWS Equation 4.3-1 and shrinkage
6. Rigidity calculated from APA Research Report 138 (the four-term equation given above)

7. Rigidity incorporating both APA Research Report 138 and shrinkage

8. Several other procedures

Only one of these approaches is represented in this design example. By using this approach, SEAOC does not intend to establish a standard procedure or indicate a standard of care for calculation of wood shear wall rigidities. It is merely employing one of the current methods.

3b. Discussion of rigidity calculation using the IBC deflection and APA Research Report equations

Because the rigidity k of a shear wall or cantilever column is based on its displacement Δ, the displacements will first be computed using the F_{tot} forces already determined in Tables 1B-2a, 1B-2b, 1B-3a, and 1B-3b.

Compute values for k

$$F = k\Delta$$

or $$k = F/\Delta$$

The basis for determining the deflection of a shear wall is the four-term equation shown below.

$$\Delta = \frac{8vh^3}{EAb} + \frac{vh}{Gt} + 0.75he_n + d_a\frac{h}{b} \qquad \text{(Eq 23-2)}$$

where

v = shear in the wall in pounds per linear foot (plf)

h = height from the bottom of the sill plate to the underside of thc framing at diaphragm level above (top plates)

A = area of the boundary element in square inches

At the roof, the boundary elements consist of two 2x4s

At the floor, the boundary elements consist of three 2x4s

b = is the shear wall length in feet

G = shear modulus values from Table 3, Plywood Design Specifications, in pounds per square inch (psi)

t = equivalent thickness values from Table 2, Plywood Design Specifications, in inches

Gt = panel rigidity through the thickness from IBC Table 2305.2.2(2) (psi)

V_n = load per fastener (nail) in pounds

e_n = nail slip values from APA Research Report 138 are for Structural-I sheathing with dry lumber = $(V_n/769)^{3.276}$

e_n = nail slip values from IBC Table 2305.2.2(1)

d_a = vertical displacement of the tiedown due to anchorage details in inches

This design example uses the G, t, and e_n values from APA Research Report 138. The designer is free to use the IBC valves that will produce nearly identical results. IBC Table 2305.2.2(2) also has values for OSB sheathing that are not in the APA report. The values in IBC Table 2305.2.2(2) combine the Gt valves, while this design example uses the separated values from the APA report.

This equation is based on a uniformly nailed, cantilever shear wall with a horizontal point load at the top and with, panel edges blocked, and reflects tests conducted by the American Plywood Association. The deflection is estimated from the contributions of four distinct parts. The first part of the equation accounts for cantilever beam action using the moment of inertia of the boundary elements. The second accounts for shear deformation of the sheathing. The third accounts for nail slippage/bending, and the fourth accounts for tiedown assembly displacement (this also should include bolt/nail slip and shrinkage). End stud elongation due to compression or tension is not considered, nor are the end rotations of the base support.

Testing on wood shear walls has indicated that the above formula is reasonably accurate for aspect ratios h/w lower than or equal to 2:1. For higher aspect ratios, the wall drift increases significantly, and testing showed that displacements were not adequately predicted. Use of the new aspect ratio requirement of 2:1 makes this formula more accurate for determining shear wall deflection/stiffness than it was in previous editions of the building codes, subject to the limitations mentioned above.

Recent testing on wood shear walls has shown that sill plate crushing under the boundary element can increase the deflection of the shear wall by as much as 20 to 30 percent. For a calculation of this crushing effect, see the deflection of wall frame at line D in Part 3c.

Fastener slip/nail deformation values (e_n).

Using the fastener slip equations from Table B-4 of Research Report 138 for 10d common nails, there are two basic equations.

When the nails are driven into *green* lumber

$$e_n = (V_n/977)^{1.894}$$ APA Research Rpt. T B-4

When the nails are driven into *dry* lumber

$$e_n = (V_n/769)^{3.276}$$ APA Research Rpt. T B-4

where

V_n is the fastener load in pounds per fastener.

These values are based on Structural-I sheathing and must be increased by 20 percent when the sheathing is not Structural-I. The language in footnote a to APA Table B-4 stating "Fabricated green/tested dry (seasoned) . . . " is very misleading. The values in the table are actually *green values*, because the lumber is fabricated when green. Don't be misled by the word "seasoned."

It is uncertain whether or not the d_a factor is intended to include wood shrinkage and crushing due to shear wall rotation because the code is not specific. This design example includes both shrinkage and crushing in the d_a factor.

Many engineers share a concern that, if the contractor installs the nails at a different spacing (too many or too few), the rigidities will then be different from those calculated. However, nominal changing of the nail spacing in a given wall does not significantly change the stiffness.

Determination of the design story drift Δ §12.8.6

For both strength and allowable stress design, the ASCE/SEI 7-05 requires building drifts to be determined by strength level forces specified in §12.8.

Wood design using the ASCE/SEI 7-05 means that the engineer must use both strength-level forces and allowable stress forces. This can create some confusion, since the code requires drift checks to be strength-level forces. However, all the design equations and tables in IBC Chapter 23 are based on allowable stress design. Drift and shear wall forces will be based on strength-level forces. Remember that the structural system factor *R* is based on using strength-level forces.

3c. Estimation of roof level rigidities

Roof design level displacements

To determine roof level wall rigidities, roof level displacements must first be established. Below is a series of calculations, in table form, to estimate the roof-level story drifts A in each shear wall connecting to the roof (Table 1B-5). Because there is a wall with openings supported by a GLB on line D, the A for this wall must also be determined. Finally, roof level wall rigidities are summarized in Table 1B-6 and a drift check is given in Table 1B-7.

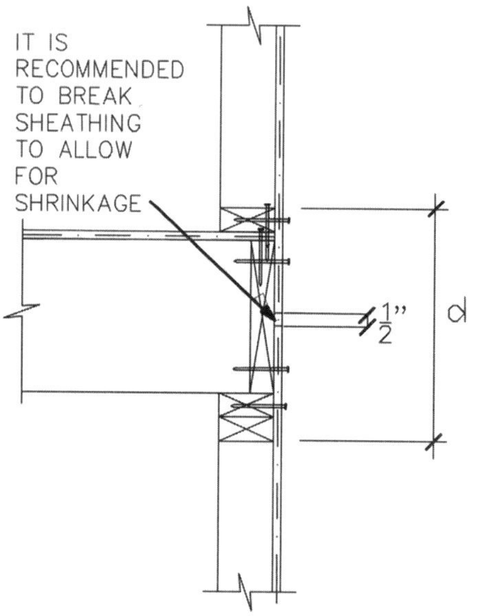

Figure 1B-10. Second floor diaphragm connection to shear wall

Table 1B-4. Determine tiedown assembly displacements for roof level shear walls[1]

Wall	ASD		Strength Design					
	Uplift (0.7)[(2)] (lb)	Tiedown[(3)] Device	Uplift (lb)	Tiedown Elongation[(4)] (in)	Tiedown Assembly Displacement			d_a[(8)] (in)
					Shrink[(5)]	Crush[(6)]	Slip[(7)]	
A1	2,450	Bolted	3,500	0.05	0.19	0.02	0.04	0.30
A2	2,450	Bolted	3,500	0.05	0.19	0.02	0.04	0.30
B	2.475	Bolted	3,535	0.05	0.19	0.02	0.04	0.30
C	3,075	Bolted	4,400	0.07	0.19	0.04	0.05	0.35
1	0	Not required	0	0	0.02	0.02	0	0.04
2	0	Not required	0	0	0.19	0.02	0	0.21
3	830	Strap	1,160	0.004	0.19	0.02	0.002	0.21
5a	0	Not required	0	0	0.19	0.02	0	0.21
5b	0	Not required	0	0	0.19	0.02	0	0.21

Notes:

1. Tiedown assembly displacement is calculated at the second floor level.
2. Uplift force is determined by using the *net* overturning force $M_{ot} - M_R$ divided by the distance between the centroid of the tiedown to the end of the shear wall where *MR* uses load combinations outlined in this design example. With 4x members at the ends of the wall, this equates to the length of the wall minus 1¾ inches for straps, or the length of the wall minus 5½ inches when using a bolted hold-down with 2-inch offset from post to anchor bolt. Using allowable stress design (ASD), tiedown devices need only be sized by using the ASD uplift force. The strength design uplift force is used to determine tiedown assembly displacement in order to determine strength-level displacements.
3. Continuous tie rod hold-down systems can also be used. See Design Example 2 for method of calculating tiedown assembly displacement.
4. Tiedown elongation is based on actual uplift force divided by tiedown capacity times tiedown elongation at capacity (from manufacturer's catalog). Example for tiedown elongation at A1: tiedown selected has a 7500-pound allowable load for a 3½-inch-thick (net) member. From the manufacturer's ICC Evaluation Report, the tiedown deflection at the highest allowable design load (7500 lb) is 0.05 inch. Since there are two tiedown devices (one above and one below the floor), the total elongation is twice the tiedown deflection of one device. Therefore, the total tiedown elongation is (3500/7500) 0.05 × 2 = 0.05 inch.
5. Wood shrinkage based on a change from 19-percent moisture content (MC) to 13-percent MC with 19-percent MC being assumed for S-Dry lumber per project specifications. The MC of 13 percent is the assumed final MC at equilibrium with ambient humidity for the project location. The final equilibrium value can be higher in coastal areas and lower in inland or desert areas. This equates to (0.002) (*d*) (19 – 13), where *d* is the dimension of the lumber (see Figure 1B-10).

   Shrinkage:

   2-2x top plate + 2x sill plate = (0.002)(3 × 1.5 in) (19 – 13) = 0.05

   2x12 floor joist = (0.002)(11.25) (19 – 13) = 0.14

   0.19

The use of pre-manufactured, dimensionally stable, wood I joists that are considered not to shrink would thereby reduce the shrinkage to 0.05 inch. (*Notes continued on next page*)

6. Per the NDS-05 §4.2.6, when compression perpendicular to grain $f_{c\perp}$ is less than $0.73F'_{c\perp}$, crushing will be approximately 0.02 inch. When $f_{c\perp} = 0.73F'_{c\perp}$, crushing is approximately 0.04 inch. The effect of sill plate crushing is the downward effect with uplift force at the opposite end of the wall and has the same rotational effect as the tiedown displacement. Short walls that have no uplift forces will still have a wood crushing effect and will contribute to rotation of the wall.
7. Per NDS-05 §10.3.6, $\gamma = (270{,}000)(1)^{1.5} = 270{,}000$ lb/in plus ¹⁄₁₆-inch oversized hole for bolts. For nails, values for e_n can be used. Example for slip at tiedown at A1 (tiedown has three ⅞-inch-diameter bolts to post)

   $$\text{Load/bolt} = 3500/3 = 1167 \text{ lb/bolt}$$

   $$\gamma = (270{,}000)\,(0.875)^{1.5} = 221{,}000 \text{ lb/in}$$

   $$\text{slip} = (1167/221{,}000) = 0.005 \text{ in}$$

   Because there are two tiedown devices (one above and one below the floor), the total slip is twice the bolt slip. Good detailing specifies that tiedown bolts must be re-tightened just prior to closing in. This accomplishes two things: it takes the slack out of the oversized bolt hole and compensates for some wood shrinkage. This design example assumes that about one-half of the bolt-hole slack is taken out.

   Therefore, total slip equals $(0.005 \times 2) + \left(\frac{1}{16}\right)\frac{1}{2} = 0.04$ in
8. Thus, d_a is the total tiedown assembly displacement. This also could include mis-cuts (short-studs) and lack of square-cut ends.

Table 1B-5. Deflections of the shear walls at the roof level[1,2,6,10,12]

Wall	ASD v (plf)	Strength v[(7)(8)] (plf)	h (ft)	A[(3)] (sq in)	E (psi)	b (ft)	G[(4)] (psi)	t[(11)] (in)	V_n (lb)	e_n[(5)] (in)	d_a (in)	Δ (in)
A1	276	394	9.0	19.25	1.7E6	5.0	90,000	0.535	131	0.0030	0.30	0.65
A2	276	394	9.0	19.25	1.7E6	5.0	90,000	0.535	131	0.0030	0.30	0.65
B	272	388	10.0	19.25	1.7E6	14.0	90,000	0.535	129	0.0029	0.30	0.32
C	308	440	10.0	12.25	1.7E6	8.5	90,000	0.535	147	0.0044	0.35	0.53
1	46	66	15.25	19.25	1.7E6	18.0	90,000	0.535	22	8.8E-6	0.04	0.06
2	105	150	9.0	12.25	1.7E6	10.0	90,000	0.535	50	0.0001	0.21	0.22
3	285	407	9.0	12.25	1.7E6	15.0	90,000	0.535	136	0.0034	0.21	0.23
5a[(9)]	190	271	9.0	12.25	1.7E6	16.0	90,000	0.535	90	0.0009	0.21	0.18
5b[(9)]	118	168	9.0	12.25	1.7E6	10.0	90,000	0.535	56	0.0002	0.21	0.23

Notes:

1. $\Delta = \frac{8vh^3}{EAb} + \frac{vh}{Gt} + 0.75he_n + d_a\frac{h}{b}$ (Eq 23-2)
2. h values are from the bottom of the sill plate to the bottom of the framing at diaphragm level (top plates).
3. A values are for 4x6 posts for walls A1, A2, B, C, and wall 1. A values are for 4x4 posts for walls 2, 3, 5a, and 5b.
4. G values are for Structural-I sheathing.
5. e_n values for Structural-I sheathing with dry lumber = $(V_n/769)^{3276}$. See APA Table B-4, Research Report 138.
6. The use of a computer spreadsheet is recommended. This will not only save time, but will also eliminate possible arithmetic errors with these repetitive calculations. (Notes continued on next page)

7. Deflection of walls Δ is based on strength level forces. The shear wall deflections must be determined using the strength design forces. The calculated deflection of a shear wall is linear up to about two times the allowable stress design values. Since there are tiedown assembly displacements, and dead loads that resist overturning, the factoring-up approach of ASD forces is not appropriate.
8. When sheathing is applied to both sides of the wall, the deflection of the shear wall is determined by using one-half the values from Tables 1B-2a and 1B-3a.
9. In-plane shears to walls 5a and 5b are proportioned based on relative lengths (not per IBC Eq 23-2). For example, wall at line 5a: $R = 16^2/(16^2 + 10^2) = 72$ percent, which is appropriate for two walls in a line, but not necessarily for three or more walls in line. Attempting to equate deflections is desirable. However, the calculations are iterative and indeterminate, and the results are very similar.
10. For deflection of shear wall at line D, Part 3c.
11. Effective thickness for shear. See Table 2, Plywood Design Specification.
12. Deflections are based on APA Research Report 138 formula with shrinkage.

Determine deflection of wall frame at line D (with force transfer around openings)

The deflection for the shear wall can be approximated by using an analysis similar to computing the stiffness for a concrete wall with an opening in it. The deflection for the solid wall is computed, then a deflection for a horizontal window strip is subtracted, and the deflection for the wall piers added back in.

Engineering judgment may be used to simplify this approximation. However, the method shown in Figure 1B-11 is one way to approximate the deflection.

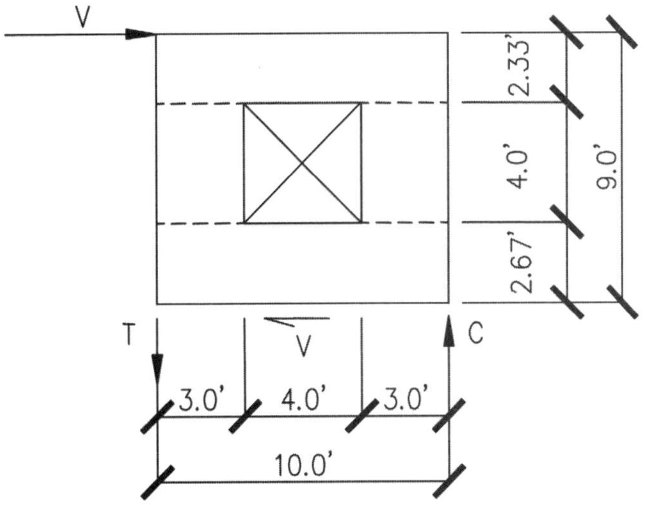

Figure 1B-11. Elevation of wall frame on line D

First, determine deflection of the entire wall, without an opening

Deflection of solid wall:

$$\Delta = \frac{8vh^3}{EAb} + \frac{vh}{Gt} + 0.75he_n + d_a\frac{h}{b} \qquad \text{(Eq 23-2)}$$

Sheathing is on one side of wall with 10d common nails @ 4 inches on center (o/c). Wall has 2x6 studs with 4x6 studs at ends

$$V = 2152\text{lb}$$

$$v = \frac{2152 \text{ lb}}{10.0 \text{ ft}} = 215 \text{ plf}$$

With edge nailing at 4 inches o/c

$$V_n = \text{load per nail} = 215(4/12) = 72 \text{ lb/nail}$$

$$e_n = (72/769)^{3.276} = 0.0004 \text{ in}$$

With a tiedown elongation of 0.05 inch, wood shrinkage of 0.13 inch, and wood crushing of 0.02 inch, it gives a tiedown assembly displacement of 0.20 inch.

For crushing: the strength level overturning moment $M_{ot} = 2152 \times 9.0 = 19{,}368$. Dividing by the distance $L = 9.7$ ft computes the seismic downward component of the 4x6 post

$$P = 19{,}368/9.7 = 1997 \text{ lb} = R_{ot}$$

$$f_c = P/A$$

$$f_c = 1997/(3.5 \times 5.5) = 104 \text{ psi} < 0.73(625) = 456 \text{ psi}$$

$$\therefore \text{crush} = 0.02 \text{ in}$$

For shrinkage of GLB fabricated to AITC specifications at 17-percent MC

$$0.002(17 - 13)\ 16.5 = 0.13 \text{ in}$$

For strap: $\frac{PL}{AE}$ + strap nail slip = 0.05 in

$$d_a = 0.05 + 0.13 + 0.02 = 0.20 \text{ in}$$

$$\Delta = \frac{8(215)9.0^3}{1.7E6(19.25)10.0} + \frac{215(9.0)}{(90{,}000)0.535} + 0.75(9.0)0.0004 + \frac{9.0(0.20)}{10.0} = 0.23 \text{ in}$$

Second, determine deflection of window strip

$$V = 2152 \text{ lb (strength)}$$

With sheathing on one side:

$$v = \frac{2152 \text{ lb}}{10.0 \text{ ft}} = 215 \text{ plf}$$

$$V_n = \text{load per nail} = 215(4/12) = 72 \text{ lb/nail}$$

$$e_n = (72/769)^{3.276} = 0.0004 \text{ in}$$

Since the boundary elements are connected to continuous posts that extend above and below the opening, the value of d_a equals the sheathing nail deformation value calculated above (boundary element "chord" elongation is neglected).

$$d_a = 0.0001 \text{ in}$$

$$-\Delta = \frac{8(215)4.0^3}{1.7E6(19.25)10.0} + \frac{215(4.0)}{(90,000)0.535} + 0.75(4.0)0.0004 + \frac{4.0(0.0001)}{10.0} = 0.02 \text{ in}$$

Note that this deflection is negative because it is subtracted from the sum of the deflections, as shown later.

Third, determine deflection of wall piers

$$V = \frac{2152 \text{ lb}}{2} = 1076 \text{ lb}$$

$$v = \frac{1076 \text{ lb}}{3.0 \text{ ft}} = 358 \text{ plf}$$

$$V_n = \text{load per nail} = 358(4/12) = 119 \text{ lb/nail}$$

$$e_n = (119/769)^{3.276} = 0.0022 \text{ in}$$

Since the boundary elements are connected to continuous posts that extend above and below the opening, the value of d_a equals the sheathing nail deformation value calculated for the wall piers.

$$d_a = 0.0004 \text{ in}$$

$$\Delta = \frac{8(358)4.0^3}{1.7E6(19.25)3.0} + \frac{358(4.0)}{(90,000)0.535} + 0.75(4.0)0.0022 + \frac{4.0(0.0004)}{3.0} = 0.04 \text{ in}$$

Last, determine the sum of the deflections

$$\Delta = 0.23 - 0.02 + 0.04 = 0.25 \text{ in}$$

Thus the stiffness of the wall is (0.23/0.25), or 92 percent of that of the solid wall.

Determine deflection of wall due to deflection of GLB (see Figure 1B-12)

Δh = Shear wall deflection due to deflection of the support beam

$$\tan\theta = \frac{\Delta V}{b} = \frac{\Delta h}{h}$$

$$\therefore \Delta h = \frac{h(\Delta V)}{b}$$

$$R_{ot} = \frac{Vh}{b}$$

$$R_{ot} = \frac{2152 \text{ lb}(9.0 \text{ ft})}{10.0 \text{ ft}} = 1936 \text{ lb (strength)}$$

For 5.125 × 16.5 GLB 24 FV4

$$E = 1{,}800{,}000 \text{ psi}$$

$$I = 1918 \text{ in}^4$$

$$\Delta V = \frac{R_{ot} a^2 b^2}{3EIL}$$

$$\Delta V = \frac{1997(8.0 \times 12)^2 (10.0 \times 12)^2}{3(1.8E6)1918(18.0 \times 12)} = 0.118 \text{ in}$$

$$\Delta h = \frac{h(\Delta V)}{b}$$

$$\Delta h = \frac{(9.0 \times 12)(0.118)}{(10.0 \times 12)} = 0.11 \text{ in}$$

Thus, total deflection of shear wall including GLB rotation and tiedown assembly displacement is

$$\Delta h = 0.25 + 0.11 = 0.36 \text{ in}$$

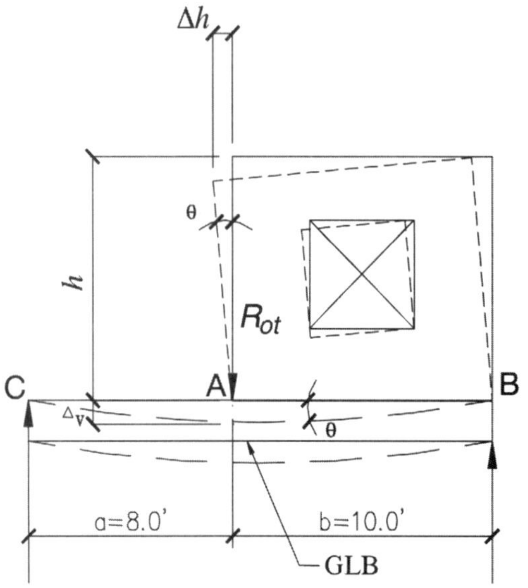

Figure 1B-12. Wall elevation at line D

Table 1B-6. Wall rigidities at roof level[1] (walls from second floor to roof)

Wall	$\Delta^{(2)}$ (in)	F_{tot} (lb)	$k = \frac{F_{tot}}{\Delta}$ (k/in)	$k = \frac{F_{tot}}{\Delta}$ (k/in)
A1	0.65	1,969	3.03	6.06
A2	0.65	1,969	3.03	
B	0.32	5,430		16.97
C	0.53	3,737		7.05
D	0.36	2,152		5.98
1	0.06	1,179		19.65
2	0.22	1,493		6.79
3	0.23	6,112		26.57
5a	0.18	4,332	24.07	31.39
5b	0.23	1,684	7.32	

Notes:

1. Deflections and forces are based on strength force levels.
2. Δ is the design level displacement from Table 1B-5 and calculations of wall frame.

Determination of Δ_a §12.8.6

Before checking drift, the story drift δ_x must be computed. This is done as follows:

$$\delta_x = \frac{C_d \delta_{xe}}{I} \qquad \text{Eq 12.8-15}$$

where

$C_d = 4.0$ for the north-south direction T 12.2-1

$C_d = 4.0$ for the east-west direction T 12.2-1

$I = 1.0$ T 11.5-1

$$\delta_x = \frac{4.0\Delta}{1.0} = 4.0\Delta$$

Determination of maximum drift T 12.12-1

The calculated story drift using δ_x shall not exceed the maximum Δ_a, which is 0.025 times the story height (h_{sx}).

Table 1B-7. Drift check at roof level

	Wall	Δ (in)	h (ft)	δ_x (in)	Max. Δ_a (1) (in)	Status
East-West	A1	0.65	9.0	2.60	2.70	ok
	A2	0.65	9.0	2.60	2.70	ok
	B	0.32	10.0	1.28	3.00	ok
	C	0.53	10.0	2.12	3.00	ok
	D	0.36	9.0	1.44	2.70	ok
North-South	1	0.06	15.25	0.24	4.57	ok
	2	0.22	9.0	0.88	2.70	ok
	3	0.23	9.0	0.92	2.70	ok
	5a	0.18	9.0	0.72	2.70	ok
	5b	0.23	9.0	0.92	2.70	ok

Note:

1. From Table 12.12-1, it should be noted that shear wall A is deflected to 96 percent of the code allowable. This is contrary to the belief of many engineers that shear wall buildings do not have drift problems.

3d. Estimation of second floor level rigidities

First floor level design displacements

First floor level rigidities are determined by first calculating tiedown displacements (Table 1B-8) and then deflections of shear walls at the second floor level (Table 1B-9). The drift check, discussed in Part 3c, is given in Table 1B-10, and wall rigidities are calculated in Table 1B-11.

Table 1B-8. Tiedown assembly displacements for first floor level walls[1]

Wall	ASD		LRFD					
	Uplift (0.7)[(2)] (lb)	Tiedown Device	Uplift (lb)	Tiedown[(3)] Elongation (in)	Tiedown Assembly Displacement			d_a (in)
					Shrink[(4)]	Crush[(5)]	Slip[(6)]	
A1	5,550	Bolted	8,925	0.06	0.01	0.02	0.04	0.12
A2	5,550	Bolted	8,925	0.06	0.01	0.02	0.04	0.12
B	5,250	Bolted	7,500	0.06	0.01	0.02	0.04	0.13
C1	4,700	Bolted	6,700	0.05	0.01	0.02	0.04	0.12
C2	1,600	Bolted	2,300	0.02	0.01	0.02	0.04	0.09
2	0	Not req'd	0	0	0.01	0.02	0	0.03
3	825	Strap	1,155	0.05	0.01	0.02	0.002	0.08
5	400	Strap	560	0.03	0.01	0.02	0.002	0.06

Notes:

1. Tiedown assembly displacement is calculated at the foundation.
2. Uplift force is determined by using the net overturning force $M_{ot} - M_R$ divided by the distance to the centroids of the boundary elements assuming 4x members at the ends of the shear wall. This equates to the length of the wall minus 3½ inches for straps, or the length of wall minus 5½ to 7¼ inches when using a bolted hold-down, which includes a 2-inch offset from post to tiedown bolt.
3. Tiedown elongation is based on actual uplift force divided by tiedown capacity times tiedown elongation at capacity (from manufacturer's catalog). Example of tiedown elongation at A1: Tiedown selected has a 15,000-pound allowable load for a 3½-inch member. From the manufacturer's Evaluation Service Approval, the tiedown deflection at the highest allowable design load (15,000 lb) is 0.09 inch, giving a tiedown elongation of (8925/15,000)0.09 = 0.05 inch. Since the tiedown device has an average ultimate strength of 55,000 pounds, the displacement can be assumed to be linear and, therefore, extrapolated.
4. Wood shrinkage is based on a change from 15-percent MC to 13-percent MC. This equates to $0.002 \times d \times (15 - 13)$ where d is 2.5 inches for a 3x sill plate. Pressure-treated lumber has an MC of less than 15 percent at completion of treatment.
5. Per NDS-05 §4.2.6, when compression perpendicular to grain $f_{c\perp}$ is less than $0.73F'_{c\perp}$, crushing will be approximately 0.02 inch; when $f_{c\perp} = F'_{c\perp}$, crushing is approximately 0.04 inch.
6. Per NDS-05 §10.3.6, γ = load/slip modulus = $(270{,}000)(D^{1.5})$ plus 1⁄16-inch oversized hole for bolts. For nails, values for e_n can be used. For example, slip at tiedown at A1 (tiedown has five 1-inch-diameter bolts to post)

   $$\text{Load/bolt} = 8{,}925/5 = 1{,}785 \text{ lb/bolt}$$

   $$\gamma = (270{,}000)\,(1)^{1.5} = 270{,}000 \text{ lb/in}$$

   $$\text{Slip} = (1785/270{,}000) = 0.007 \text{ in}$$

   Good detailing specifies that tiedown bolts must be re-tightened just prior to closing in. This accomplishes two things: it takes the slack out of the oversized bolt hole and compensates for some wood shrinkage. This design example assumes that about one-half of the bolt-hole slack is taken out.

   Therefore, total slip $= (0.007) + \left(\dfrac{1}{16}\right)\dfrac{1}{2} = 0.04$ in

Table 1B-9. Deflections of the shear walls at the second floor level[1,2,3,4]

Wall	ASD v (plf)	Strength v (plf)	h (ft)	A (sq in)	E (psi)	b (ft)	G (psi)	t (in)	V_n (lb)	e_n (in)	d_a (in)	Δ (in)
A1	308	441	9.0	19.25	1.7E6	5.0	90,000	0.535	147	0.0044	0.12	0.34
A2	308	441	9.0	19.25	1.7E6	5.0	90,000	0.535	147	0.0044	0.12	0.34
B	328	469	9.0	19.25	1.7E6	14.0	90,000	0.535	156	0.0054	0.13	0.23
C1(5)	215	308	9.0	19.25	1.7E6	10.0	90,000	0.535	103	0.0014	0.12	0.19
C2(5)	215	308	9.0	19.25	1.7E6	9.0	90,000	0.535	103	0.0014	0.09	0.16
2	111	159	9.0	12.25	1.7E6	10.0	90,000	0.535	53	0.0001	0.03	0.06
3	251	359	9.0	12.25	1.7E6	22.0	90,000	0.535	120	0.0023	0.08	0.12
5	405	578	9.0	12.25	1.7E6	14.0	90,000	0.535	192	0.0106	0.06	0.23

Notes:

1. h values are from bottom of sill plate to bottom of framing at diaphragm level (top plates).
2. $\Delta = \dfrac{8vh^3}{EAb} + \dfrac{vh}{Gt} + 0.75he_n + d_a\dfrac{h}{b}$ (Eq 23-2)
3. G values are for Structural-I sheathing.
4. e_n values for Structural-I sheathing with dry lumber = $(V_n/769)^{3.276}$.
5. Shear distributed to walls C1 and C2 is proportioned based on relative lengths. Attempting to equate deflections is desirable; however, the calculations are iterative and indeterminate, and the results are very similar. The average Δ for walls A, B, and C at the second floor level is 0.30 inch. For deformation compatibility, it has been decided to size the cantilever column elements at line E for the deflections nearest shear wall at C, where the average is $\Delta = 0.18$ inch. Another approach would be to use a weighted average that includes the force in the wall. For example, if 99 percent of the load is carried by a stiff wall with $\Delta = 0.10$ inch and 1 percent is carried by a wall with $\Delta = 1.00$ inch, then the weighted average approach is appropriate. $\Delta = 0.10 \times 0.99 + 1.0 \times 0.01 = 0.11$ inch assumes no rotation and a rigid diaphragm. If the diaphragm is flexible, then deflection compatibility is not an issue. The engineer should exercise good judgment in determining deformation compatibility.

Table 1B-10. Drift check at second floor level

	Wall	Δ (in)	h (ft)	δ_x (in)	Max. Δ_a (in)	Status
East-West	A1	0.34	9.0	1.36	2.70	ok
	A2	0.34	9.0	1.36	2.70	ok
	B	0.23	9.0	0.92	2.70	ok
	C1	0.19	9.0	0.76	2.70	ok
	C2	0.16	9.0	0.64	2.70	ok
North-South	2	0.06	9.0	0.24	2.70	ok
	3	0.12	9.0	0.48	2.70	ok
	5	0.23	9.0	0.92	2.70	ok

Drift for cantilever columns at line E

The cantilever column is assumed to be fixed at the base. This can be accomplished by setting the column on a footing and then casting the grade beam around the column. With this type of connection, the stresses in the flange of the column caused by concrete bearing at the top of the grade beam should be checked. Another approach is to provide a specially detailed base plate with anchor bolts that are bolted to the top of the grade beam. The bolts and base plate will allow for some rotation, which should be considered when computing the column deflections. The grade beam should have a stiffness at least 10 times greater than that of the column for the column to be considered fixed at the base. It is common for columns of this type to have drift control the size of the column rather than bending. In addition, IBC §12.2.5.2 requires the grade beam to be designed for load combinations with overstrength factor (Ω_O).

$$\Delta = \frac{PL^3}{3EI}$$

It should be noted that if the steel columns were not needed to resist lateral forces (gravity columns only), and all lateral forces were resisted by the wood shear walls, then only relative rigidities of the wood shear walls would need to be calculated.

When designing the cantilever columns, deformation compatibility with the other lateral-resisting elements needs to be considered.

From Figure 1B-7 at line E, the force to each of the three cantilever columns becomes

$$E = (2340 \text{ lb} + 168\text{lb})/3 = 836 \text{ lb/column}$$

$$I_{req'd} = \frac{836(9 \times 12)^3}{3(29 \times 10^6)0.18} = 67 \text{ in}^4$$

where a $\Delta = 0.18$ inch is used, which is the average drift of the nearest shear wall at line C.

Try *HSS* 8 × 4 × ½

$$I_x = 71.8 \text{ in}^4$$

$$\Delta_{TS} = \left(\frac{67}{71.8}\right) 0.18 = 0.17 \text{ in}$$

Check for cantilever column limitations §12.2.5.2

Cantilever columns have a very low buiding *R* value, and their past performance in earthquakes has been poor. The first publication of the Structural Seismic Design Manual illustrated the severe penalty placed on a building where cantilever columns were used. The 1997 UBC did not allow for a mixed R value system in one direction without penalizing the entire direction. The code now allows mixed R value systems in one direction (see Section 1 of this design example for elaboration on this issue).

To safeguard against a P-delta buckling type of failure, the code limits the amount of axial load placed on cantilever columns. Section 12.2.5.2 places a limit of 15 percent of the permissible axial stress. For allowable stress design, the axial load on the cantilever columns is to be in accordance with the load combinations of §2.4.

For cantilever columns the recommended k factor is 2.1

$$\frac{k\ell}{r_x} = \frac{2.1 \times 9.0 \times 12}{2.71} = 83.7 \quad \frac{k\ell}{r_y} = \frac{2.1 \times 9.0 \times 12}{1.56} = 145.4 \text{ (controls)}$$

Allowable axial stress $F_a = 7.11$ ksi

Deadload of column (critical for center column)

$$19.5 \text{ psf} \times \frac{(9.0 \text{ ft})^2}{2 \times 7.0 \text{ ft}} \left(\frac{10.0 \text{ ft}}{2} + \frac{18.0 \text{ ft}}{2} \right) = 1.58k$$

Live load on column

$$20.0 \text{ psf} \times \frac{(9.0 \text{ ft})^2}{2 \times 7.0 \text{ ft}} \left(\frac{10.0 \text{ ft}}{2} + \frac{18.0 \text{ ft}}{2} \right) = 1.62k$$

From §2.4, the critical load combination is

$$D + 0.525E + 0.75\, L_r$$

$$P = 1.58 + 0 + 0.75\,(1.62) = 2.80k$$

$$f_a = \frac{2.80}{9.74} = 0.30 \text{ ksi} < 0.15 \times 7.11 = 1.07 \text{ ksi} \ldots o.k.$$

As mentioned in Example 1, the seismic modification factor R of 6.5 was used for the building. The seismic modification factor for the cantilever columns is 1.25. Therefore, the seismic force to the cantilever columns needs to be factored up by the factor of (6.5/1.25) = 5.20.

$$E_{column} = 836 \times 5.20 = 4347 \text{ lb}$$

$$M = E \times h = 4347 \times 9.0 = 39.1 \text{ ft-k}$$

$$f_b = \frac{39.1 \times 12}{17.9} = 26.2 \text{ ksi} < 0.66 \times 46 = 30.4 \text{ ksi}$$

Combined:

$$\frac{f_a}{2F_a} + \frac{f_b}{F_b} = \frac{0.30}{2 \times 7.07} + \frac{26.2}{30.4} = 0.02 + 0.86 = 0.88 < 1.0 \ldots o.k. \quad \text{AISC Eq (H1-16)}$$

Drift check for cantilever columns

$$\delta_x = \frac{C_d \Delta}{I} \qquad \text{Eq 12.8-15}$$

$$\delta_x = \frac{1.25 \times 0.16}{1.0} = 0.20 \text{ in}$$

$$\Delta_{max} = 0.025 \times \text{the story height } h_{sx}$$

$$\Delta_{max} = 0.025 \times 9 \times 12 = 2.7 \text{ in} > 0.20 \text{ in} \ldots o.k.$$

Use *HSS* 8 × 4 × ½ steel columns (3 locations).

Table 1B-11. Wall rigidities at second floor level (walls from first to second floor)[(1)]

Wall	$\Delta^{(2)}$ (in)	F_{tot} (lb)	$k = \frac{F_{tot}}{\Delta}$ (k/m)	$k = \frac{F_{tot}}{\Delta}$ (k/m)
A1	0.34	2,024	5.953	11.91
A2	0.34	2,024	5.953	
B	0.23	6,550		28.48
C1	0.19	3,082	16.22	33.56
C2	0.16	2,774	17.34	
E	0.16	2,508		15.67
2	0.06	1,592		26.53
3	0.12	7,891		65.76
5	0.23	8,090		35.17

Notes:

1. Deflections and forces are based on strength force levels.
2. Δ is the design level displacement from Table 1B-9.

4. Determine centers of mass and rigidity of diaphragms (Figure 1-13)

It has been common strategy for practicing engineers to assume flexible diaphragms and distribute loads to shear walls based on tributary areas, and this is a well-established conventional design assumption. In this design example, the rigid diaphragm assumption will be used. This is not intended to imply that seismic design of residential construction in the past should have been necessarily performed in this manner. However, recent earthquakes and testing of wood panel shear walls have indicated that expected drifts are considerably higher than what was known or assumed in the past. This knowledge of the increased drifts of short wood panel shear walls has increased the need for the engineer to consider the relative rigidities of shear walls. This, and the fact that diaphragms tend to be much more rigid than the shear walls, has necessitated consideration of diaphragm rigidities. Therefore, in this part, the diaphragms are assumed to be rigid. See Part 6 for confirmation of this assumption.

4a. For roof diaphragm

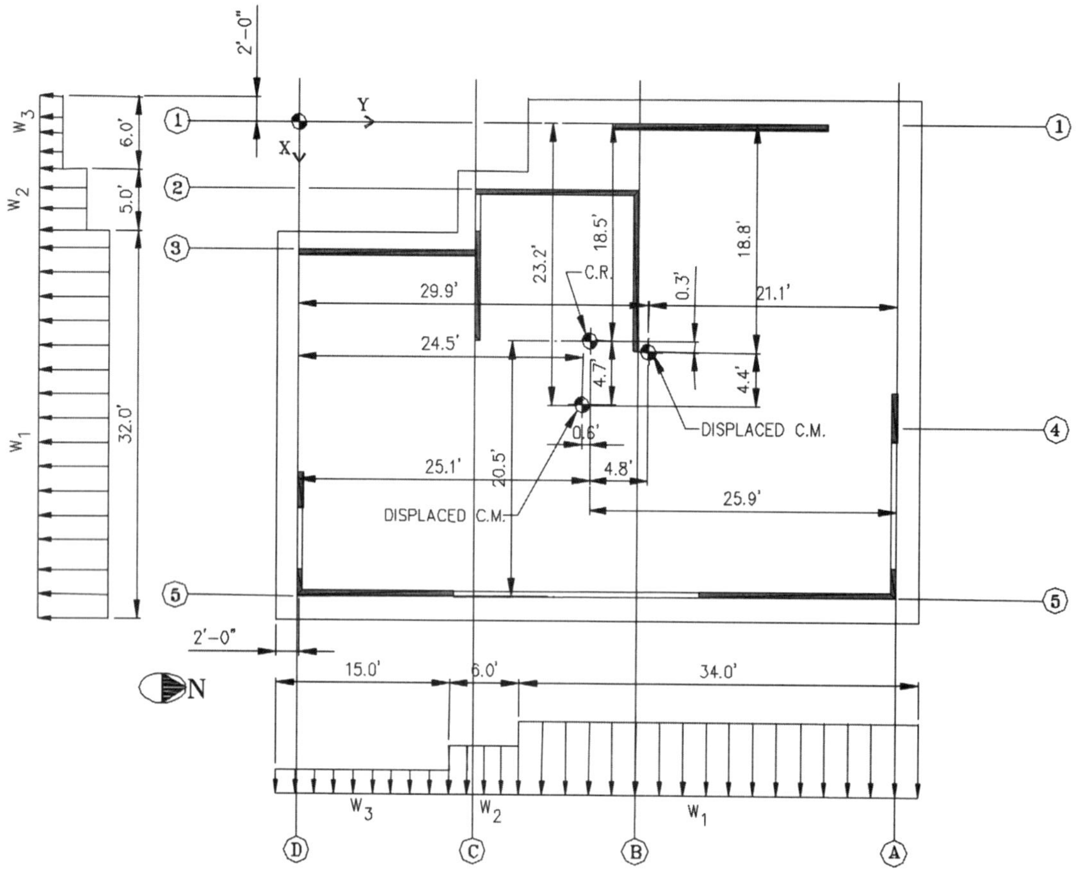

Figure 1B-13. Roof diaphragm centers of rigidity and mass

Determine center-of-mass (CM) of roof diaphragm from wall loads

Using diaphragm loading from flexible diaphragm analysis for east-west direction (see Figure 1B-6) and summing forces about line D:

$$303 \text{ plf } (34.0 \text{ ft}) = 10{,}302 \text{ lb} \times \left(\frac{34}{2} + 6 + (15-2)\right) \text{ft} = 370{,}872 \text{ ft-lb}$$

$$261 \text{ plf } (6.0 \text{ ft}) = 1{,}566 \text{ lb} \times \left(\frac{6}{2} + (15-2)\right) \text{ft} = 25{,}056 \text{ ft-lb}$$

$$226 \text{ plf } (15.0 \text{ ft}) = 3{,}390 \text{ lb} \times \left(\frac{15}{2} - 2\right) \text{ft} = 18{,}645 \text{ ft-lb}$$

$$15{,}258 \text{ lb} \qquad 414{,}573 \text{ ft-lb}$$

$$\therefore \bar{y}_m = \frac{\Sigma wx}{\Sigma w} = \frac{414{,}573 \text{ ft-lb}}{15{,}258 \text{ lb}} = 27.2 \text{ ft @ roof}$$

Using diaphragm loading from flexible diaphragm analysis for north-south direction (see Figure 1B-8) and summing forces about line 1

$$376 \text{ plf } (32.0 \text{ ft}) = 12{,}032 \text{ lb} \times 25.0 \text{ ft} = 300{,}800 \text{ ft-lb}$$

$$274 \text{ plf } (5.0 \text{ ft}) = 1{,}370 \text{ lb} \times 6.5 \text{ ft} = 8{,}905 \text{ ft-lb}$$

$$233 \text{ plf } (6.0 \text{ ft}) = \underline{1{,}398 \text{ lb}} \times 1.0 \text{ ft} = \underline{1{,}398 \text{ ft-lb}}$$

$$14{,}800 \text{ lb} \qquad 311{,}103 \text{ ft-lb}$$

$$\therefore \bar{x}_m = \frac{\Sigma wy}{\Sigma w} = \frac{311{,}103 \text{ ft-lb}}{14{,}800 \text{ lb}} = 21.0 \text{ ft @ roof}$$

Determine center-of-rigidity (CR) for roof diaphragm

Using the rigidity values k from Table 1B-6 and the distance y from line D to the shear wall

$$\bar{y} = \frac{\Sigma(k_{xx} y)}{\Sigma k_{xx}} \quad or \quad \bar{y}\,\Sigma k_{xx} = \Sigma k_{xx}\, y$$

$$\bar{y}(5.98 + 7.05 + 16.97 + 6.06) = 5.98(0) + 7.08(15.0) + 16.97(29.0) + 6.06(51.0)$$

$$\therefore \bar{y}_r = \frac{907}{36.1} = 25.1 \text{ ft @ roof}$$

Using the rigidity values k from Table 1B-6 and the distance x from line 1 to the shear wall

$$\bar{x} = \frac{\Sigma(k_{yy} x)}{\Sigma k_{yy}} \quad or \quad \bar{x}\,\Sigma k_{yy} = \Sigma k_{yy} x$$

$$\bar{x}(19.65 + 6.79 + 26.57 + 31.39) = 19.65(0) + 6.79(6.0) + 26.57(11.0) + 31.39(39.0)$$

$$\therefore \bar{x}_r = \frac{1557.2}{84.40} = 18.5 \text{ ft @ roof}$$

4b. For second floor diaphragm

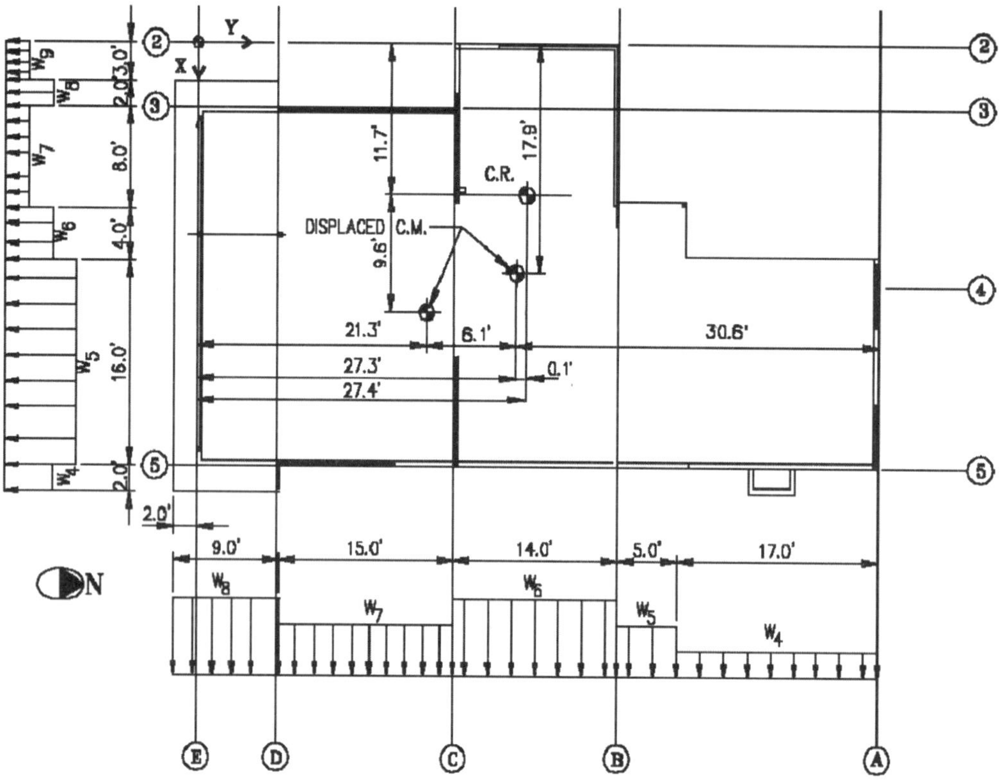

Figure 1B-14. Second floor diaphragm centers of rigidity and mass

Determine CM of floor diaphragm from wall loads.

Using diaphragm loading from flexible diaphragm analysis for east-west direction (See Figure 1B-7) and summing forces about line E

$$
\begin{aligned}
42 \text{ plf } (17.0 \text{ ft}) &= 714 \text{ lb} \times 49.5 \text{ ft} = 35{,}343 \text{ ft-lb} \\
53 \text{ plf } (5.0 \text{ ft}) &= 265 \text{ lb} \times 38.5 \text{ ft} = 10{,}202 \text{ ft-lb} \\
87 \text{ plf } (14.0 \text{ ft}) &= 1218 \text{ lb} \times 29.0 \text{ ft} = 35{,}322 \text{ ft-lb} \\
74 \text{ plf } (15.0 \text{ ft}) &= 1110 \text{ lb} \times 14.5 \text{ ft} = 16{,}095 \text{ ft-lb} \\
84 \text{ plf } (9.0 \text{ ft}) &= \underline{756 \text{ lb}} \times 2.5 \text{ ft} = \underline{1{,}890 \text{ ft-lb}} \\
&\quad 4063 \text{ lb} \qquad\qquad\quad 98{,}852 \text{ ft-lb}
\end{aligned}
$$

$$\therefore \bar{y}_m = \frac{98{,}852 \text{ ft-lb}}{4063 \text{ lb}} = 24.3 \text{ ft @ second floor}$$

Using diaphragm loading from flexible diaphragm analysis for north-south direction (see Figure 1B-9) and summing forces about line 2

$$
\begin{aligned}
23 \text{ plf } (2.0 \text{ ft}) &= 46 \text{ lb} \times 34.0 \text{ ft} = 5{,}564 \text{ ft-lb} \\
154 \text{ plf } (16.0 \text{ ft}) &= 2464 \text{ lb} \times 25.0 \text{ ft} = 61{,}600 \text{ ft-lb} \\
110 \text{ plf } (4.0 \text{ ft}) &= 440 \text{ lb} \times 15.0 \text{ ft} = 6{,}600 \text{ ft-lb} \\
97.2 \text{ plf } (8.0 \text{ ft}) &= 778 \text{ lb} \times 9.0 \text{ ft} = 7{,}002 \text{ ft-lb} \\
58.9 \text{ plf } (2.0 \text{ ft}) &= 118 \text{ lb} \times 4.0 \text{ ft} = 471 \text{ ft-lb} \\
35.8 \text{ plf } (3.0 \text{ ft}) &= 107 \text{ lb} \times 1.5 \text{ ft} = 160 \text{ ft-lb} \\
&\quad 3953 \text{ lb} \qquad\qquad 77{,}397 \text{ ft-lb}
\end{aligned}
$$

$$\therefore \bar{x}_m = \frac{77{,}397 \text{ ft-lb}}{3953 \text{ lb}} = 19.6 \text{ ft @ second floor}$$

Determine CR for floor diaphragm

Using the rigidity values k from Table 1B-11 and the distance y from line E to the shear wall

$$\bar{y}_r(15.67 + 33.56 + 28.48 + 11.91) = 15.67(0) + 33.56(22.0) + 28.48(36.0) + 11.91(58.0)$$

$$\therefore \bar{y}_r = \frac{2454}{89.62} = 27.4 \text{ ft @ second floor}$$

Using the rigidity values k from Table 1B-11 and the distance x from line 2 to the shear wall

$$\bar{x}_r(26.53 + 65.76 + 35.17) = 26.53(0.0) + 65.76(5.0) + 35.17(33.0)$$

$$\therefore \bar{x}_r = \frac{1489.4}{127.5} = 11.7 \text{ ft @ second floor}$$

5. Distribution of lateral forces to the shear walls with rigid diaphragms §12.8.4

Using the rigid diaphragm assumption, the base shear was distributed to the two levels in Part 1. In this part, the story forces are distributed to the shear walls that support each level.

The code requires in §12.8.4.2 that the story force at the CM be displaced from the calculated CM a distance of 5 percent of the building dimension at that level perpendicular to the direction of force. This is to account for accidental torsion. The code requires the most severe load combination to be examined and also permits the negative torsional shear to be subtracted from the direct load shear. However, lateral forces must be considered to act in each direction of the two principal axes. This design example does not reflect on eccentricities between the CMs between levels. In this example, these eccentricities are small and are therefore deemed insignificant. The engineer must exercise good judgment in determining when these effects need to be considered.

5a. For the roof diaphragm (see Figure 1B-13)

Forces in the east-west (x) direction

Distance to the calculated CM: $\bar{y}_m$ = 27.2 ft

Displaced $e_y = (0.05 \times 55 \text{ ft})$ = 2.7 ft

New $\bar{y}$ to displace CM = 27.2 ft + 2.7 ft = 29.9 ft *or* 24.5 ft

Distance to the calculated CR: $\bar{y}_r$ = 25.1 ft

$$e_y = 29.9 - 25.1 = 4.8 \text{ ft}$$

or

$$e_y = 25.1 - 24.5 = 0.6 \text{ ft}$$

Note that displacing the CM by 5 percent can result in the CM being on either side of the CR and can produce added torsional shears to all walls.

$$T_x = F_x e_y = 15{,}230 \text{ lb } (4.8 \text{ ft}) = 73{,}104 \text{ ft-lb}$$

or

$$T_x = F_x e_y = 15{,}230 \text{ lb } (0.6 \text{ ft}) = 9138 \text{ ft-lb}$$

Forces in the north-south (y) direction

Distance to the calculated CM: $\bar{x}_m$ = 21.0 ft

Displaced $e_x = (0.05 \times 43 \text{ ft})$ = 2.2 ft

New $\bar{x}$ to displace CM = 21.0 ft ± 2.2 ft = 23.2 ft *or* 18.8 ft

Distance to the calculated CR: $\bar{x}_r$ = 18.5 ft

$$e_x = 23.2 - 18.5 = 4.7 \text{ ft}$$

or

$$e_x = 18.8 - 18.5 = 0.3 \text{ ft}$$

$$T_y = F_y e_x = 14{,}800 \text{ lb } (4.7 \text{ ft}) = 69{,}560 \text{ ft-lb}$$

or

$$T_y = F_y e_x = 14{,}800 \text{ lb } (0.3 \text{ It}) = 4440 \text{ ft-lb}$$

$F_{e\text{-}w}$ = 15,230 lb (Table 1B-1b)

$F_{n\text{-}s}$ = 14,800 lb (Table 1B-1a)

T_x = 73,104 ft-lb for walls A and B

T_x = 9138 ft-lb for walls C and D

T_y = 69,560 ft-lb for wall 5

T_y = 4440 ft-lb for walls 1, 2, and 3

5b. For roof wall forces

The direct shear force F_v is determined from

$$F_v = \frac{R}{\Sigma R}$$

and the torsional shear force F_t is determined from

$$F_T = T\frac{Rd}{J}$$

where

$J = \Sigma Rd_x^2 + \Sigma Rd_y^2$

R = rigidity of lateral resisting element

d = distance from lateral resisting element to the CR

$T = F_e$

Table 1B-12. Distribution of forces to shear walls below the roof level

	Wall	R_x	R_y	d_x	d_y	Rd	Rd^2	Direct Force F_v	Torsional Force F_t	Total Force $F_v + F_t$
East-West	A	6.06			25.9	157	4,065	2,559	367	2,926
	B	16.97			3.9	66.2	258	7,167	155	7,322
	C	7.05			10.1	71.2	719	2,978	21	2,999
	D	5.98			25.1	150	3,749	2,526	44	2,570
	Σ	36.06					8,791	15,230		
North-South	1		19.65	−18.5		−363.5	6,725	3,446	−52	3,394
	2		6.79	−12.5		−84.8	1,061	1,191	−12	1,179
	3		26.57	−7.5		−199.3	1,495	4,659	−29	4,630
	5		31.39	20.5		643.5	13,192	5,504	1,432	6,936
	Σ		84.40				22,473	14,800		
	Σ						31,264			

For simplicity, many engineers will add 5 or 10 percent of the direct force shears to account for torsional effects. The average torsional force added to the shears walls in this design example is 8 percent of the direct force. Adding only 5 percent of the wall shears would not be conservative.

Torsional forces are subtracted from direct forces for this design example as is now allowed by the code. This only occurs when both displaced CMs are on the same side of the CR for a given direction. When the CR occurs between the two displaced CMs, then torsional forces cannot be subtracted (which occurs at the roof in the east-west direction). Many engineers still neglect these negative forces.

5c. For the floor diaphragm (see Figure 1B-14)

Forces in the east-west (x) direction

Distance to the calculated CM: $\bar{y}_m$ = 24.3 ft

Displaced $e_y = (0.05 \times 60 \text{ ft})$ = 3.0 ft

New $\bar{y}$ to displaced CM = 24.3 ft ± 3.0 ft = 27.3 ft *or* 21.3 ft

Distance to the calculated CR: $\bar{y}_r$ = 27.4 ft

$e_y = 27.3 - 27.4 = -0.1$ ft

or

$e_y = 27.4 - 21.3 = 6.1$ ft

$T_x = F_x e_y = 19{,}275$ lb (0.1 ft) = 1,930 ft-lb

or

$T_x = F_x e_y = 19{,}275$ lb (6.1 ft) = 117,580 ft-lb

Forces in the north-south (y) direction

Distance to the calculated CM: $\bar{x}_m$ = 19.6 ft

Displaced $e_x = (0.05 \times 35 \text{ ft})$ = 1.7 ft

New $\bar{x}$ to displaced CM = 19.6 ft ± 1.7 ft = 21.3 ft *or* 17.9 ft

Distance to the calculated CR: $\bar{x}_r$ = 11.7 ft

$e_x = 21.3 - 11.7 = 9.6$ ft

or

$e_x = 17.9 - 11.7 = 6.2$

$T_y = F_y e_x = 18{,}750$ lb (9.6 ft) = 180,000 ft-lb

or

$T_y = F_y e_x = 18{,}750$ lb (6.2 ft = 116,250 ft-lb

$F_{e\text{-}w} = (15{,}230 + 4045)$ = 19,275 lb (adding forces from roof and floor from Table 1B-1b)

T_x = 1,930 ft-lb for walls A and B

T_x = 117,580 ft-lb for walls C and E

$F_{n\text{-}s} = (14{,}800 + 3950)$ = 18,750 lb (adding forces from roof and floor from Table 1B-1a)

T_y = 116,250 ft-lb for walls 2 and 3

T_y = 180,000 ft-lb for wall 5

Table 1B-13. Distribution of forces to shear walls below the second floor level

	Wall	R_x	R_y	d_x	d_y	Rd	Rd^2	Direct Force F_v	Torsional Force F_t	Total Force $F_v + F_t$
East-West	A	11.91			30.6	364.4	11,152	2,562	14	2,576
	B	28.48			8.6	244.9	2,106	6,125	10	6,135
	C	33.56			5.4	181.2	979	7,218	439	7,657
	E	15.67			27.4	429.3	11,764	3,370	1,040	4,410
	Σ	89.62					26,001	19,275		
North-South	2		26.53	−11.7		−310	3,632	3,903	−1,427	2,476
	3		65.76	−6.7		−440	2,952	9,674	−2,025	7,649
	5		35.17	21.3		749	15,956	5,173	5,337	10,510
	Σ		127.46				22,540	18,750		
	Σ						48,541			

Table 1B-14. Comparison of loads on shear walls using flexible vs. rigid diaphragm analysis and recheck of nailing in walls

Wall	$F_{flexible}$	F_{rigid}	Rigid/ Flexible Ratio	b (ft)	$V = \frac{F_{max}(0.7)}{(b)}$	Sheathing 1 or 2 sides	Allowable Shear (plf)	Edge Nail Spacing (in)
					Roof Level			
A	3,939	2,926	0.74	10.0	276	One	510	4
B	5,430	7,322	1.38	14.0	366	One	510	4
C	3,737	2,999	0.80	8.5	308	One	510	4
D	2,152	2,570	1.19	6.0	300	One	510	4
1	1,179	3,394	2.88	18.0	132	One	510	4
2	1,493	1,179	0.79	10.0	105	One	510	4
3	6,112	4,630	0.76	15.0	285	One	510	4
5	6,016	6,936	1.15	26.0	187	One	510	4
					Floor Level			
A	4,407	2,576	0.58	10.0	308	One	510	4
B	6,550	6,135	0.94	14.0	328	One	510	4
C	5,856	7,657	1.28	19.0	282	One	510	4
E	2,508	4,410	1.75	—	—	—	—	—
2	1,592	2,476	1.56	10.0	173	One	510	4
3	7,891	7,649	0.97	22.0	251	One	510	4
5	8,090	10,510	1.30	14.0	526	One	510	4

Note that the nailing for the shear wall at line 5 at the first floor level will need to be decreased.

In SDC D, E, or F, shear walls with shears that exceed allowable shears of 350 plf will require 3x framing at abutting panel edges with staggered nails. See also notes at bottom of Table 1B-2a.

Where rigid diaphragm analysis shows seismic forces to the shear walls are higher than those from flexible diaphragm analysis, the wall stability and anchorage must be re-evaluated. Engineering judgment should be used to determine if a rigid diaphragm analysis should be repeated because of changes in wall rigidity.

If rigid diaphragm loads are used, the diaphragm shears should be rechecked for total load divided by diaphragm length along the individual wall lines.

6. Diaphragm deflections and whether diaphragms are flexible or rigid

This step is shown only as a reference for how to calculate horizontal diaphragm deflections. Since the shear wall forces were determined using both flexible and rigid diaphragm assumptions, there is no requirement to verify that the diaphragm is either rigid or flexible.

The design seismic force in the roof and floor diaphragms using Equation 12.10-1 must first be found. The design seismic force is then divided by the diaphragm area to determine the horizontal loading in pounds per square foot (refer to Figures 1B-13 and 1B-14). The design seismic force shall not be less than $0.20 S_{DS} I w_{px}$ nor greater than $0.4 S_{DS} I w_{px}$.

The basic equation for determining seismic forces on a diaphragm is shown below. The following will compute the seismic forces in the east-west direction

$$F_{px} = \frac{\sum_{i=x}^{n} F_i}{\sum_{i=x}^{n} w_i} w_{px} \qquad \text{Eq 12.10-1}$$

Note that the forces in the east-west direction are higher.

$$F_{p\,roof} = \frac{(15{,}230 \times 65{,}400)}{65{,}400} = 15{,}230 \text{ lb}$$

$$F_{p\,roof} = \frac{15{,}230 \text{ lb}}{2164 \text{ sf}} = 7.04 \text{ psf}$$

For simplicity, the wall weights parallel to the direction of loading are included in w_{px}.

For the uppermost computed in level, the above calculation will always produce the same force as that Equation (12.8-11).

$$F_{p\,floor} = \frac{(15{,}230 + 4045) \times 39{,}900}{(39{,}900 + 65{,}400)} = 7304 \text{ lb}$$

$$F_{p\,min} = 0.2 S_{DS} I_w w_{px} = 0.20(1.19)(1.0) w_{px} = 0.238(39{,}900) = 9496 \text{ (governs)} \qquad \text{§12.10.1.1}$$

$$F_{p\,max} = 0.4 S_{DS} I_w w_{px} = 0.4(1.19)(1.0) w_{px} = 0.476(39{,}900) = 18{,}992 \text{ lb}$$

$$F_{p\,floor} = \frac{9496 \text{ lb}}{1542 \text{ sf}} = 6.16 \text{ psf} \qquad \text{§12.10.1.1}$$

The designer is required to check diaphragm forces in each direction. This design example performs the check in one direction only.

In this example, the roof and floor diaphragms spanning between line A and line B will be used to illustrate the method. The basic equation for determining the deflection of a diaphragm is

$$\Delta = \frac{5vL^3}{8EAb} + \frac{vL}{4Gt} + 0.188Le_n + \frac{\Sigma(\Delta_c X)}{2b} \qquad \text{(Eq 23-1)}$$

The above equation is based on a uniformly loaded, uniformly nailed, simple span diaphragm with blocked panel edges, and monotonic tests conducted by the American Plywood Association (APA). The equation has four separate parts: the first accounts for beam bending; the second, for shear deformation; the third, for nail slippage/bending; and the fourth part accounts for chord slippage.

The SDPWS has changed the traditional 4-term expression to a simplified 3-term expression for diaphragm deflection. This new equation is shown below:

$$\Delta = \frac{5vL^3}{8EAW} + \frac{0.25vL}{1000G_a} + \frac{\Sigma(x\Delta_c)}{2W} \qquad \text{SDPWS Eq 4.2-1}$$

The new simplified 3-term equation combines the second and third terms (of the 4-term equation) into one term. Computed deflections by using either the 4-term equation or the 3-term equation produces nearly identical results at the critical strength level (1.4 times the allowable shear values for seismic).

For the purpose of this calculation, assume the diaphragm is a simple span supported at A and B (refer to Figures 1B-13 and 1B-14). In reality, with continuity at B, the actual deflection will be less.

6a. Roof diaphragm

Check diaphragm shear.

Based on the $F_{p\,roof} = 7.04$ psf as computed above, find roof shear to line A for the east-west direction:

1. Area of roof including overhangs: 22 × 43 ft
2. Wall length: 39 ft
3. Diaphragm shears are converted to allowable stress design by multiplying by 0.7

$$v = \frac{(7.04)43.0(22.0)0.7}{(39.0)2} = 60 \text{ plf} < 190 \text{ plf allowable}$$

From IBC Table 2306.3.1, the allowable shear of 190 plf is based on 15/32-inch APA or TECO performance-rated wood structural panels (DOC PS1 or PS2) with unblocked edges and 10d nails with 1½-inch penetration spaced at 6 inches o/c at boundaries and panel edges. APA or TECO performance-rated wood structural panels may be either plywood or oriented strand board (OSB).

Check diaphragm deflection using the IBC equation.

The code specifies that the deflection be calculated on an equivalent tributary lateral load basis. In other words, the diaphragm deflection should be based on the *same load* as the load used for the lateral resisting elements, not F_{px} total force at the level considered. Since the code requires building drifts to be determined by strength level forces as specified in §12.8, determine strength loads on building diaphragm.

$$F_{p\,roof} = \frac{15{,}230 \text{ lb}}{2164 \text{ sf}} = 7.04 \text{ psf}$$

$$v = \frac{(7.04 \text{ psf})43.0 \text{ ft}(22.0 \text{ ft})}{2(39.0 \text{ ft})} = 85 \text{ plf}$$

With nails at 6 inches o/c, the load per nail is 85(6/12) = 43 lb/nail = V_n

L = 22.0 ft

b = 39.0 ft

G = 90,000 psi — Plywood Design Specifications (PDS) T 3

E = 1,700,000 psi

$A_{2x4\,chords}$ = 5.25 sq in × 2 = 10.50 sq in

Sum of individual chord-splice slip.

Note that the area for two 2x4 top plates (chord) has been used. All top plates are connected with metal straps. If a metal strap is not used, then use of the area for one top plate is recommended. Also note that the top plates at line 1 are two 2x6s. The deflection calculation will conservatively use the chord area of the two 2x4s at line 5.

Fastener slip/nail deformation values (e_n)

$e_n = 1.20(43/769)^{3.276} = 0.0001$

t = 0.298 in (for CDX or Standard Grade) — PDS T 1

The chord-splice of the diaphragm will be spliced with a 12-gage metal strap using 10d nails. Assume a chord splice of the diaphragm at mid-span. The slippage for both the diaphragm chords is to be included. The nail slip value from APA Research Report 138 can be used

$$e_n = (V_n/769)^{3.276} = (127/769)^{3.276} = 0.003 \text{ in}$$

where

The allowable load is 127 pounds per nail (from NDS Table 11P for a 10d nail in a 12-gage strap).

V_n = 127 lb/nail in the strap. The elongation of the metal strap is assumed to be 0.03 inch.

Therefore, the chord slip is

$$\Delta_c = 0.003 + 0.03$$

$$\Delta_c = 0.033 \text{ in}$$

$$\Sigma(\Delta_c X) = (0.033)11.0 \text{ ft } (2) = 0.73 \text{ in-ft}$$

where the distance to the nearest support is 11 feet, and to get the sum for both chords, multiply by 2.

$$\Delta = \frac{5(85)22.0^3}{8(1.7E6)10.50(39.0)} + \frac{85(22.0)}{4(90{,}000)0.298} + 0.188(22.0)0.0001 + \frac{0.70}{2(39.0)} = 0.03 \text{ in}$$

This deflection is based on a *blocked* diaphragm. The IBC does not have a formula for an *unblocked* diaphragm.

As a comparison, determine diaphragm deflection for the unblocked roof diaphragm using the SDPWS 3-term expression.

$$\Delta = \frac{5vL^3}{8EAW} + \frac{0.25vL}{1000G_a} + \frac{\Sigma(x\Delta_c)}{2W} \qquad \text{SDPWS Eq 4.2-1}$$

v = 85 plf

L = 22.0 ft

E = 1,700,000 psi

A = 10.5 sq in 2x4 chords

W = 39.0 ft

G_a = 15.0 k/in — SDPWS T 4.2B

$\Sigma(x\Delta_c)$ = 0.73 in-ft

$$\Delta = \frac{5(85)22.0^3}{8(1.7E6)10.5(39.0)} + \frac{0.25(85)22.0}{1000(15.0)} + \frac{0.73}{2(39.0)} = 0.04 \text{ in}$$

This deflection is based on an unblocked diaphragm. Comparing this to the IBC deflection of 0.03 for a blocked diaphragm, deflection is about 40 percent higher than the blocked diaphragm deflection.

Note that at gable-ended roofs, when the chord is in the plane of the roof (pitched), the chord connection at the ridge should be carefully detailed to accommodate the uplift component of the chord.

6b. Floor diaphragm

Check diaphragm shear.

Based on the $F_{p\,floor}$ = 6.16 psf as computed in Part 6 above, find floor shear to line A for the east-west direction (floor area is 22 × 16).

Diaphragm shears are converted to allowable stress design by multiplying by 0.7

where

$$v = \frac{(6.16\text{ psf})16.0\text{ ft}(22.0\text{ ft})0.7}{2(16.0)} = 47\text{ plf} < 190\text{ plf} \qquad \text{T 2306.3.1}$$

Allowable shear of 190 plf is based on 15/32-inch APA or TECO performance-rated sheathing with unblocked edges and 10d nails with 1½-inch penetration spaced at 6 inches o/c at boundaries and panel edges supported on framing. APA or TECO performance-rated wood structural panels may be either plywood or OSB.

Check diaphragm deflection.

$$f_{p\,floor} = \frac{4045}{1542} = 2.62\text{ psf}$$

$$v = \frac{(2.62\text{ psf})16.0\text{ ft}(22.0\text{ ft})}{2(16.0)} = 29\text{ plf}$$

With nails at 6 inches o/c the load per nail is 29(6/12) = 15 lb/nail = V_n

L = 22.0 ft

b = 16.0 ft

G = 90,000 psi PDS T 3

E = 1,700,000 psi

$A_{2x4\ chords}$ = 5.25 sq in × 2 = 10.50 sq in

$e_n = 1.2(15/769)^{3.276} = 3.0E-06$

t = 0.298 in PDS T 2

Using an assumed single chord-splice slip of 0.033 inch at the mid-span of the diaphragm

$$\Sigma\Delta_c X = (0.033)11.0\text{ ft }(2) = 0.73\text{ in-ft}$$

$$\Delta = \frac{5(29)22.0^3}{8(1.7E6)10.50(16.0)} + \frac{29(22.0)}{4(90{,}000)0.298} + 0.188(22.0)3.0E\text{-}06 + \frac{0.73}{2(16.0)} = 0.03\text{ in}$$

Converting to an unblocked diaphragm

$$\Delta = 0.03(1.4) = 0.04\text{ in}$$

6c. Flexible versus rigid diaphragms §12.3.1.3

The maximum diaphragm deflection is 0.04 inch, assuming a simple span for the diaphragm. The average story drift is on the order of 0.50 inch (see Part 4, Tables 1B-7 and 1B-10 for the computed deflections of the shear walls). For the diaphragms to be considered flexible, the maximum diaphragm deflection will have to be more than 2 times the average story drift, or 1.0 inch. This would be 25 times the computed "simple span" deflections of the diaphragms. As defined by the code, the diaphragms are considered rigid. Because some amount of diaphragm deformation will occur, the analysis is highly complex and beyond the scope of what is normally done for this type of construction.

Diaphragm deflection analysis and testing to date has been performed on level/flat diaphragms. There has not been any testing of sloped (e.g., roof) and complicated diaphragms as found in the typical wood-framed single-family residence. Consequently, some engineers create their design based on the roof diaphragm being flexible and the floor diaphragm being rigid.

In this procedure, the engineer should exercise good judgment in determining if the higher load of the two methodologies is actually required. In other words, if the load to two walls by rigidity analysis is found to be 5 percent to line A and 95 percent to line B, but by flexible analysis it is found to be 50 percent to line A and 50 percent to line B, the engineer should probably design for the larger of the two loads for the individual walls. Note that the same definition of a flexible diaphragm has been in the UBC since the 1988 edition. However, it generally has not been enforced by building officials for Type V construction. ASCE/SEI 7-05 has repeated this same definition. For further discussion, see the commentary at the end of this example.

Commentary

Following are some issues and topics related to the seismic design of wood frame residences that can be used to improve design practices and/or understanding of important aspects of design.

Rigid versus flexible diaphragm.

This design example illustrates seismic design using both flexible and rigid diaphragms. It also illustrates that most one- and two-family dwellings have rigid diaphragms as defined by code. This being the case, a design needs to be based on either a rigid or flexible diaphragm assumption, not both. Using the common approach of basing wall rigidities on deflections of shear walls and other vertical elements, the engineer first needs to know or assume how the shear walls will be constructed (e.g., nail size and spacing). Without performing a preliminary analysis, the procedure of just doing a design based on rigid diaphragms may be subject to a trial and error process. One method (as used in this design example) to avoid this process is to first perform an analysis based on flexible diaphragms, then use the construction required from the flexible diaphragms for determining the wall rigidities.

Part 2 of this design example uses flexible diaphragms to determine shear wall construction. Parts 3, 4, and 5 of this design example use rigid diaphragms per ASCE/SEI 7-05 requirements without the exception for one- and two-family dwellings. The shear wall deflections employed in this design example use IBC and SDPWS equations. This needs to be viewed as one possible approach that is substantiated by the code. However, other approaches can also be used. Two of these are given below.

1. The rigidities of the shear walls can be based on the length of the wall times the allowable shear capacity. This method can be appropriate, provided the tiedown assembly displacements are kept to a minimum, which may involve using specific types of tiedown devices that limit displacements to less than ⅛ inch.

2. Shear wall rigidities can be based on graphs of the four-term shear wall code deflection equation (see Part 3b). A chart of these (Figure 1B-15) is included in this section and is also considered appropriate in determining wall rigidities.

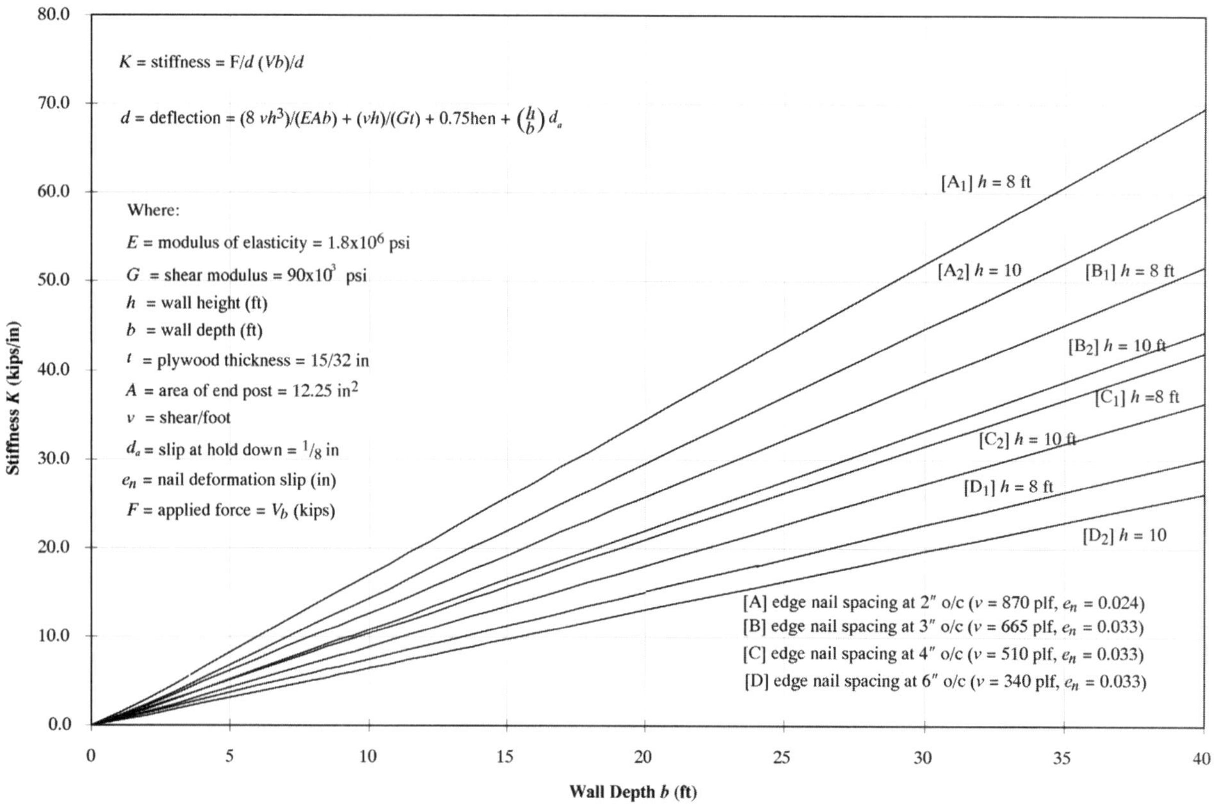

Figure 1B-15. Stiffness of one-story 1/2-inch Structural-I plywood shear walls

References

American Forest and Paper Association, 2005, AF&PA Supplement Special Design Provisions for Wind and Seismic (SDPWS-05), American Forest and Paper Association, Washington D.C.

American Forest and Paper Association, 2005, *National Design Specification for Wood Construction including Supplements, NDS-05*. American Forest and Paper Association, Washington D.C.

American Forest and Paper Association, 1996, *Wood Construction Manual*. American Forest and Paper Association, Washington D.C.

American Plywood Association, 1997, *Design/Construction Guide—Diaphragms and Shear Walls*. Form L350, Engineered Wood Association, Tacoma, Washington.

American Plywood Association, 1997, *Diaphragms and Shear Walls*. Engineered Wood Association, Tacoma, Washington.

American Plywood Association, 1993, revised, *Wood Structural Panel Shear Walls*. Report 154, Engineered Wood Association, Tacoma, Washington.

American Plywood Association, 1994, *Northridge, California Earthquake*. Report T-94-5. Engineered Wood Association, Tacoma, Washington.

American Plywood Association, *Performance Standards and Policies for Structural-Use Panels [Sheathing Standard, Sec. 2.3.3]*. Standard PRP-108. Engineered Wood Association, Tacoma, Washington.

American Plywood Association, 1997, *Plywood Design Specifications*, Form Y510, Engineered Wood Association, Tacoma, Washington.

American Plywood Association, 1988, *Plywood Diaphragms*, Research Report 138. American Plywood Association, Tacoma, Washington.

American Plywood Association, 2002, Effect of Green Lumber Framing on Wood Structural Panel Shear Wall Performance. APA Report T2002-53, American Plywood Association, Tacoma, Washington.

American Plywood Association, 2005, Using Narrow Pieces of Wood Structural Panel Sheathing in Wood Shear Walls, APA T2005-08, The Engineered Wood Association, Tacoma, Washington.

Applied Technology Council, 1995, *Cyclic Testing of Narrow Plywood Shear Walls ATCR-1*. Applied Technology Council, Redwood City, California.

Applied Technology Council, 1981, *Guidelines for Design of Horizontal Wood Diaphragms*, ATC-7. Applied Technology Council, Redwood City, California.

Applied Technology Council, 1980, *Proceedings of a Workshop on Design of Horizontal Wood Diaphragms*, ATC-7-1. Applied Technology Council, Redwood City, California.

Building Seismic Safety Council, 1997, *National Earthquake Hazard Reduction Program, Recommended Provisions for Seismic Regulations for New Buildings*. Building Seismic Safety Council, Washington D.C.

Breyer, Donald E., Kenneth J. Fridley, David G. Pollock and Kelly E. Cobeen, 2007, *Design of Wood Structures ASD*. McGraw-Hill Book Co., New York, New York

Bugni, David A., 1999, "A Linear Elastic Dynamic Analysis of a Timber Framed Structure." *Building Standards*, International Conference of Building Officials, Whittier, California

Cobeen, K. E., 1996, "Performance Based Design of Wood Structures." *Proceeding: Annual SEAOC Convention*. Structural Engineers Association of California, Sacramento, California.

Coil, J., 1999, "Seismic Retrofit of an Existing Multi-Story Wood Frame Structure," *Proceedings: Annual SEAOC Convention*. Structural Engineers Association of California, Sacramento, California.

Commins, A. and Gregg, R., 1996, *Effect of Hold Downs and Stud-Frame Systems on the Cyclic Behavior of Wood Shear Walls*, Simpson Strong-Tie Co., Pleasanton, California.

Countryman, D., and Col Benson, 1954, *1954 Horizontal Plywood Diaphragm Tests*. Laboratory Report 63, Douglas Fir Plywood Association, Tacoma Washington.

CUREe, 1999, *Proceedings of the Workshop on Seismic Testing, Analysis, and Design of Wood Frame Construction*. California Universities for Research in Earthquake Engineering.

Dolan, J. D., 1996, *Experimental Results from Cyclic Racking Tests of Wood Shear Walls with Openings*. Timber Engineering Report No. TE- 1996-001. Virginia Polytechnic Institute and State University, Blacksburg, Virginia.

Dolan, J. D. and Heine, C. P., 1997a, *Monotonic Tests of Wood Frame Shear Walls with Various Openings and Base Restraint Configurations*. Timber Engineering Report No. TE-1997-001, Virginia Polytechnic Institute and State University, Blacksburg, Virginia.

Dolan, J. D. and Heine, C. P., 1997b, *Sequential Phased Displacement Cyclic Tests of Wood Frame Shear Walls with Various Openings and Base Restrain Configurations*. Timber Engineering Report No. TE-1997-002, Virginia Polytechnic Institute and State University, Blacksburg, Virginia.

Dolan, J. D., and Heine, C. P., 1997c, *Sequential Phased Displacement Test of Wood Frame Shear Walls with Corners*. Timber Engineering Report No. TE-1997-003, Virginia Polytechnic Institute and State University, Blacksburg, Virginia.

Earthquake Engineering Research Institute, 1996, "Reconnaissance Report: Northridge Earthquake of January 17, 1994," *Earthquake Spectra*. Vol. 11, Supplement C. Earthquake Engineering Research Institute, Oakland, California.

Faherty, Keith F., and Williamson, Thomas G., 1995, *Wood Engineering Construction Handbook*. McGraw Hill, Washington D.C.

Federal Emergency Management Agency, 2003, *National Earthquake Hazard Reduction Program, Recommended Provisions for Seismic Regulations for New Buildings and Other Structures and Commentary*. Federal Emergency Management Agency, Washington D.C.

Ficcadenti, S. K., T. A. Castle, D. A. Sandercock, and R. K. Kazanjy, 1996, "Laboratory Testing to Investigate Pneumatically Driven Box Nails for the Edge Nailing of 3/8" Plywood Shear Walls," *Proceedings: Annual SEAOC Convention*. Structural Engineers Association of California, Sacramento, California.

Foliente, Greg C., 1994,*Analysis, Design and Testing of Timber Structures Under Seismic Loads*. University of California Forest Products Laboratory, Richmond, California.

Foliente, Greg C., 1997, *Earthquake Performance and Safety of Timber Structures*. Forest Products Society, Madison Wisconsin.

Forest Products Laboratory, 1999, *Wood Handbook Publication FPL—GTR -113*. Madison, Wisconsin.

Ghosh, S. K., Dowty, S., 2006, "Section 1602.1 Definitions (Diaphragm, flexible) and Section 1617.5.3 Horizontal Distribution" Structural Engineer Vol. 7, No. 7 Awelg White, Chicago, Illinois.

Goers R. and Associates, 1976, *A Methodology for Seismic Design and Construction of Single-Family Dwellings*. Applied Technology Council, Redwood City, California.

Ju, S. and Lin, M., 1999, "Comparison of Building Analysis Assuming Rigid or Flexible Floors," *Journal of Structural Engineering*. American Society of Civil Engineers, Washington, D.C.

Matteson, Thor, 2004, Wood-framed Shear Wall Construction. International Code Council, Country Club Hills, Illinois.

Mendes, S., 1987, "Rigid versus Flexible: Inappropriate Assumptions Can Cause Shear Wall Failures!" *Proceedings: Annual SEAOC Convention*. Structural Engineers Association of California, Sacramento, California.

Mendes, S., 1995, "Lessons Learned From Four Earthquake Damaged Multi-Story Type V Structures," *Proceedings: Annual SEAOC Convention*. Structural Engineers Association of California, Sacramento, California.

Rose, J. D., 1998, *Preliminary Testing of Wood Structural Panel Shear Walls Under Cyclic (Reversed) Loading*. Research Report 158, APA—Engineered Wood Association, Tacoma, Washington.

Rose, J. D., and E. L. Keith, 1996, *Wood Structural Panel Shear Walls with Gypsum Wallboard and Window [Sheathing Standard, Sec. 2.3.3]*. Research Report 158. APA—The Engineered Wood Association, Tacoma Washington.

SEAOC, 1999, *Acceptable Diaphragm-Rigidity Assumptions for Distribution of Horizontal Forces in Light-Frame Construction*. Structural Engineers Association of California, Sacramento, California.

SEAOC, 1999, *Cantilever Column Opinion*. Structural Engineers Association of California, Sacramento, California.

SEAOC, 1997, *Seismic Detailing Examples for Engineered Light Frame Timber Construction*. Structural Engineers Association of California, Sacramento, California.

SEAOC, 1999, *Guidelines for Diaphragms and Shear Walls*. Structural Engineers Association of California, Sacramento, California.

SEAOC, 1999, *Plan Review—Codes and Practice*. Structural Engineers Association of California, Sacramento, California.

Shipp, J., and D. Thompson, 2001, *Timber Design I, II, and III*. Volumes VIII, IX, and X, Professional Engineering Development Publications, Inc., Irvine, California.

Steinbrugge, J., 1994, "Standard of Care in Structural Engineering Wood Frame Multiple Housing," *Proceedings: Annual SEAOC Convention*. Structural Engineers Association of California, Sacramento, California.

Design Example 2
Wood Light-frame Three-story Structure

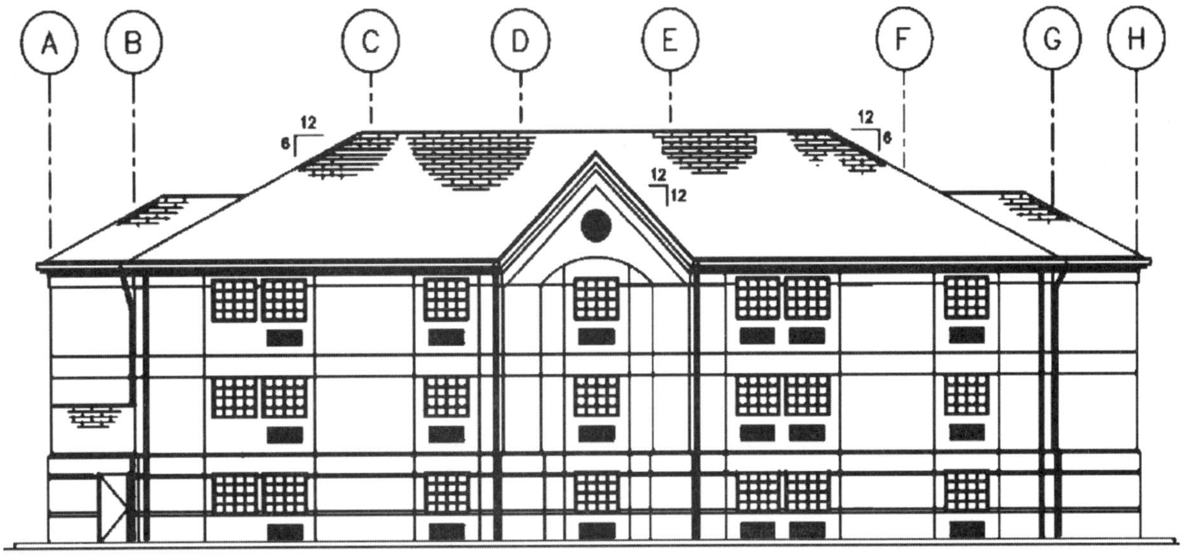

Figure 2-1. Wood light-frame three-story structure elevation

Foreword

After careful consideration and extensive discussion, SEAOC is recommending that large wood frame structures, such as the three-story building in this example (see Figure 2-1), be designed for seismic forces considering *both* rigid and flexible diaphragm assumptions (i.e., semirigid modeling assumption). This method represents a significant change from the current practice that almost exclusively uses the flexible diaphragm assumption for determining distribution of story shears to shear walls. There are two principal reasons for considering both rigid and flexible diaphragms.

First, since adoption of the 1988 UBC, there has been a *definition* of diaphragm flexibility in the code (§12.3.1.1 of ASCE/SEI 7-05). This same definition has been used in the IBC and ASCE/SEI 7-05. Arguably, when introduced in 1988, this definition may not have been intended to apply to wood framed diaphragms. After considerable discussion and re-evaluation, it is the joint opinion of the SEAOC Code and Seismology Committees that this definition should be considered in wood framed diaphragms. The application of this definition in wood construction often requires the use of the rigid diaphragm assumption, and subsequent calculation of shear wall rigidities, for distribution of story shears to shear walls. In fact, this definition results in many, if not most, diaphragms in wood frame construction being considered rigid.

ASCE/SEI 7-05 exempts one- and two-family residential buildings of light-frame construction from a rigid or semi-rigid structural analysis (§12.3.1.1), while the 2006 IBC exempts nearly all diaphragms of light-frame construction from a rigid or semi-rigid structural analysis (IBC §1613.6.1). This design example will follow the ASCE/SEI 7-05 requirements.

Many engineers feel that exclusive use of the flexible diaphragm assumption results in underestimation of forces on some shear walls. For example, a rigid diaphragm analysis is judged more appropriate when the shear walls are more flexible compared to the diaphragm, particularly where one or more lines of shear walls (or other vertical resisting elements) are more flexible than the others.

Second, in some instances, the use of flexible diaphragm assumptions can actually force the engineer to provide a more favorable lateral-force-resisting system than would occur by using only rigid diaphragm assumptions. Flexible diaphragm assumptions encourage the placement of shear walls around the perimeter of the floor and roof area, thus minimizing the need to have wood diaphragms to resist torsional forces.

In this design example, the floor diaphragms are constructed using screw shank nails, sheathing is glued to the framing members (to reduce floor squeaks), and lightweight concrete fill is placed over the floor sheathing (for sound insulation). Additionally, gyp board is applied to the framing underside for ceiling finish. These materials in combination provide significantly stiffer diaphragms than those represented by the IBC or NDS supplement diaphragm deflection equations.

For the part of the analysis that assumes a rigid diaphragm, the engineer must also select a method to estimate shear wall rigidities (and rigidities of other vertical resisting elements). This requires using careful judgment because at the present time there is no consensus method for estimating rigidities. In the commentary of Design Example 1, several alternatives are discussed.

Prior to commencing design of a wood light-frame structure, users of this document should check with the local jurisdiction regarding both the level of analysis required and acceptable methodologies.

Overview

This design example illustrates the seismic design of a three-story, 30-unit hotel structure. The light-frame structure, shown in Figures 2-1, 2-2, 2-3, and 2-4, has wood structural panel shear walls, and roof and floor diaphragms. The roofs have composite shingles and are framed with plated trusses. The floors have a 1½-inch lightweight concrete topping framed with engineered I-joists. The primary tiedowns for the shear walls use a continuous tiedown system.

This structure cannot be built using conventional construction methods for reasons shown in Part 6 of this design example. The following sections illustrate a detailed analysis for some of the important seismic requirements of the ASCE/SEI 7-05. This design example is not a complete building design, and many aspects of a complete design, including wind design (see ASCE/SEI 7-05, Chapter 6), are not included. Only selected items of the seismic design are illustrated.

In general, the ASCE/SEI 7-05 recognizes only two diaphragm categories: flexible and rigid. However, the diaphragms in this design example are considered semi-rigid. Hence, the analysis will use the envelope method, which considers the worst loading condition from the flexible and rigid diaphragm analyses for each vertical shear resisting element. It should be noted that the envelope method, although not explicitly required by code, is deemed necessary and is good engineering practice for this design example.

Initially, the shear wall nailing and tiedown requirements are determined using the flexible diaphragm assumption. Next, these shear wall forces are used to determine shear wall rigidities for the rigid diaphragm analysis. Finally, further iterations with significant stiffness redistributions may be required.

The method of determining shear wall rigidities used in this design example is far more rigorous than normal practice but is *not* the only method available to determine shear wall rigidities. The commentary following Design Example 1 illustrates two other simplified approaches that would also be appropriate.

Outline

This example will illustrate the following parts of the design process

1. Design base shear and vertical distributions of seismic forces

2. Lateral forces on shear walls and shear wall nailing assuming flexible diaphragms

3. Rigidities of shear walls

4. Distribution of lateral forces to the shear walls

5. Redundancy factor ρ

6. Does structure meet requirements of conventional construction provisions?

7. Diaphragm deflections to determine if diaphragm is flexible or rigid

8. Tiedown forces for shear wall on line C

9. Tiedown connection at the third floor for the shear wall on line C

10. Tiedown connection at the second floor for the shear wall on line C

11. Anchor bolt spacing and tiedown anchor embedment for shear wall on line C

12. Detail of tiedown connection at the third floor for shear wall on line C (Figure 2-9)

13. Detail of tiedown connection at the second floor for shear wall at line C. (Figure 2-10)

14. **Detail of wall intersection at exterior shear walls (Figure 2-11)**

15. **Detail of tiedown connection at foundation (Figure 2-12)**

16. **Detail of shear transfer at interior shear wall at roof (Figure 2-13)**

17. **Detail of shear transfer at interior shear walls at floors (Figure 2-14)**

18. **Detail of shear transfer at interior shear walls at foundation (Figure 2-15)**

19. **Detail of sill plate at foundation edge (Figure 2-16)**

20. **Detail of shear transfer at exterior wall at roof (Figure 2-17)**

21. **Detail of shear transfer at exterior wall at floor (Figure 2-18)**

Given Information

Roof weights (slope 6:12):

Roofing	3.5 psf
½-inch sheathing	1.5
Trusses	3.5
Insulation	1.5
Miscellaneous	0.7
Gyp ceiling	2.8
DL (along slope)	13.5 psf

Floor weights:

Flooring	1.0 psf
Lt. wt concrete	14.0
⅝-inch sheathing	1.8
Floor framing	5.0
Miscellaneous	0.4
Gyp ceiling	2.8
	25.0 psf

DL (horiz. proj.) = 13.5 (13.41/12) = 15.1 psf

Stair landings do not have lightweight concrete fill

Area of floor plan is 5288 sq ft

Weights of respective diaphragm levels, including tributary exterior and interior walls

W_{roof} = 134,250 lb

$W_{3rd\ floor}$ = 228,750 lb

$W_{2nd\ floor}$ = 228,750 lb

W = 591,750 lb

Weights of diaphragms are typically determined by taking one-half the height of walls at the third floor to the roof and (with equal story heights) the full height of walls for the third and second floor diaphragms.

Framing lumber is Douglas Fir-Larch (DF-L) grade stamped No. 1 S-Dry.
Note: The designer must recognize the increased potential for shrinkage problems when green lumber is used. The shrinkage of lumber can affect the architectural and mechanical systems as well as the structural system. The potential for wood shrinkage problems increases proportionally with the number of stories in the structure.

Foundation sill plates are pressure-treated Hem-Fir.

DOC PS1 or PS2 rated wood structural panels with a trademark of an approved testing and grading agency for shear walls will be $^{15}/_{32}$-inch-thick Structural-I, 32/16 panel index span rating, 5-ply with Exposure I glue is specified. However, 4-ply is also acceptable.

Three-ply $^{15}/_{32}$-inch sheathing has lower allowable shears in some local jurisdictions and the inner ply voids can cause nailing problems.

The roof is $^{15}/_{32}$-inch-thick DOC PS-1 or PS 2 (APA or TECO performance-rated) sheathing, 32/16 span rating with Exposure I glue.

The floor is $^{19}/_{32}$-inch-thick DOC PS-1 or PS 2 (APA or TECO performance-rated) Sturd-I-Floor 24-in o/c rating (or APA or TECO performance-rated sheathing, 48/24 span rating) with Exposure I glue.

Common wire nails are used for diaphragms, shear walls, and straps. Sinker nails will be used for design of the shear wall sill plate nailing at the second and third floors. (Note: Many nailing guns use the smaller diameter box and sinker nails instead of common nails. Closer nail spacing may be required if the smaller diameter nails are used).

Seismic and site data:

$S_S = 1.78$	F22-1
$S_1 = 0.55$	F22-2
$S_{DS} = 1.19$	Eq 11.4-3
Seismic Design Category D	T 11.6-1
$I = 1.0$	T 11.5-1

Site Class C has been determined by geotechnical investigation. Without a geotechnical investigation, Site Class D shall be used as a default value.

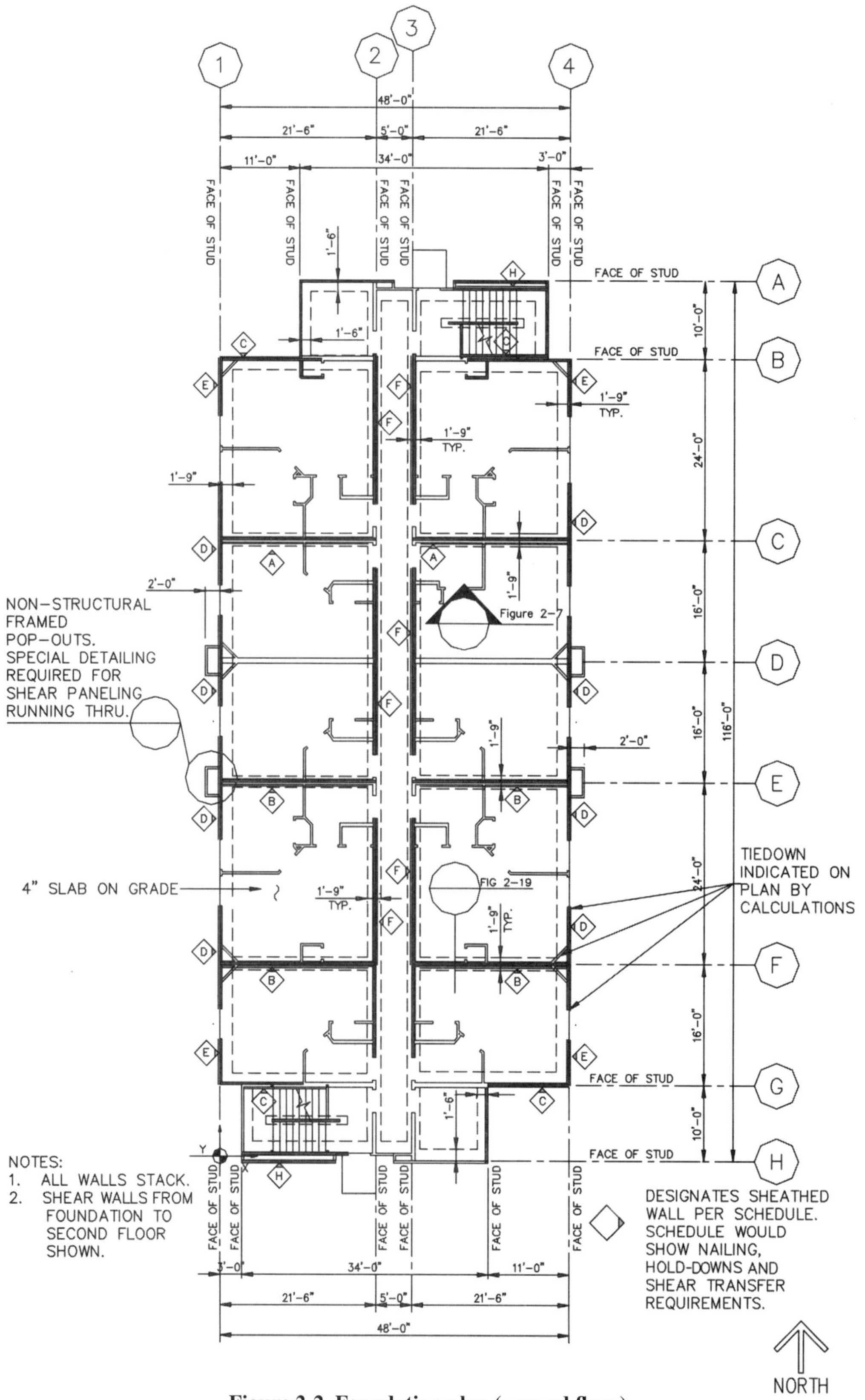

Figure 2-2. Foundation plan (ground floor)

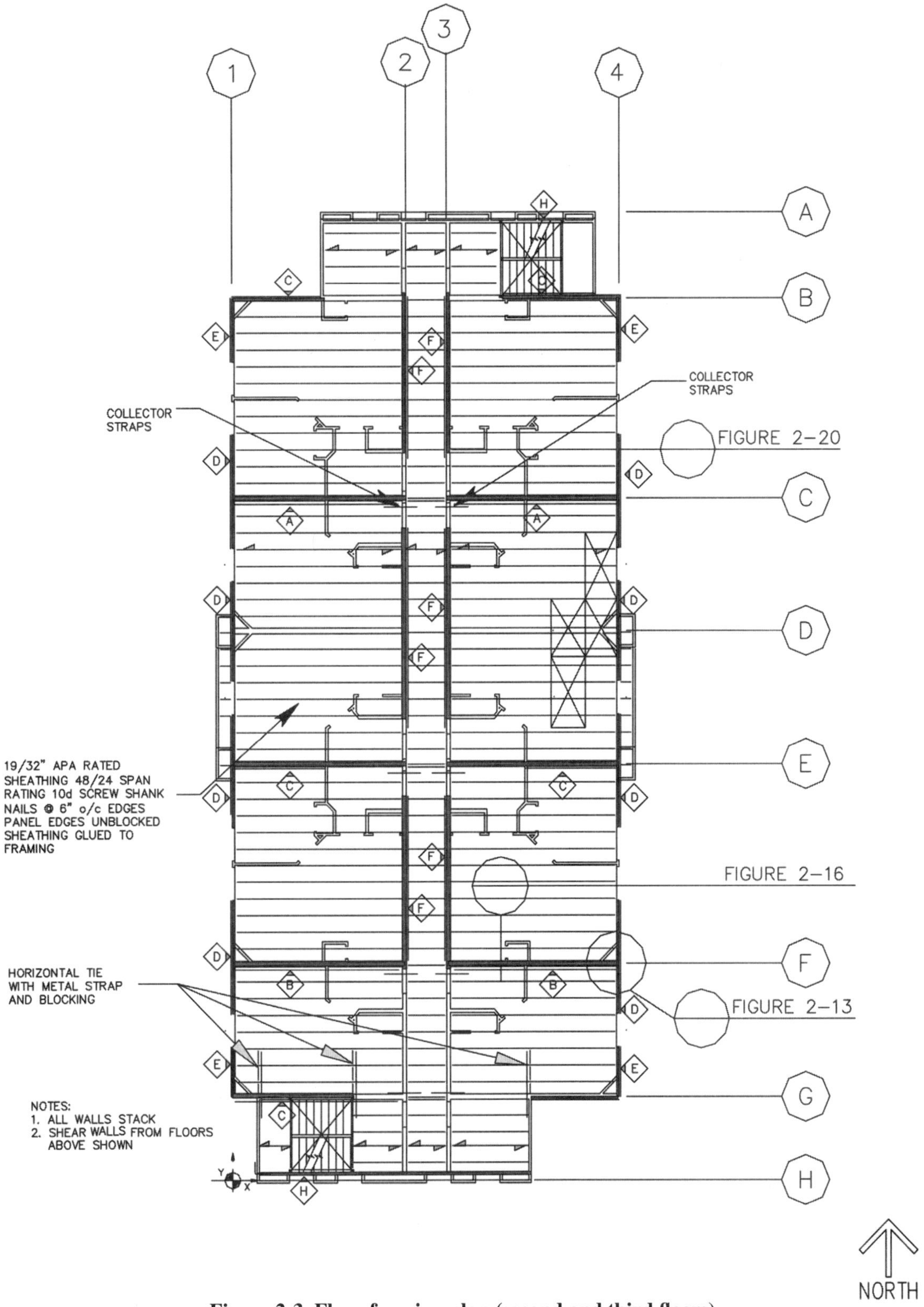

Figure 2-3. Floor framing plan (second and third floors)

Note: Shear walls on lines 2 and 3 do not extend from the third floor to the roof.

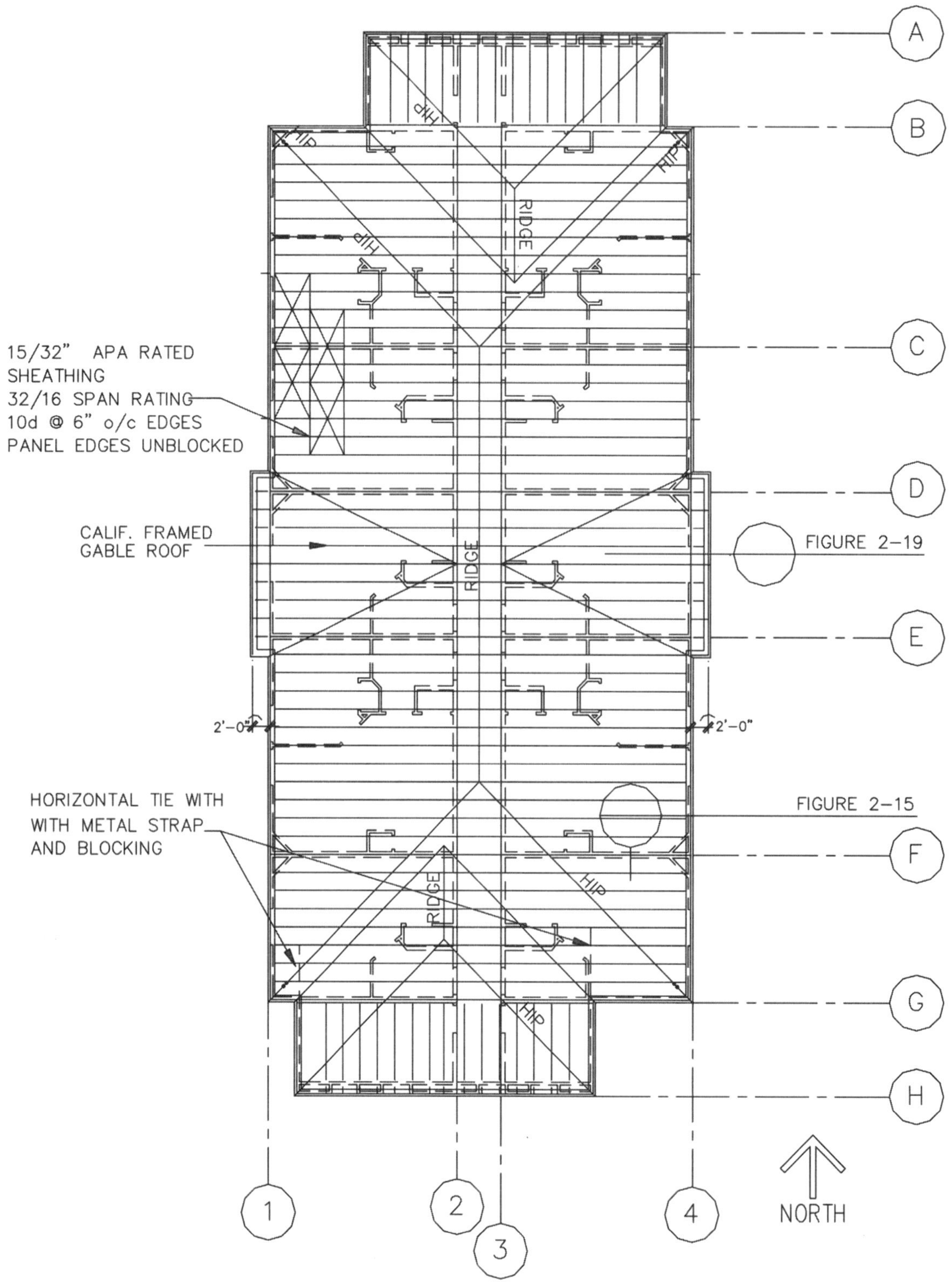

Figure 2-4. Roof framing plan

Factors That Influence Design

Before starting the example, four important related aspects of the design will be discussed. These are the effect of moisture content on lumber, the use of pre-manufactured roof trusses, proper detailing of shear walls at building pop-outs, and effects of box nails on wood structural panel shear walls.

Moisture content in lumber connections — NDS-05 T 10.3.3

This design example is based on dry lumber. Project specifications typically call for lumber to be grade-stamped S-Dry (surfaced dry). Dry lumber has a moisture content (MC) less than or equal to 19 percent. Partially seasoned or green lumber grade stamped S-GRN (surfaced green) has an MC between 19 and 30 percent. Wet lumber has an MC greater than 30 percent. Construction of structures using lumber with MC greater than 19 percent can produce shrinkage problems. Also, many engineers and building officials are not aware of the reduction requirements or wet service factors related to installation of nails, screws, and bolts (fasteners) into lumber with MC greater than 19 percent. For fasteners installed in lumber with MC greater than 19 percent, the wet service factors $C_M = 0.7$ for dowel-type fasteners such as nails, bolts, lags, and screws (NDS-05 Table 10.3.3) are used.

For construction using lumber with MC greater than 19 percent, there is a 30-percent reduction in the strength of connections, that is permanent. The effect of green lumber (mc > 19 percent) on diaphragms and shear walls is different. Recently, the APA conducted tests (APA Report T2002-53) with green lumber and dry lumber, and the tests showed that stiffness is greatly affected but the strength is for the most part, not affected. Strength is not affected because the diaphragm and shear wall allowable shears/nominal capacities are based on their ultimate capacity and not by limit state as are connections. Therefore, it is recommended that diaphragm and shear wall deflections be modified when green lumber is used, but the shear wall and diaphragm capacities need no modification.

The engineer needs to exercise good judgment in determining whether it is prudent to base the structural design on dry or green lumber. Other concerns are the geographical area and the time of year the structure is built. It is possible for green lumber (or dry lumber that has been exposed to rain) to dry out to an MC below 19 percent on the construction site. For 2x framing, this generally takes about 2 to 3 weeks of exposure to dry air, 4x lumber takes even longer. Drying occurs when the surfaces are exposed to air on all sides, not while stacked on pallets (unless shimmed with stickers). Moisture content can easily be verified by a hand-held moisture meter.

Use of pre-manufactured roof trusses to transfer lateral forces — (§2303.4)

The structural design in this design example uses the pre-manufactured wood roof trusses. Under seismic forces, these must transfer the lateral forces from the roof diaphragm to the tops of the interior shear walls. To accomplish this, special considerations must be made in the design and detailed on the plans. In particular, any trusses that are to be used as collectors or lateral drag struts should be clearly indicated on the structural framing plan. The magnitude of the forces, the means by which the forces are applied to the trusses and transferred from the trusses to the shear walls, must be shown on the plans. In addition, if the roof sheathing at the hip ends breaks above the joint between the end jack trusses and the supporting girder truss, the lateral forces to be resisted by the end jacks should be specified so that an appropriate connection can be provided to resist these forces. The drawings also must specify the load combinations and whether or not a stress increase is permitted. If ridge vents are being used, special detailing for shear transfers must be included because normal diaphragm continuity is disrupted.

Proper detailing of shear walls at building pop-outs

The structure for this design example has doubled-framed walls for party walls and exterior "planted-on" box columns (pop-outs). The designer should not consider these walls as shear walls unless special detailing and analysis is provided to substantiate that there is a viable lateral force path to that wall and the wall is adequately braced.

Effects of box nails on wood structural panel shear walls

This design example uses common nails for fastening wood structural panels. Based on cyclic testing of shear walls and performance in past earthquakes, the use of common nails is preferred. SDPW Table 4.3A lists nominal unit shear capacities for wood structural panel shear walls for common or galvanized box nails. IBC Table 2306.4.1 lists allowable shears for wood structural panel shear walls for common or galvanized box nails. Footnote j of Table 2306.4.1, states that the galvanized nails shall be "hot-dipped" or tumbled (these nails are not gun nails). Most contractors use gun nails for diaphragm and shear wall installations. The IBC does not have a table for allowable shears for wood structural panel shear walls or diaphragms using box nails.

Box nails have a smaller diameter shank and a smaller head size than common nails. Using 10d box nails would result in a 19-percent reduction in allowable load for diaphragms and shear walls as compared to 10d common nails. Using 8d box nails would result in a 22-percent reduction in allowable load for diaphragms and shear walls as compared to 8d common nails. This is based on comparing allowable shear values listed in Table 11Q in the NDS-05 for $^{15}/_{32}$-inch side member thickness, t_s, and Douglas Fir-Larch framing. In addition to the reduction of the shear wall and diaphragm capacities, when box nails are used, the walls will also drift more than when common nails are used.

A contributor to the problem is that when contractors buy large quantities of nails (for nail guns), the words *box* or *common* do not appear on the carton label. Nail length and diameters are the most common listing on the labels. Thus, it is **extremely important** to list the required nail lengths and diameters on the structural drawings for all diaphragms and shear walls. Another problem is that contractors prefer box nails because their use reduces splitting, eases driving, and costs less.

Just to illustrate a point, if an engineer designs for dry lumber (as discussed above) and common nails, and subsequently green lumber and box nails are used in the construction, the result is a compounding of the reductions. For example, for 10d nails installed into green lumber, the reduction would be 0.81 times 0.7 or a 43-percent reduction in capacity.

Calculations and Discussion	Code Reference

1. Design base shear and vertical distributions of seismic forces §12.8.1

1a. Design base shear

Determine period using approximate fundamental period (see Figure 2-5 for section through structure):

$$T_a = C_t(h_n)^x = 0.020(33.63)^{3/4} = 0.28 \text{ sec}$$ Eq 12.8-7

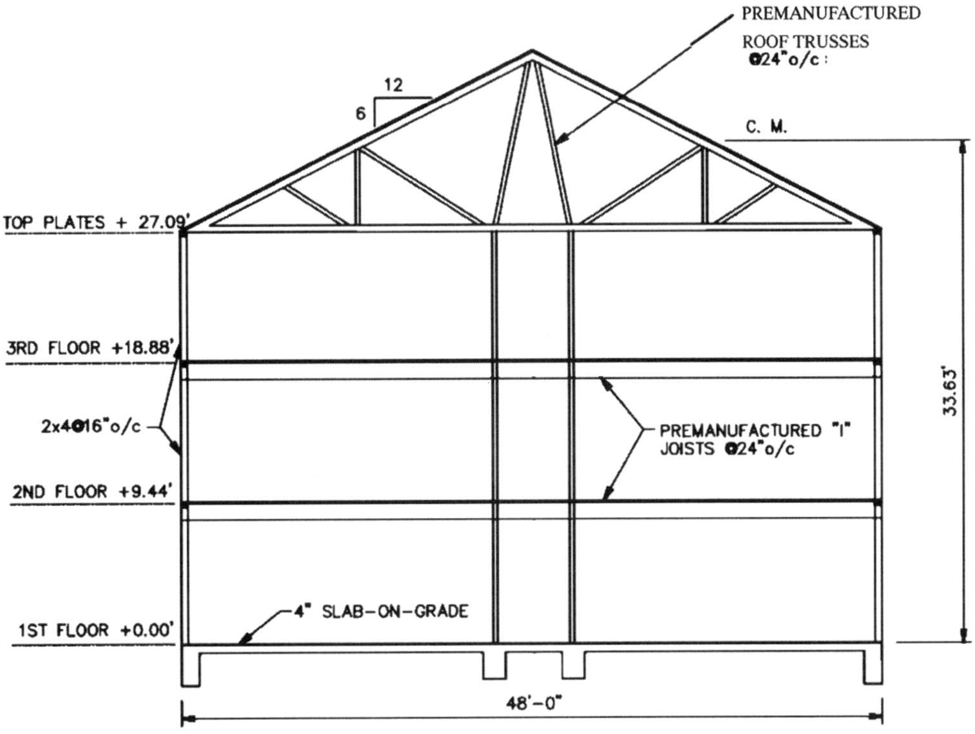

Figure 2-5. Typical cross-section through building

Because the stud walls are both wood structural panel shear walls and bearing walls

$$R = 6.5$$ T 12.1-1

Design base shear is

$$V = C_s W$$ Eq 12.8-1

Where

$$C_s = \frac{S_{DS}}{\left(\frac{R}{I}\right)}$$ Eq 12.8-2

Note: design base shear is on a strength design basis.

All the tables in the IBC for wood diaphragms and shear walls are based on allowable loads.

$$I = 1.0$$

$$R = 6.5$$

$$S_{DS} = ⅔\, S_{MS} = ⅔\,(1.78) = 1.19$$

$$S_{MS} = F_a S_S = 1.0 \times 1.78 = 1.78$$

where:

$$F_a = 1.0$$

$$S_S = 1.78$$

Site Class C

$$C_s = \frac{1.19}{\left(\frac{6.5}{1.0}\right)} = 0.183$$

but need not exceed

$$C_s = \frac{S_{D1}}{T\left(\frac{R}{I}\right)} \qquad \text{Eq 12.8-3}$$

$$S_1 = 0.55$$

$$S_{M1} = F_v S_1 = 1.3 \times 0.55 = 0.72$$

$$S_{D1} = ⅔\, S_{ML} = ⅔\,(0.72) = 0.48$$

$$F_v = 1.3$$

$$C_s = \frac{0.48}{\left(\frac{6.5}{1.0}\right)0.28} = 0.264 > 0.183 \text{ . . . therefore, does not control}$$

$$C_s = 0.183$$

but shall not be less than

$$C_s = 0.01$$

$$V = 0.183\, W$$

Comparison of the above result with the simplified static method permitted under §12.14 shows that it is more advantageous to use the standard method of determining the design base shear.

$$V = \frac{FS_{DS}}{R}\, W = \frac{1.2(1.0)}{6.5}\, W = 0.185W > 0.183W \qquad \text{Eq 12.14-11}$$

where

$F\ = 1.2$ for three-story buildings

$S_{DS} = (\tfrac{2}{3}) \times 1.0 \times 1.5 = 1.0$ §12.14.8.1

$S_a\ = 1.5$ max

$F_a\ = 1.0$ for $S_S \geq 1.25$

Check for maximum S_a value: §12.8.1.3

Since the structure does not have any Horizontal or Vertical Irregularities listed in Tables 12.3-1 and 12.3-2, the structure is classified as "regular." Also, since the structure is five stories or fewer in height and the period, *T*, is less than 0.5 seconds, C_S is permitted to be calculated using a value of 1.5 for S_S.

$$S_{MS} = F_a\,S_S = 1.0 \times 1.5 = 1.5$$

$$S_{DS} = \tfrac{2}{3}\,S_{MS} = \tfrac{2}{3}(1.5) = 1.0$$

$$C_S = \frac{1.0}{\left(\dfrac{6.5}{1.0}\right)} = 0.154$$

$$V = 0.154W < 0.183W$$

The engineer can choose which method to use. This design example illustrates the Equivalent Lateral Force Procedure using the full value of $S_S = 1.78$.

$$V = 0.183W$$

$$\therefore\ V = 0.183\ (591{,}750) = 108{,}290 \text{ lb}$$

In addition, equation (12.8-6) requires an additional check for C_S minimum for structures that are located where S_1 is equal to or greater than $0.6g$:

$$C_S = \frac{0.5S_1}{\left(\dfrac{R}{I}\right)} = \frac{0.5(1.0)}{\left(\dfrac{6.5}{1.0}\right)} = 0.07$$

However, this condition does not control.

It is desirable to use the strength level forces throughout the design of the structure for two reasons:

1. Errors in calculations can occur and which load is being used—strength design or allowable stress design—may be confused. This design example will use the following format.

   $V_{base\ shear}$ = strength

   F_{px} = strength

   F_x = force-to-wall (strength)

   v = wall shear at element level (ASD)

   $$v = \frac{0.7\,\rho Q_E}{b}(ASD)$$

2. Future editions of the code will use only strength design.

Seismic load effect E:

Where the effects of gravity and the seisimic ground motion are additive, the seismic load E is defined as

$$E = \rho Q_E + 0.2\ S_{DS}\ D$$ Eq 12.4-1

Where the effects of gravity and the seisimic ground motion counteract, the seismic load E is defined as

$$E = \rho Q_E - 0.2\ S_{DS}\ D$$ Eq 12.4-2

The redundancy ρ will be assumed to be 1.0 (in most cases it is 1.0 for Type V construction with interior shear walls). The assumed value for ρ will have to be verified. (This will be shown in Part 5.)

The basic load combinations for allowable stress design for horizontal forces is

$$(1.0 + 0.14\ S_{DS})\ D + 0.7\ \rho Q_E$$

and §12.4.2.3

$$(1.0 + 0.105\ S_{DS})\ D + 0.525\ \rho Q_E + 0.75L + 0.75L_r$$

For vertical downward loads, it is

$$(1.0 + 0.105\ S_{DS})\ D + 0.75\ L + 0.75\ S + 0.525\ \rho Q_E$$ §12.4.2.3

where

$$(1.0 + 0.105\ S_{DS})\ D = (1.0 + 0.105 \times 1.19)\ D = 1.125\ D$$ §12.4.2.3

and for vertical uplift

$$(0.6 - 0.14\ S_{DS})\ D + 0.7\ \rho Q_E$$ §12.4.2.3

where

$$(0.6 - 0.14\ S_{DS})\ D = (0.6 - 0.14 \times 1.19)\ D = 0.43\ D$$ Eq 12.8-1

1b. Vertical distributions of forces

The base shear must be distributed to each level. This is done as follows:

$$F_x = C_{vx} V$$ Eq 12.8-11

$$C_{vx} = \frac{w_x h_x^k}{\sum_{i=1}^{n} w_i h_i^k}$$ Eq 12.8-12

where h_x is the height from the base to level x, in feet.

k is a distribution exponent related to the building period

Since $T = 0.28$ seconds < 0.5 seconds, $k = 1$

Determination of F_x is shown in Table 2-1.

Note: Although not shown here, designers must also check wind loading. In this example, wind loading may control the design in the east-west direction.

Table 2-1. Vertical distribution of seismic forces

Level	w_x (k)	h_x	$w_x h_x$ (k-ft)	$\frac{w_x h_x}{\Sigma w_i h_i}$ (%)	F_x (k)	$\frac{F_x}{w_x}$	F_{tot} (k)
Roof	134.25	33.6	4,511	41.1	44.5	0.330	44.5
3rd Floor	228.75	18.9	4,323	39.4	42.7	0.186	87.2
2nd Floor	228.75	9.4	2,150	19.5	21.1	0.092	108.3
Σ	591.75		10,984		108.3		

2. Lateral forces on shear walls and shear wall nailing assuming flexible diaphragms

In this step, forces on shear walls due to seismic forces will be determined. As was customary in the past, this portion of the example assumes flexible diaphragms. The ASCE/SEI 7-05 does not require torsional effects to be considered for flexible diaphragms. The effects of torsion and wall rigidities will be considered in Part 4 of this design example.

Under the flexible diaphragm assumptions, loads to shear walls are determined based on tributary areas with simple spans between supports. Another method of determining loads to shear walls can assume a continuous beam. This design example uses the total building weight *W* applied to each respective direction. The results shown will be slightly conservative, since the building weight *W* includes the wall weights for the direction of load, which can be subtracted out. This example converts the story forces into seismic forces per square foot of floor or roof area. This may result in losing a certain amount of precision, but also results in much simpler calculations. This approach is generally considered acceptable unless there appears to be a concentration of dead load in a particular area (e.g., a mechanical penthouse).

A detailed analysis will include the derivation of these tributary weights, which include the tributary exterior and interior wall weights.

Using forces from Table 2-2 and the area of the floor plan = 5288 square feet, calculate tributary weights.

For roof diaphragm

Roof area = 5288 sq ft

$$f_{p\,roof} = \frac{44.5 \times 1000}{5288} = 8.415 \text{ psf}$$

For third floor diaphragm

Floor area = 5288 sq ft

$$f_{p\,3rd} = \frac{42.7 \times 1000}{5288} = 8.075 \text{ psf}$$

For second floor diaphragm

Floor area = 5288 sq ft

$$f_{p\,2nd} = \frac{21.1 \times 1000}{5288} = 3.990 \text{ psf}$$

Table 2-2. Forces to walls and required panel nailing or east-west direction[1, 2,3]

Wall (grid line)	Trib Area (sq ft)	ΣF_{above} (lb)	ΣF_x	F_{tot} (lb)	b^4 (ft)	$v = \frac{F_{tot}(0.7)^5}{b}$ (plf)	Sheathed 1 or 2 sides	Allowable Shear[6] (plf)	Edge Nail Spacing (in)
Shear Walls at Roof Level[7]									
A	170	0	1,430	1,430	12.5	80	1	340	6
B	746	0	6,280	6,280	22.0	200	1	340	6
C	1,344	0	11,310	11,310	43.0	185	1	340	6
E	1,344	0	11,310	11,310	43.0	185	1	340	6
F	960	0	8,080	8,080	43.0	135	1	340	6
G	554	0	4,660	4,660	22.0	150	1	340	6
H	170	0	1,430	1,430	12.5	80	1	340	6
Σ	5,288	0	44,500	44,500	198				
Shear Walls at Third Floor Level									
A	170	1,430	1,375	2,805	12.5	160	1	340	6
B	746	6,280	6,025	12,305	22.0	395	1	510	4
C	1,344	11,310	10,850	22,160	43.0	365	1	510	4
E	1,344	11,310	10,850	22,160	43.0	365	1	510	4
F	960	8,080	7,750	15,830	43.0	260	1	510	4
G	554	4,660	4,475	9,135	22.0	290	1	510	4
H	170	1,430	1,375	2,805	12.5	160	1	340	6
Σ	5,288	44,500	42,700	87,200	198				
Shear Walls at Second Floor Level									
A	170	2,805	680	3,485	12.5	195	1	340	6
B	746	12,305	2,975	15,280	22.0	490	1	665	3
C	1,344	22,160	5,365	27,525	43.0	450	1	665	3
E	1,344	22,160	5,365	27,525	43.0	450	1	665	3
F	960	15,830	3,830	19,660	43.0	320	1	665	3
G	554	9,135	2,210	11,345	22.0	365	1	665	3
H	170	2,805	680	3,485	12.5	195	1	340	6
Σ	5,288	87,200	21,100	108,300	198				

Notes

1. In SDC D, E, or F, the 2006 IBC (Table 2306.4.1 Footnote i and §2305.3.11) requires 3x nominal thickness stud framing at abutting panels and at foundation sill plates when the allowable shear values exceed 350 pounds per foot.
2. Sill bolt washers: For SDC D, E, or F, Section 2305.3.11 of the 2006 IBC requires that a minimum of 3-inch-square by 0.229-inch-thick plate washers be used for each foundation sill bolt (regardless of allowable shear values in the wall). These changes resulted from the splitting of framing studs and sill plates observed in the Northridge earthquake and in cyclic testing of shear walls. The plate washers are intended to help resist uplift forces on shear walls. Because of observed vertical displacements of tiedowns, these plate washers are required even if the wall has tiedowns designed to take uplift forces at the wall boundaries. The washer edges shall be parallel/perpendicular to the sill plate. Section 2305.3.11 has an exception to the 3x foundation sill plates by allowing 2x foundation sill plates when the allowable shear values are less than 600 pounds per foot, provided that sill bolts are designed for 50 percent of allowable values. Sill bolt plate washers are not required in SDCs A, B and C.
3. The 1999 SEAOC Blue Book recommends special inspection when the nail spacing is closer than 4 inches o/c.
4. The shear wall length used for wall shears is the "out-to-out" wall length.
5. Note that forces are strength level and that shear in wall is multiplied by 0.7 to convert to allowable stress design.
6. DOC PS-1 or PS-2 (APA or TECO performance-rated) Structural-I-rated wood structural panels may be either plywood or OSB. Allowable shear from IBC Table 2306.4.1 using 10d common nails with a minimum 1½-inch penetration.
7. Shear walls at lines C, E, and F extend to the bottom of the prefabricated wood trusses at the roof level. Shear transfer is obtained by framing clips from the bottom chord of the trusses to the top plates of the shear walls. Project plans call for trusses at these lines to be designed for these horizontal forces (see also comments in Part 8). Roof shear forces are also transferred to lines A, B, G, and H.

Table 2-3. Forces to walls and required panel nailing for north-south direction[1, 2,3]

Wall	Trib Area (sq ft)	ΣF_{above} (lb)	ΣF_x	F_{tot} (lb)	b^4 (ft)	$v = \frac{F_{tot}(0.7)^5}{b}$ (plf)	Sheathed 1 or 2 sides	Allowable Shear[6] (plf)	Edge Nail Spacing (in)
Shear Walls at Roof Level[7]									
1	2,644	0	22,250	22,250	64.5	245	1	340	6
2	0	0	0	0	0	0			
3	0	0	0	0	0	0			
4	2,644	0	22,250	22,250	64.5	245	1	340	6
Σ	5,288	0	44,500	44,500	129.0				
Shear Walls at Third Floor Level									
1	1,202	22,250	9,705	31,955	64.5	350	1	510	4
2	1,442	0	11,645	11,645	60.0	140	1	340	6
3	1,442	0	11,645	11,645	60.0	140	1	340	6
4	1,202	22,250	9,705	31,955	64.5	350	1	510	4
Σ	5,288	44,500	42,700	87,200	249.0				
Shear Walls at Second Floor Level									
1	1,202	31,955	4,795	36,750	64.5	400	1	510	4
2	1,442	11,645	5,755	17,400	60.0	205	1	340	6
3	1,442	11,645	5,755	17,400	60.0	205	1	340	6
4	1,202	31,955	4,795	36,750	64.5	400	1	510	4
Σ	5,288	87,200	21,100	108,300	249.0				

Notes

1. In SDC D, E, or F, the 2006 IBC (Table 2306.4.1 Footnote i and §2305.3.11) requires 3x nominal thickness stud framing at abutting panels and at foundation sill plates when the allowable shear values exceed 350 pounds per foot.
2. Sill bolt washers: In SDC D, E, or F, Section 2305.3.11 of the 2006 IBC requires that a minimum of 3-inch-square by 0.229-inch-thick plate washers be used for each foundation sill bolt (regardless of allowable shear values in the wall). These changes resulted from the splitting of framing studs and sill plates observed in the Northridge earthquake and in cyclic testing of shear walls. The plate washers are intended to help resist uplift forces on shear walls. Because of observed vertical displacements of tiedowns, these plate washers are required even if the wall has tiedowns designed to take uplift forces at the wall boundaries. The washer edges shall be parallel/perpendicular to the sill plate. Section 2305.3.11 has an exception to the 3x foundation sill plates by allowing 2x foundation sill plates when the allowable shear values are less than 600 pounds per foot, provided that sill bolts are designed for 50 percent of allowable values. Sill bolt plate washers are not required in SDCs A, B, and C.
3. The 1999 SEAOC Blue Book recommends special inspection when the nail spacing is closer than 4 inches o/c.
4. The shear wall length used for wall shears is the "out-to-out" wall length.
5. Note that forces are strength level and that shear in wall is multiplied by 0.7 to convert to allowable stress design.
6. DOC PS-1 or PS-2 (APA or TECO performance-rated) Structural-I-rated wood structural panels may be either plywood or oriented strand board (OSB). Allowable shear from IBC Table 2306.4.1 using 10d common nails with a minimum 1½-inch penetration.
7. The interior shear walls at lines 2 and 3 were not used to brace the roof diaphragm. This is because installing wall sheathing (blocking panels) perpendicular to plated trusses is labor intensive. Often it is not installed correctly, and occasionally, because of contractor error, it is not even installed. This approach will increase the third floor diaphragm transfer (redistribution) forces. With rigid diaphragms, you must carefully follow the load paths.

3. Rigidities of shear walls

3a. Rigidity calculation using the IBC, APA Research Report, and SDPWS deflection equations

Determination of wood shear wall rigidities is not a simple task. In practice, approximate methods are often used. The method illustrated in this example is by far the most rigorous method used. There are other, more simplified methods, and their use is often appropriate. The alternate methods are briefly discussed in Design Example 1B.

It must be emphasized that, at the present time, *every* method is approximate, particularly for multistory structures such as those in this example. Until more definite general procedures are established through further testing and research, the designer must exercise judgment in selecting the appropriate method to be used for a given structure. When in doubt, consult with the local building official regarding methods acceptable to the jurisdiction. At the time of this publication, the type of seismic design required for a project of this type varies greatly from one jurisdiction to another.

Wall rigidities are approximate. The initial rigidity, R, of the structure can be significantly higher because of stucco, drywall, brick and stone veneers, stiffening effects of walls not considered, and areas over doors and windows. During an earthquake, some low-stressed walls may maintain their stiffness while others degrade in stiffness. Some walls and their collectors may attract significantly more lateral load than anticipated in flexible or rigid diaphragm analysis. It must be understood that the method of analyzing a structure using rigid diaphragms takes significantly more engineering effort. However, use of the rigid diaphragm method indicates that some lateral resisting elements can attract significantly higher seismic demands than tributary area (i.e., flexible diaphragm) analysis methods.

In this example, shear wall rigidities k are computed using the basic stiffness equation

$$F = k\Delta$$

or

$$k = \frac{F}{\Delta}$$

The well-known 4-term equation to determine the shear wall deflections is shown to be

$$\Delta = \frac{8vh^3}{EAb} + \frac{vh}{Gt} + 0.75he_n + d_a\frac{h}{b} \qquad \text{(Eq 23-2)}$$

The above equation is based on tests conducted by the American Plywood Association and on a uniformly nailed, cantilever shear wall with fixed base and free top, a horizontal point load at top, and panel edges blocked, and deflection is estimated from the contributions of four distinct parts. The first part of the equation accounts for cantilever beam action using the moment of inertia of the boundary elements. The second term accounts for shear deformation of the sheathing. The third term accounts for nail slippage/bending, and the fourth term accounts for tiedown assembly displacement (this also should include bolt/nail slip and shrinkage).

The engineer should be cautioned to use the units as listed in IBC §2305.3.2. Do not attempt to change the units.

Testing on wood shear walls has indicated that the above deflection formula is reasonably accurate for wall aspect ratios, h/w, lower than or equal to 2:1. For higher aspect ratios, the wall drift increases significantly, and displacements were not adequately predicted by the formula. Using the aspect ratio requirement of 2:1 makes this formula more accurate for determining shear wall deflection/stiffness than it was in previous editions of the building code, subject to the limitations mentioned above.

Recent testing on wood shear walls has shown that sill plate crushing under the boundary element can increase the shear wall deflection by as much as 20 to 30 percent. For a calculation of this crushing effect, see the deflection of wall frame at line D later in part 11c of this example.

This design example uses the G, t, and e_n values from the APA Research Report 138. The designer is free to use the IBC values, which will produce nearly identical results. IBC Table 2305.2.2(2) also has values for OSB sheathing that are not in the APA Report. The values in IBC Table 2305.2.2(2) combine the Gt values, where this design example uses separated values from the APA report.

The American Forest and Paper Association (AF&PA) has recently published a new supplement to the National Design Specification, the Special Design Provisions for Wind and Seismic (SDPWS). In this publication, the SDPWS has changed the traditional 4-term expression to a new simplified 3-term expression for shear wall deflection. This new equation is expressed

$$\Delta = \frac{8vh^3}{EAb} + \frac{vh}{1000\,G_a} + d_a \frac{h}{b}$$ SDPWS Eq 4. 3-1

The new simplified 3-term equation combines the second and third terms (of the 4-term equation) into one term. Computed deflections by using either the 4-term equation or the 3-term equation produces nearly identical results.

The small differences in computed shear wall deflections from using the 4-term equation versus the 3-term equation and the method of determining the tiedown displacement will impact shear wall rigidities and load distribution. It is recommended that the designer use consistent equations and tiedown displacement computations on a particular project.

Faster slip/nail deformation values (e_n)

Using the fastener slip equations from APA Research Report 138, Table B-4 for 10d common nails used in this example, there are two basic equations:

When nails are driven into green lumber: $e_n = (V_n/977)^{1.894}$ APA T B-4

When nails are driven into dry lumber: $e_n = (V_n/769)^{3.276}$ APA T B-4

where

V_n = fastener load in pounds per fastener

These values from the above formulas are based on Structural-I sheathing and must be increased by 20 percent when the sheathing is not Structural-I. The language in Footnote A in Research Report 138, Table B-4, which states "Fabricated green/tested dry (seasoned) . . ." is potentially misleading. The values in the table are actually *green values*, because the assembly is fabricated when green. Don't be misled by the word "seasoned."

It is uncertain whether or not the d_a factor is intended to include wood shrinkage and crushing due to shear wall rotation, because the code is not specific. This design example includes shrinkage and crushing in the d_a factor.

Many engineers are concerned that if the contractor installs the nails at a different spacing (too many or too few), the rigidities will be different from those calculated. However, nominal changing of the nail spacing in a given wall does not significantly change the stiffness.

3b. Calculation of shear wall rigidities

In this example, shear wall rigidities are calculated using the four-term code deflection equation in IBC §2305.3.2. These calculations are facilitated by the use of a spreadsheet program, which eliminates possible arithmetic errors from the many repetitive computations that must be made.

The first step is to calculate the displacement (i.e., vertical elongation) of the tiedown assemblies and the crushing effect of the boundary element. This is the term d_a. The force considered to act on the tiedown assembly is the net uplift force determined from the flexible diaphragm analyses of Part 2. These forces are summarized in Tables 2-4, 2-9, and 2-13 for the roof, the third floor, and the second floor, respectively.

After the tiedown assembly displacements are determined, the four-term deflection equation is used to determine the deflection, Δ, of each shear wall. These are summarized in Tables 2-5 and 2-6 for the roof level, in Tables 2-10 and 2-11 for the third floor level, and in Tables 2-14 and 2-15 for the second floor level.

Finally, the rigidities of the shear walls are summarized in Tables 2-7, 2-12, and 2-16 for the roof, third floor, and second floor, respectively.

For both strength and allowable stress design, the ASCE/SEI 7-05 now requires building drifts to be determined by earthquake forces without multiplying by 0.7.

Using strength level forces for wood design utilizing the ASCE/SEI 7-05 and 2006 IBC means that the engineer will use both strength-level forces and allowable stress forces. This can create some confusion because the code requires drift checks to be strength-level forces. However, all the design equations and tables in IBC Chapter 23 are based on allowable stress design. Drift and shear wall rigidities should be calculated from the strength-level forces. Remember that the structural system factor R is based on using strength-level forces.

3c. Estimation of roof level rigidities

To estimate roof level wall rigidities, roof level displacements must first be determined. Given below are a series of calculations in table form, to estimate the roof level drifts, Δ, in each shear wall. First, the shear wall tiedown assembly displacements are determined (Table 2-4). These, and the parameters given in Table 2-5, are used to arrive at the drifts, Δ, for each shear wall at the roof level (Tables 2-5 and 2-6). Rigidities are estimated in Table 2-7 for walls in both directions. Once the Δ drifts are known, a drift check is performed. This is summarized in Table 2-8.

Table 2-4. Determine tiedown assembly displacements at roof level[1]

	ASD		Strength Design					
					Tiedown Assembly Displacement			
Wall	Uplift $(0.7)^2$ (lb)	Tiedown Device	Uplift[2] (lb)	Tiedown[3] Elongation (in)	Shrink[4]	Crush[5]	Slip[6]	d_a[7] (in)
A	0	Not required	0	0	0.05	0.02	0	0.07
B1	840	Strap	1,175	0.04	0.05	0.02	0.002	0.11
B2	840	Strap	1,175	0.04	0.05	0.02	0.002	0.11
C1	100	Strap	140	0.02	0.05	0.02	0.002	0.09
C2	100	Strap	140	0.02	0.05	0.02	0.002	0.09
E1	100	Strap	140	0.02	0.05	0.02	0.002	0.09
E2	100	Strap	140	0.02	0.05	0.02	0.002	0.09
F1	0	Not required	0	0	0.05	0.02	0	0.07
F2	0	Not required	0	0	0.05	0.02	0	0.07
G1	500	Strap	700	0.02	0.05	0.02	0.002	0.09
G2	500	Strap	700	0.02	0.05	0.02	0.002	0.09
H	0	Not required	0	0	0.05	0.02	0	0.07
1a, 4a	120	Strap	170	0.02	0.05	0.02	0.002	0.09
1b, 4b	0	Not required	0	0	0.05	0.02	0	0.07
1c, 4c	0	Not required	0	0	0.05	0.02	0	0.07
1d, 4d	0	Not required	0	0	0.05	0.02	0	0.07
1e, 4e	0	Not required	0	0	0.05	0.02	0	0.07
1f, 4f	120	Strap	170	0.02	0.05	0.02	0.002	0.09

Notes:

1. Tiedown assembly displacements for the roof level are calculated for the tiedowns at the third floor level.
2. Uplift force is determined by using the *net* overturning moment ($M_{ot} - M_R$) divided by the distance between the *centroids* of the boundary elements with 4x members at the ends of the shear wall where M_R uses load combinations outlined in part 1a of this design example. This equates to the length of the wall minus 3½ inches for straps or the length of wall minus 7¼ inches when using a bolted tiedown with 2-inch offset from post to anchor bolt. Using allowable stress design, tiedown devices need only be sized by using the ASD uplift force. The strength design uplift force is used to determine tiedown assembly displacement, and then to determine strength-level displacements.
3. The continuous tiedown (rod) system selected for this structure will have a "shrinkage compensating" system. Most of these systems have shrinkage compensation by either pre-tensioning of cables or a "self-ratcheting" hardware connector and are proprietary. The device selected in this design example has adjusting grooves at 1⁄10-inch increments, meaning the most the "system" will not have compensated for in shrinkage and crushing will be 1⁄10 inch. If the selected device does not have a shrinkage compensating device, then shrinkage of floor framing, sill plates, compression bridges, crushing of bridge support studs, and collector studs will need to be considered. See Design Example 1, Part 3c for an example calculation for a bolted connection. The tiedown rod at line B will elongate as follows:

for ⅝-inch rod: $\Delta = \dfrac{PL}{AE} = 6090\text{ lb}(4.5)(12)/0.31(29E6) = 0.04\text{ in}$ *(Notes continued on next page)*

Note that the rod length is 4.5 feet (Figure 2-12). The elongation for the portion of the rod at the level below will be considered at the level below.
For level below (Table 2-13) rod length is 9.44 feet (Figure 2-12)

$$\text{for } \tfrac{5}{8}\text{-inch rod: } \Delta = \frac{PL}{AE} = 12{,}040 \text{ lb}(9.44)(12)/0.31(29E6) = 0.15 \text{ in}$$

4. Wood shrinkage is based on a change in moisture content (MC) from 19 percent to 15 percent, with 19-percent MC being assumed for S-Dry lumber per project specifications. The MC of 15 percent is the assumed final MC at equilibrium with ambient humidity for the project location. The final equilibrium value can be higher in coastal areas and lower in inland or desert areas. This equates to (0.002) (*d*) (19 – 15), where *d* is the dimension of the lumber (see Figure 2-11). Pressure-treated lumber has a moisture content of less than 16 percent at treatment completion. Shrinkage of 2x DBL top plate + 2x DBL sill plate = (0.002) (4 × 1.5 in)(19 – 15) = 0.05 in
5. Per NDS-05 §4.2.6, when compression perpendicular to grain $f_{c\perp}$ is less than $0.73F'_{c\perp}$, crushing will be approximately 0.02 inches. When $f_{c\perp} = 0.73F'_{c\perp}$, crushing is approximately 0.04 inches. The effect of sill plate crushing is the downward effect at the opposite end of the wall with uplift force and has the same rotational effect as the tiedown displacement. Short walls that have no uplift forces will still have a crushing effect and contribute to rotation of the wall.
6. Per NDS-05 §10.3.6 load/slip modulus $\gamma = (270{,}000)(D^{1.5})$, plus an additional 1/16 inch for the oversized hole for bolts. For nails, values for e_n can be used.
7. d_a is the total tiedown assembly displacement. This also could include mis-cuts (short studs) and lack of square-cut ends.

Table 2-5. Deflections of shear walls at the roof level in east-west direction

Wall	ASD *v* (plf)	Strength *v* (plf)	*h* (ft)	*A* (in^2)	*E* (psi)	*b* (ft)	*G* (psi)	*t* (in)	Nail Spacing (in)	V_n (lb)	e_n (in)	d_a (in)	Δ (in)
A	80	114	8.21	10.5	1.7E6	12.5	90,000	0.535	6	60	0.0002	0.07	0.07
B1	200	285	8.21	10.5	1.7E6	11.0	90,000	0.535	6	144	0.0041	0.11	0.16
B2	200	285	8.21	10.5	1.7E6	11.0	90,000	0.535	6	144	0.0041	0.11	0.16
B						22.0							
C1	185	263	8.21	10.5	1.7E6	21.5	90,000	0.535	6	133	0.0032	0.09	0.10
C2	185	263	8.21	10.5	1.7E6	21.5	90,000	0.535	6	133	0.0032	0.09	0.10
C						43.0							
E1	185	263	8.21	10.5	1.7E6	21.5	90,000	0.535	6	133	0.0032	0.09	0.10
E2	185	263	8.21	10.5	1.7E6	21.5	90,000	0.535	6	133	0.0032	0.09	0.10
E						43.0							
F1	135	188	8.21	10.5	1.7E6	21.5	90,000	0.535	6	95	0.0011	0.07	0.07
F2	135	188	8.21	10.5	1.7E6	21.5	90,000	0.535	6	95	0.0011	0.07	0.07
F						43.0							
G1	150	212	8.21	10.5	1.7E6	11.0	90,000	0.535	6	109	0.0017	0.09	0.12
G2	150	212	8.21	10.5	1.7E6	11.0	90,000	0.535	6	109	0.0017	0.09	0.12
G						22.0							
H	80	114	8.21	10.5	1.7E6	12.5	90,000	0.535	6	60	0.0002	0.07	0.07

Table 2-6. Deflections of shear walls at the roof level in north-south direction

Wall	ASD v (plf)	Strength v (plf)	h (ft)	A (in^2)	E (psi)	b (ft)	G (psi)	t (in)	Nail Spacing (in)	V_n (lb)	e_n (in)	d_a (in)	Δ (in)
1a, 4a	245	345	8.21	10.5	1.7E6	8.0	90,000	0.535	6	175	0.0078	0.09	0.21
1b, 4b	245	345	8.21	10.5	1.7E6	14.0	90,000	0.535	6	175	0.0078	0.07	0.16
1c, 4c	245	345	8.21	10.5	1.7E6	11.5	90,000	0.535	6	175	0.0078	0.07	0.17
1d, 4d	245	345	8.21	10.5	1.7E6	11.5	90,000	0.535	6	175	0.0078	0.07	0.17
1e, 4e	245	345	8.21	10.5	1.7E6	11.5	90,000	0.535	6	175	0.0078	0.07	0.17
1f, 4f	245	345	8.21	10.5	1.7E6	8.0	90,000	0.535	6	175	0.0078	0.09	0.21
1, 4						64.5							

Table 2-7. Shear wall rigidities at roof level[1]

Wall	Δ^2 (in)	F (lb)	$k_i = \frac{F}{\Delta}$ (k/in)	k_{total} (k/in)
A	0.07	1,430	20.43	20.43
B1	0.16	3,140	19.62	
B2	0.16	3,140	19.62	
B		6,280	39.24	39.24
C1	0.10	5,655	56.55	
C2	0.10	5,655	56.55	
C		11,310	113.1	113.1
E1	0.10	5,655	56.55	
E2	0.10	5,655	56.55	
E		11,310	113.1	113.1
F1	0.07	4,040	57.71	
F2	0.07	4,040	57.71	
F		8,080	115.4	115.4
G1	0.12	2,330	19.42	
G2	0.12	2,330	19.42	
G		4,660	38.84	38.84
H	0.07	1,430	20.42	20.42
1a, 4a	0.21	2,760	13.14	
1b, 4b	0.16	4,830	30.19	
1c, 4c	0.17	3,965	23.32	
1d, 4d	0.17	3,970	23.35	
1e, 4e	0.17	3,965	23.32	
1f, 4f	0.21	2,760	13.14	
1, 4		22,250	126.5	126.5

Notes:

1. Deflections and forces are based on strength force levels.
2. Δ are the design level drifts from Tables 2-5 and 2-6.

3d. Drift check at roof level §12.12

To establish drift, the story drift, δ_x, must be determined as follows:

$$\delta_x = \frac{C_d \delta_{xe}}{I} \qquad \text{Eq 12.8-15}$$

where

$C_d = 4.0$ T 12.1-1

$I = 1.0$ T 11.5-1

$$\delta_x = \frac{4.0\Delta}{1.0} = 4.0\Delta$$

The calculated story drift using δ_x shall not exceed the maximum Δ_a, which is 0.025 times the story height. The drift check is summarized in Table 2-8.

Table 2-8. Drift check at roof level

	Wall	Δ (in)	Height (ft)	δ_x (in)	Max Δ_a (in)	Status
East-West	A	0.07	8.21	0.28	2.46	ok
	B	0.16	8.21	0.64	2.46	ok
	C	0.10	8.21	0.40	2.46	ok
	E	0.10	8.21	0.40	2.46	ok
	F	0.07	8.21	0.28	2.46	ok
	G	0.12	8.21	0.48	2.46	ok
	H	0.07	8.21	0.28	2.46	ok
North-South	1a, 4a	0.21	8.21	0.84	2.46	ok
	1b, 4b	0.16	8.21	0.64	2.46	ok
	1c, 4c	0.17	8.21	0.68	2.46	ok
	1d, 4d	0.17	8.21	0.68	2.46	ok
	1e, 4e	0.17	8.21	0.68	2.46	ok
	1f, 4f	0.21	8.21	0.84	2.46	ok

3e. Estimation of third floor level rigidities

Shear wall rigidities at the third floor are estimated in the same manner as those at the roof. The calculations are summarized in Tables 2-9, 2-10, 2-11, and 2-12. A drift check is not shown.

Table 2-9. Tiedown assembly displacements at third floor level[1]

	ASD		Strength Design					
					Tiedown Assembly Displacement			
Wall	Uplift $(0.7)^2$ (lb)	Tiedown Device	Uplift[2] (lb)	Tiedown Elongation[3] (in)	Shrink[4]	Crush[5]	Slip[6]	d_a[7] (in)
A	135	Strap	190	0.02	0.05	0.02	0.002	0.09
B1	4,350	Rod	6,090	0.04	0	0	0.10	0.14
B2	4,350	Rod	6,090	0.04	0	0	0.10	0.14
C1	2,000	Strap	2,800	0.02	0.05	0.02	0.002	0.09
C2	2,000	Strap	2,800	0.02	0.05	0.02	0.002	0.09
E1	2,000	Strap	2,800	0.02	0.05	0.02	0.002	0.09
E2	2,000	Strap	2,800	0.02	0.05	0.02	0.002	0.09
F1	550	Strap	770	0.02	0.05	0.02	0.002	0.09
F2	550	Strap	770	0.02	0.05	0.02	0.002	0.09
G1	2,800	Rod	3,920	0.02	0	0	0.10	0.12
G2	2,800	Rod	3,920	0.02	0	0	0.10	0.12
H	135	Strap	190	0.02	0.05	0.02	0.002	0.09
1a, 4a	2,275	Strap	3,185	0.02	0.05	0.02	0.002	0.09
1b, 4b	0	Not req'd	0	0	0.05	0.02	0	0.07
1c, 4c	0	Not req'd	0	0	0.05	0.02	0	0.07
1d, 4d	0	Not req'd	0	0	0.05	0.02	0	0.07
1e, 4e	0	Not req'd	0	0	0.05	0.02	0	0.07
1f, 4f	2,275	Strap	3,185	0.02	0.05	0.02	0.002	0.09
2a, 3a	0	Not req'd	0	0	0.05	0.02	0	0.07
2b, 3b	0	Not req'd	0	0	0.05	0.02	0	0.07
2c, 3c	0	Not req'd	0	0	0.05	0.02	0	0.07

Notes:

1. Tiedown assembly displacements for the third floor level are calculated for the tiedowns at the second floor level.
2. Footnotes 2-6, see Table 2-4.

Table 2-10. Deflections of shear walls at third floor level in east-west direction

Wall	ASD v (plf)	Strength v (plf)	h (ft)	A (in^2)	E (psi)	b (ft)	G (psi)	t (in)	Space (in)	V_n (lb)	e_n (in)	d_a (in)	Δ (in)
A	160	224	9.43	15.7	1.7E6	12.5	90,000	0.535	6	112	0.0018	0.09	0.13
B1	395	560	9.43	15.7	1.7E6	11.0	90,000	0.535	4	187	0.0097	0.14	0.31
B2	395	560	9.43	15.7	1.7E6	11.0	90,000	0.535	4	187	0.0097	0.14	0.31
B						22.0							
C1	365	518	9.43	15.7	1.7E6	21.5	90,000	0.535	4	173	0.0075	0.09	0.20
C2	365	518	9.43	15.7	1.7E6	21.5	90,000	0.535	4	173	0.0075	0.09	0.20
C						43.0							
E1	365	518	9.43	15.7	1.7E6	21.5	90,000	0.535	4	173	0.0075	0.09	0.20
E2	365	518	9.43	15.7	1.7E6	21.5	90,000	0.535	4	173	0.0075	0.09	0.20
E						43.0							
F1	260	371	9.43	15.7	1.7E6	21.5	90,000	0.535	4	124	0.0025	0.09	0.13
F2	260	371	9.43	15.7	1.7E6	21.5	90,000	0.535	4	124	0.0025	0.09	0.13
F						43.0							
G1	290	415	9.43	15.7	1.7E6	11.0	90,000	0.535	4	140	0.0038	0.12	0.22
G2	290	415	9.43	15.7	1.7E6	11.0	90,000	0.535	4	140	0.0038	0.12	0.22
G						22.0							
H	160	224	9.43	15.7	1.7E6	12.5	90,000	0.535	6	112	0.0018	0.09	0.13

Table 2-11. Deflections of shear walls at the third floor level in north-south direction

Wall	ASD v (plf)	Strength v (plf)	h (ft)	A (in^2)	E (psi)	b (ft)	G (psi)	t (in)	Space (in)	V_n (lb)	e_n (in)	d_a (in)	Δ (in)
1a, 4a	350	497	9.43	15.7	1.7E6	8.0	90,000	0.535	4	166	0.0066	0.09	0.27
1b, 4b	350	497	9.43	15.7	1.7E6	14.0	90,000	0.535	4	166	0.0066	0.07	0.20
1c, 4c	350	497	9.43	15.7	1.7E6	11.5	90,000	0.535	4	166	0.0066	0.07	0.21
1d, 4d	350	497	9.43	15.7	1.7E6	11.5	90,000	0.535	4	166	0.0066	0.07	0.21
1e, 4e	350	497	9.43	15.7	1.7E6	11.5	90,000	0.535	4	166	0.0066	0.07	0.21
1f, 4f	350	497	9.43	15.7	1.7E6	8.0	90,000	0.535	4	166	0.0066	0.09	0.27
1, 4						64.5							
2a, 3a	140	196	9.43	15.7	1.7E6	18.0	90,000	0.535	6	98	0.0012	0.07	0.09
2b, 3b	140	196	9.43	15.7	1.7E6	24.0	90,000	0.535	6	98	0.0012	0.07	0.08
2c, 3c	140	196	9.43	15.7	1.7E6	18.0	90,000	0.535	6	98	0.0012	0.07	0.09
2, 3						60.0							

Table 2-12. Shear wall rigidities at third floor[1]

Wall	Δ^2 (in)	F (lb)	$k_i = \frac{F}{\Delta}$ (k/in)	k_{total} (k/in)
A	0.13	2,805	21.58	21.58
B1	0.31	6,152	19.84	
B2	0.31	6,153	19.84	
B		12,305	39.68	39.68
C1	0.20	11,080	55.40	
C2	0.20	11,080	55.40	
C		22,160	110.80	110.80
E1	0.20	11,080	55.40	
E2	0.20	11,080	55.40	
E		22,160	110.80	110.80
F1	0.13	7,915	60.88	
F2	0.13	7,915	60.88	
F		15,830	121.70	121.70
G1	0.22	4,568	20.76	
G2	0.22	4,567	20.76	
G		9,135	41.52	41.52
H	0.13	2,805	21.58	21.58
1a, 4a	0.27	3,965	14.68	
1b, 4b	0.20	6,936	34.68	
1c, 4c	0.21	5,696	27.12	
1d, 4d	0.21	5,696	27.12	
1e, 4e	0.21	5,696	27.12	
1f, 4f	0.27	3,966	14.68	
1, 4		31,955	145.40	145.40
2a, 3a	0.09	3,494	38.82	
2b, 3b	0.08	4,657	58.21	
2c, 3c	0.09	3,494	38.82	
2, 3		11,645	135.80	135.80

Notes:

1. Deflections and forces are based on strength levels.
2. Δ are the design level displacements from Tables 2-10 and 2-11.

3f. Estimation of second floor level rigidities

Shear wall rigidities at the second floor level are estimated in the same manner as those for the roof and third floor. The calculations are summarized in Tables 2-13, 2-14, 2-15, and 2-16. A drift check is not shown.

Table 2-13. Tiedown assembly displacements at second floor level[1]

	ASD		Strength Design					
					Tiedown Assembly Displacement			
Wall	Uplift $(0.7)^2$ (lb)	Tiedown Device	Uplift[2] (lb)	Tiedown Elongation[3] (in)	Shrink[4]	Crush[5]	Slip[6]	d_a[7] (in)
A	1,090	Strap	1,525	0.02	0.01	0.02	0.002	0.05
B1	8,600	Rod	12,040	0.15	0	0	0.10	0.25
B2	8,600	Rod	12,040	0.15	0	0	0.10	0.25
C1	4,380	Rod	6,130	0.08	0	0	0.10	0.18
C2	4,380	Rod	6,130	0.08	0	0	0.10	0.18
E1	4,380	Rod	6,130	0.08	0	0	0.10	0.18
E2	4,380	Rod	6,130	0.08	0	0	0.10	0.18
F1	1,565	Rod	2,200	0.03	0	0	0.10	0.13
F2	1,565	Rod	2,200	0.03	0	0	0.10	0.13
G1	5,700	Rod	7,980	0.10	0	0	0.10	0.20
G2	5,700	Rod	7,980	0.10	0	0	0.10	0.20
H	1,090	Strap	1,525	0.02	0.01	0.02	0.002	0.05
1a, 4a	5,240	Rod	7,340	0.10	0	0	0.10	0.20
1b, 4b	0	Not req'd	0	0	0.01	0.02	0	0.03
1c, 4c	1,000	Strap	1,400	0.02	0.01	0.02	0.002	0.05
1d, 4d	1,000	Strap	1,400	0.02	0.01	0.02	0.002	0.05
1e, 4e	1,000	Strap	1,400	0.02	0.01	0.02	0.002	0.05
1f, 4f	5,240	Rod	7,340	0.10	0	0	0.10	0.20
2a, 3a	0	Not req'd	0	0	0.01	0.02	0	0.03
2b, 3b	0	Not req'd	0	0	0.01	0.02	0	0.03
2c, 3c	0	Not req'd	0	0	0.01	0.02	0	0.03

Notes:

1. Tiedown assembly displacements for the second floor level are calculated for the tiedowns at the first floor level

2. Uplift force is determined by using the *net* overturning moment ($M_{ot} - M_R$) divided by the distance between the *centroids* of the boundary elements with 4x members at the ends of the shear wall where M_R uses load combinations outlined in Part 1a of this design example. This equates to the length of the wall minus 3½ inches for straps or the length of wall minus 7¼ inches when using a bolted tiedown with a 2-inch offset from post to anchor bolt. Using allowable stress design, tiedown devices need only be sized by using the ASD uplift force. The strength design uplift force is used to determine tiedown assembly displacement, and then to determine strength-level displacements.
3. The continuous tiedown (rod) system selected for this structure will have a "shrinkage compensating" system. Most of these systems have shrinkage compensation by either pre-tensioning of cables or a "self-ratcheting" hardware connector and are proprietary. The device selected in this design example has adjusting grooves at ⅒-inch increments, meaning the most the "system" will not have compensated for in shrinkage and crushing will be ⅒ inch. If the selected device does not have a shrinkage compensating device, then shrinkage of floor framing, sill plates, compression bridges, crushing of bridge support studs, and collector studs will need to be considered. See Design Example 1, Part 3c for an example calculation for a bolted connection. The tiedown rod at line B will elongate as follows

   for ⅝-inch rod: $\Delta = \dfrac{PL}{AE}$ = 6090 lb(4.5)(12)/0.31(29E6) = 0.04 in

   Note that the rod length is 4.5 feet (Figure 2-12). The elongation for the portion of the rod at the level below will be considered at the level below.
   For level below (Table 2-13) rod length is 9.44 feet (Figure 2-12):

   for ⅝-inch rod: $\Delta = \dfrac{PL}{AE}$ = 12,040 lb(9.44)(12)/0.31(29E6) = 0.15 in

4. Wood shrinkage is based on a change in moisture content (MC) from 19 percent to 15 percent, with 19-percent MC being assumed for S-Dry lumber per project specifications. The MC of 15 percent is the assumed final MC at equilibrium with ambient humidity for the project location. The final equilibrium value can be higher in coastal areas and lower in inland or desert areas. This equates to (0.002) (d) (19 – 15), where d is the dimension of the lumber (see Figure 2-11). Pressure-treated lumber has a moisture content of less than 16 percent at treatment completion.

   Shrinkage of 2x DBL Top Plate + 2x DBL sill plate = (0.002) (4 × 1.5 in)(19 – 15) = 0.05 in

5. Per NDS-05 §4.2.6, when compression perpendicular to grain $f_{c\perp}$ is less than $0.73F'_{c\perp}$, crushing will be approximately 0.02 inches. When $f_{c\perp} = F'_{c\perp}$, crushing is approximately 0.04 inches. The effect of sill plate crushing is the downward effect at the opposite end of the wall with uplift force and has the same rotational effect as the tiedown displacement. Short walls that have no uplift forces will still have a crushing effect and contribute to rotation of the wall.
6. Per NDS-05 §7.3.6 load/slip modulus $\gamma = (270{,}000)(D^{1.5})$, plus an additional 1/16 inch for the oversized hole for bolts. For nails, values for e_n can be used.
7. d_a is the total tiedown assembly displacement. This also could include mis-cuts (short studs) and lack of square-cut ends.

Table 2-14. Deflections of shear walls at the second floor level in east-west direction

Wall	ASD v (plf)	Strength v (plf)	h (ft)	A (in^2)	E (psi)	b (ft)	G (psi)	t (in)	Space (in)	V_n (lb)	e_n (in)	d_a (in)	Δ (in)
A	195	280	9.43	26.2	1.7E6	12.5	90,000	0.535	6	140	0.0038	0.05	0.12
B1	490	700	9.43	26.2	1.7E6	11.0	90,000	0.535	3	175	0.0078	0.25	0.42
B2	490	700	9.43	26.2	1.7E6	11.0	90,000	0.535	3	175	0.0078	0.25	0.42
B						22.0							
C1	450	644	9.43	26.2	1.7E6	21.5	90,000	0.535	3	161	0.0060	0.18	0.25
C2	450	644	9.43	26.2	1.7E6	21.5	90,000	0.535	3	161	0.0060	0.18	0.25
C						43.0							
E1	450	644	9.43	26.2	1.7E6	21.5	90,000	0.535	3	161	0.0060	0.18	0.25
E2	450	644	9.43	26.2	1.7E6	21.5	90,000	0.535	3	161	0.0060	0.18	0.25
E						43.0							
F1	320	462	9.43	26.2	1.7E6	21.5	90,000	0.535	3	115	0.0020	0.13	0.16
F2	320	462	9.43	26.2	1.7E6	21.5	90,000	0.535	3	115	0.0020	0.13	0.16
F						43.0							
G1	365	518	9.43	26.2	1.7E6	11.0	90,000	0.535	3	130	0.0030	0.20	0.30
G2	365	518	9.43	26.2	1.7E6	11.0	90,000	0.535	3	130	0.0030	0.20	0.30
G						22.0							
H	195	280	9.43	26.2	1.7E6	12.5	90,000	0.535	6	140	0.0038	0.05	0.12

Table 2-15. Deflections of shear walls at the second floor level in north-south direction

Wall	ASD v (plf)	Strength v (plf)	h (ft)	A (in^2)	E (psi)	b (ft)	G (psi)	t (in)	Space (in)	V_n (lb)	e_n (in)	d_a (in)	Δ (in)
1a, 4a	400	574	9.43	26.2	1.7E6	8.0	90,000	0.535	4	191	0.0104	0.20	0.43
1b, 4b	400	574	9.43	26.2	1.7E6	14.0	90,000	0.535	4	191	0.0104	0.03	0.21
1c, 4c	400	574	9.43	26.2	1.7E6	11.5	90,000	0.535	4	191	0.0104	0.05	0.23
1d, 4d	400	574	9.43	26.2	1.7E6	11.5	90,000	0.535	4	191	0.0104	0.05	0.23
1e, 4e	400	574	9.43	26.2	1.7E6	11.5	90,000	0.535	4	191	0.0104	0.05	0.23
1f, 4f	400	574	9.43	26.2	1.7E6	8.0	90,000	0.535	4	191	0.0104	0.20	0.43
1, 4						64.5							
2a, 3a	205	294	9.43	26.2	1.7E6	18.0	90,000	0.535	6	147	0.0044	0.03	0.10
2b, 3b	205	294	9.43	26.2	1.7E6	24.0	90,000	0.535	6	147	0.0044	0.03	0.10
2c, 3c	205	294	9.43	26.2	1.7E6	18.0	90,000	0.535	6	147	0.0044	0.03	0.10
2, 3						60.0							

Table 2-16. Wall rigidities at second floor[1]

Wall	Δ^2 (in)	F (lb)	$k_i = \frac{F}{\Delta}$ (k/in)	k_{total} (k/in)
A	0.12	3,485	29.04	29.04
B1	0.42	7,640	18.19	
B2	0.42	7,640	18.19	
B		15,280	36.38	36.38
C1	0.25	13,762	55.05	
C2	0.25	13,763	55.05	
C		27,525	110.1	110.1
E1	0.25	13,762	55.05	
E2	0.25	13,763	55.05	
E		27,525	110.1	110.1
F1	0.16	9,830	61.44	
F2	0.16	9,830	61.44	
F		19,660	122.8	122.8
G1	0.30	5,672	18.91	
G2	0.30	5,673	18.91	
G		11,345	37.82	37.82
H	0.12	3,485	29.04	29.04
1a, 4a	0.43	4,558	10.60	
1b, 4b	0.21	7,978	37.99	
1c, 4c	0.23	6,552	28.48	
1d, 4d	0.23	6,552	28.48	
1e, 4e	0.23	6,552	28.48	
1f, 4f	0.43	4,558	10.60	
1, 4		36,750	144.6	144.6
2a, 3a	0.10	5,221	52.21	
2b, 3b	0.10	6,958	69.58	
2c, 3c	0.10	5,221	52.21	
2, 3		17,400	174.0	174.0

Notes:

1. Deflections and forces are based on strength force levels.
2. Δ are the design level displacements from Tables 2-14 and 2-15.

4. Distribution of lateral forces to the shear walls. §12.8.4

The base shear was distributed to the three levels in Part 2. In this step, the story forces are distributed to the shear walls supporting each level using the rigid diaphragm assumption. See Part 7 for a confirmation of this assumption.

For many years it has been common engineering practice to assume flexible diaphragms and to distribute loads to shear walls based on tributary areas. This has become a well-established conventional design assumption. In this design example, the rigid diaphragm assumption will be used. This is not intended to imply that seismic design of wood light-frame construction in the past should have been performed in this manner. However, recent earthquakes and testing of wood panel shear walls have indicated that drifts can be considerably higher than what was known or assumed in the past. This knowledge of the increased drifts of short wood panel shear walls and the fact that the diaphragms tend to be much more rigid than the shear walls has increased the need for the engineer to consider the relative rigidities of shear walls.

The code requires that the story force at the center of mass be displaced from the calculated center of mass (CM) a distance of 5 percent of the building dimension at that level perpendicular to the direction of force. This is to account for accidental torsion. The code requires the most severe load combination to be considered and also permits the negative torsional shear to be subtracted from the direct load shear. The net effect of this is to add 5-percent accidental eccentricity to the calculated eccentricity.

However, lateral forces must be considered to act in each direction of the two principal axes. This design example does not consider eccentricities between the centers of mass between levels. In this design example, these eccentricities are small and are therefore deemed insignificant. The engineer must exercise good judgment in determining when those effects need to be considered.

The direct shear force F_v is determined from

$$F = F\frac{R}{\Sigma R}$$

and the torsional shear force F_t is determined from

$$F_t = T\frac{Rd}{J}$$

where

$J = \Sigma R d_x^2 + \Sigma R d_y^2$

R = shear wall rigidity

d = distance from the lateral resisting element (e.g., shear wall) to the center of rigidity (CR)

$T = Fe$

F = 44,500 lb (for roof diaphragm)

e = eccentricity

4a. Determine center of rigidity, center of mass, and eccentricities for roof diaphragm

Forces in the east-west (x) direction

$$\bar{y}_r = \frac{\Sigma k_{xx} y}{\Sigma k_{xx}} \quad or \quad \bar{y}_r \, \Sigma k_{xx} = \Sigma k_{xx}\, y$$

Using the rigidity values k from Table 2-7 and the distance y from line H to the shear wall:

$$\bar{y}_r \, (20.43 + 39.24 + 113.1 + 113.1 + 115.4 + 38.84 + 20.42)$$
$$= 20.43(116) + 39.24(106) + 113.1(82.0) + 113.1(50.0)$$
$$+ 115.4(26.0) + 38.84(10.0) + 20.42(0)$$

Distance to calculated CR $\bar{y}_r = \dfrac{24{,}847.3}{460.53} = 53.9 \text{ ft}$

The building is symmetrical about the x-axis (Figure 2-6) and the CM is determined to be

$$\bar{y}_m = \frac{116.0}{2} = 58.0 \text{ ft}$$

The minimum 5-percent accidental eccentricity for east-west forces, e_y, is computed from the length of the structure perpendicular to the applied story force

$$e_y = (0.05 \times 116 \text{ ft}) = \pm 5.8$$

The new $\bar{y}_m$ to the displaced CM = 58 ft ± 5.8 ft = 63.8 ft or 52.2 ft

The total eccentricity is the distance between the displaced CM and the CR

$$y_r \quad = 53.9 \text{ ft}$$
$$\therefore\; e_y = 63.8 - 53.9 = 9.9 \text{ ft} \quad or \quad 52.2 - 53.9 = -1.7 \text{ ft}$$

Note that displacing the CM by 5 percent can result in the CM being on either side of the CR and can produce added torsional shears to all walls.

Note that the 5 percent may not be conservative. The contents-to-structure weight ratio can be higher in wood framing than in heavier types of construction. Also, the location of the calculated CR is less reliable than in other structural systems. Use good judgment when selecting the eccentricity *e*.

Amplification of accidental torsional moment required in §12.8.4.3 of the code is exempted for structures of light-frame construction. Therefore, code check is not required.

Forces in the north-south (y) direction

The building is symmetrical about the y-axis (Figure 2-6). Therefore, the distance to the CM and CR is

$$\bar{x}_m = \frac{48.0}{2} = 24.0 \text{ ft}$$

$$e'_x = (0.05)(48 \text{ ft}) = \pm 2.4 \text{ ft}$$

Because, the CM and CR locations coincide

$$e_x = e'_x$$

$$\therefore e_x = 2.4 \quad or \quad -2.4 \text{ ft}$$

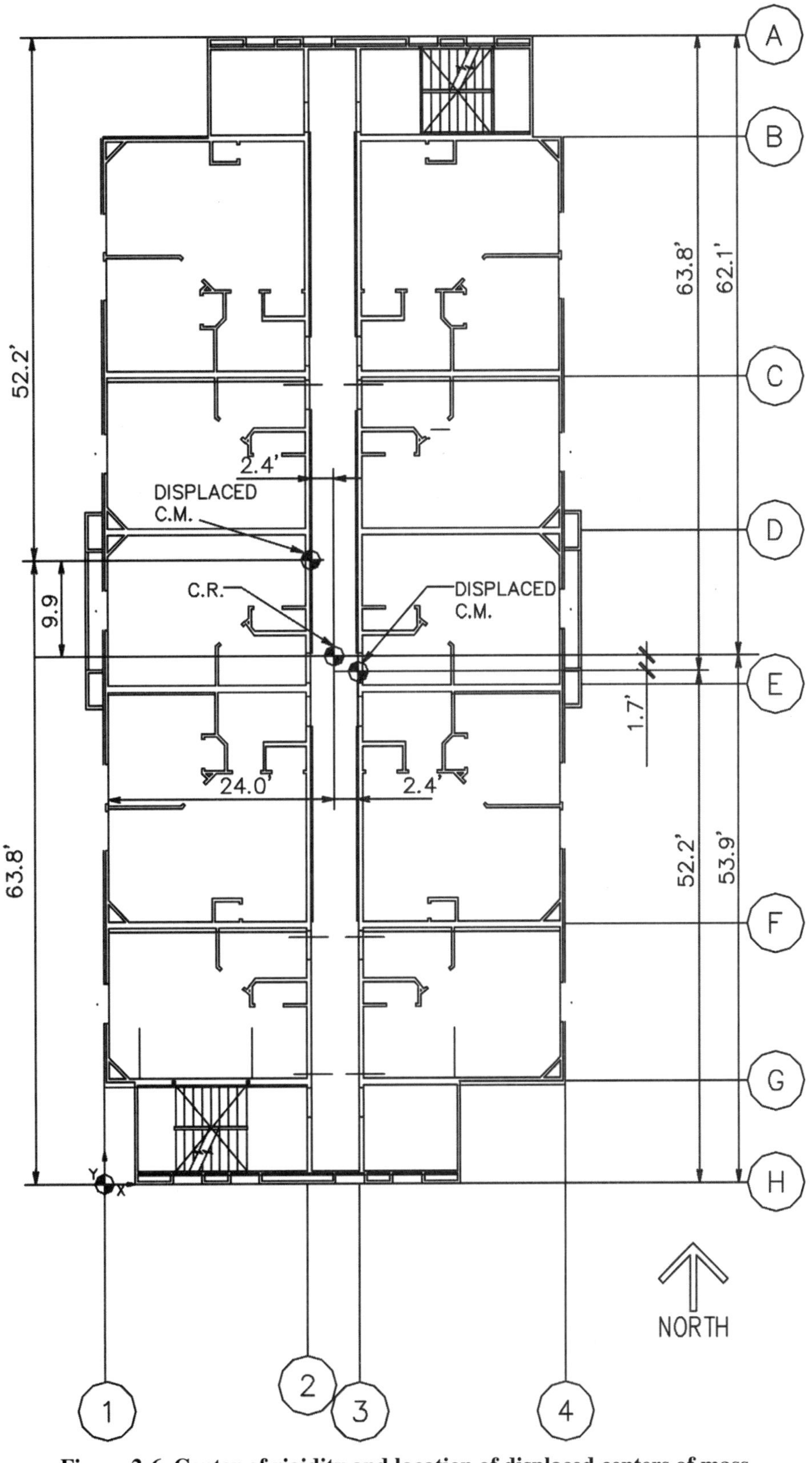

Figure 2-6. Center of rigidity and location of displaced centers of mass for second and third floor diaphragms

4b. Determine total shears on walls at roof level

The total shears on the walls at the roof level are the direct shears, F_v, and the shears due to torsion (combined actual torsion and accidental torsion), F_t. Torsion on the roof diaphragm is computed as

$T_x = Fe_y = 44{,}500$ lb (9.9 ft) = 440,550 ft-lb for walls A, B, and C

or $T_x = 44{,}500$ lb (1.7 ft) = 75,650 ft-lb for walls E, F, G, and H

$T_y = Fe_x = 44{,}500$ lb (2.4 ft) = 106,800 ft-lb for walls 1 and 4

Because the building is symmetrical for forces in the north-south direction, the torsional forces can be subtracted for those walls located on the opposite side from the displaced CM. The critical force will then be used for the design of these walls. Table 2-17 summarizes the spreadsheet for determining combined forces on the roof level walls.

4c. Determine the center of rigidity, center of mass, and eccentricities for the third and second floor diaphragms

Since the walls stack with uniform nailing, it can be assumed that the CR for the third floor and the second floor diaphragms will coincide with the CR of the roof diaphragm.

Torsion on the third floor diaphragms

$F = (44{,}500 + 42{,}700)$ = 87,200 lb

$T_x = Fe_y = 87{,}200$ lb (9.9 ft) = 863,280 ft-lb for walls A, B, and C

or 87,200 lb (1.7 ft) = 148,240 ft-lb for walls E, F, G, and H

$T_y = Fe_x = 87{,}200$ lb (2.4 ft) = 209,280 ft-lb for walls 1 and 4

Results for the third floor are summarized in Table 2-18.

Torsion on the second floor diaphragms

$F = (44{,}500 + 42{,}700 + 21{,}100)$ = 108,300 lb

$T_x = Fe_y = 108{,}300$ lb (9.9 ft) = 1,072,170 ft-lb for walls A, B, and C

or 108,300 lb (1.7 ft) = 184,110 ft-lb for walls E, F, G, and H

$T_y = Fe_x = 108{,}300$ lb (2.4 ft) = 259,920 ft-lb for walls 1 and 4

Results for the second floor are summarized in Table 2-19.

4d. Comparison of flexible vs. rigid diaphragm results

Table 2-20 summarizes wall forces determined under the separate flexible and rigid diaphragm analysis. Since nailing requirements were established in the flexible diaphragm analyses of Part 2, they must be checked for results of the rigid diaphragm analysis and adjusted if necessary (also given in Table 2-20).

Table 2-17. Distribution of forces to shear walls below the roof level

	Wall	R_x	R_y	d_x	d_y	Rd	Rd^2	Direct Force F_v	Torsional Force F_t	Total Force $F_v + F_t$
East-West	A	20.43			62.1	1,269	78,786	1,970	865	2,835
	B	39.24			52.1	2,044	106,513	3,791	1394	5,185
	C	113.10			28.1	3,178	89,305	10,932	2167	13,099
	E	113.10			3.9	441	1,720	10,932	52	10,984
	F	115.40			27.9	3,220	89,829	11,153	377	11,530
	G	38.84			43.9	1,705	74,853	3,752	200	3,952
	H	20.42			53.9	1,101	59,324	1,970	129	2,099
	Σ	460.53					500,330	44,500		
North-South	1		126.5	24.0		3,036	72,864	22,250	502	22,752
	4		126.5	−24.0		−3,036	72,864	22,250	−502	21,748
	Σ		253.0				145,728	44,500		
	Σ						646,058			

Table 2-18. Distribution of forces to shear walls below the third floor level

	Wall	R_x	R_y	d_x	d_y	Rd	Rd^2	Direct Force F_v	Torsional Force F_t	Total Force $F_v + F_t$
East-West	A	21.58			62.1	1,340	83,221	4,024	1,685	5,709
	B	39.68			52.1	2,067	107,708	7,399	2,559	9,998
	C	110.8			28.1	3,113	87,489	20,660	3,914	24,574
	E	110.8			3.9	432	1,685	20,660	93	20,693
	F	121.7			27.9	3,395	94,732	22,692	733	23,425
	G	41.52			43.9	1,823	80,018	7,741	393	8,134
	H	21.58			53.9	1,163	62,694	4,024	251	4,275
	Σ	467.66					517,547	87,200		
North-South	1		145.4	24.0		3,490	83,750	22,544	1,064	23,608
	2		135.8	2.5		340	849	21,056	259	21,315
	3		135.8	−2.5		−340	849	21,056	−259	20,797
	4		145.4	−24.0		−3,490	83,750	22,544	−1,064	21,480
	Σ		562.4				169,198	87,200		
	Σ						686,745			

Table 2-19. Distribution of forces to shear walls below second floor level

	Wall	R_x	R_y	d_x	d_y	Rd	Rd^2	Direct Force F_v	Torsional Force F_t	Total Force $F_v + F_t$
East-West	A	29.04			62.1	1,803	111,990	6,617	2,682	9,299
	B	36.38			52.1	1,911	98,750	8,290	2,843	11,133
	C	110.1			28.1	3,094	86,936	25,088	4,602	29,690
	E	110.1			3.9	429	1,675	25,088	109	25,197
	F	122.8			27.9	3,426	95,589	27,982	875	28,857
	G	37.82			43.9	1,660	72,887	8,618	424	9,042
	H	29.04			53.9	1,565	84,367	6,617	400	7,017
	Σ	475.28					552,194	108,300		
North-South	1		144.6	24.0		3,470	83,290	24,576	1,251	25,827
	2		174.0	2.5		435	1,088	29,574	157	29,731
	3		174.0	−2.5		−435	1,088	29,574	−157	29,417
	4		144.6	−24.0		−3,470	83,290	24,576	−1,251	23,325
	Σ		637.2				168,756	108,300		
	Σ						720,950			

Table 2-20. Comparison of loads on shear walls using flexible versus rigid diaphragm analysis and recheck of nailing in walls

Wall	$F_{flexible}$	F_{rigid}	Rigid/ Flexible ratio %	b (ft)	$v = \frac{F_{max}(0.7)}{b}$ (plf)	Plywood 1 or 2 sides	Allowable Shear (plf)[1,2]	Edge Nail Spacing (in)
Roof Level								
A	1,430	2,835	98	12.5	160	1	340	6
B	6,280	5,185	−17	22.0	200	1	340	6
C	11,310	13,099	15	43.0	215	1	340	6
E	11,310	10,984	−3	43.0	185	1	340	6
F	8,080	11,530	43	43.0	190	1	340	6
G	4,660	3,952	−15	22.0	150	1	340	6
H	1,430	2,099	46	12.5	120	1	340	6
1	22,250	22,752	2	64.5	250	1	340	6
4	22,250	22,752[3]	2	64.5	250	1	340	6
Third Floor								
A	2,805	5,709	103	12.5	320	1	340	6
B	12,305	9,998	−18	22.0	395	1	510	4[2]
C	22,160	24,574	11	43.0	400	1	510	4
E	22,160	20,693	−7	43.0	365	1	510	4
F	15,830	23,425	48	43.0	385	1	510	4
G	9,135	8,134	−11	22.0	295	1	510	4
H	2,805	4,275	52	12.5	240	1	340	6
1	31,955	23,608	−26	64.5	350	1	510	4
2	11,645	21,315	83	60.0	250	1	340	6
3	11,645	21,315[3]	83	60.0	250	1	340	6
4	31,955	23,608[3]	−26	64.5	350	1	510	4
Second Floor								
A	3,485	9,299	167	12.5	520	1	510	4[4]
B	15,280	11,133	−27	22.0	490	1	665	3
C	27,525	29,690	7	43.0	485	1	665	3
E	27,525	25,197	−9	43.0	450	1	665	3
F	19,660	28,857	47	43.0	470	1	665	3
G	11,345	9,042	−20	22.0	365	1	665	3
H	3,485	7,017	100	12.5	395	1	510	4
1	36,750	25,827	−30	64.5	400	1	510	4
2	17,400	29,731	70	60.0	350	1	340	6[5]
3	17,400	29,731[3]	70	60.0	350	1	340	6[5]
4	36,750	25,827[3]	−30	64.5	400	1	510	4

Notes:

1. Allowable shears from IBC Table 2306.4.1
2. Shear walls with shears that exceed 350 plf will require 3x framing at abutting panel edges with staggered nails. See also notes at bottom of Table 1-3.
3. Designates the force used was the higher force for the same wall at the opposite side of the structure.
4. The shear of 520 plf exceeds allowable of 510 plf, therefore the nail spacing will need to be decreased to 3-inch spacing. A redesign will not be necessary.
5. The shear of 350 plf exceeds allowable of 340 plf, therefore the nail spacing will need to be decreased to 4-inch spacing. A redesign will not be necessary.

Where forces from rigid diaphragm analysis are higher than those from the flexible diaphragm analysis, wall stability and anchorage must be re-evaluated. However, engineering judgment may be used to determine if a complete rigid diaphragm analysis should be repeated because of changes in wall rigidity.

If rigid diaphragm loads are used, the diaphragm shears should be rechecked for total load divided by diaphragm length along the individual wall lines.

5. Determine redundancy coefficient ρ §12.3.4.2

The redundancy coefficient penalizes lateral-force-resisting systems that do not have adequate redundancy. In Part 1 of this example, the reliability/redundancy factor was previously assumed to be $\rho = 1.0$. This will now be checked.

The method for determining the redundancy factor p is different in the ASCE/SEI 7-05. The code now requires structures in Seismic Design Categories D, E, or F to use a $\rho = 1.3$, unless one of two exceptions is met, in which case $\rho = 1.0$.

Step 1: Determine if one of the exceptions is met.

a) Each story resisting more than 35 percent of the base shear in the direction under consideration complies with Table 12.3-3. For shear walls with a height-to-length ratio of more than 1.0, the removal of that wall would not result in more than a 33-percent reduction in the overall story strength. From Table 2-1 of this design example all three levels resist more than 35 percent of the base shear. However, all of the shear walls for this structure have a height-to-length ratio less than 1.0. Therefore, this exception is met.

b) Structure is regular in plan, and there are at least two bays of lateral-load-resisting perimeter framing on each side of the structure. Therefore, this exception is also met.

Therefore, for both directions and all levels, no increase in base shear is required due to lack of redundancy.

6. Determine if structure meets requirements of conventional construction provisions in the 2006 IBC

While SEAOC is not encouraging the use of conventional construction methods, this step is included because conventional construction is allowed by the IBC (however, it is often misused) and can lead to poor structure performance.

The structure must be checked against the individual requirements of IBC §2308, and because it is in Seismic Category D, it must also be checked against IBC §2308.12. Results of these checks are shown below.

6a. Floor dead loads (§2308.2 Item 3)

The dead load weight of the floor exceeds the limit of 15 psf and, therefore, the structure requires an engineering design for vertical and lateral forces.

6b. Braced wall lines (§2308.12.3)

The spacing of braced wall lines exceeds 25 feet o/c and, therefore, the entire lateral system requires an engineering design.

Thus, the hotel structure requires an engineering design for both vertical and lateral loads.

7. Diaphragm deflections to determine if the diaphragm is flexible or rigid

This step is shown only as a reference for how to calculate horizontal diaphragm deflections. Since the shear wall forces were determined using both flexible and rigid diaphragm assumptions, there is no requirement to verify that the diaphragm is actually rigid or flexible.

The roof diaphragm has been selected to illustrate the methodology. The design seismic force in the roof diaphragm using Equation 12.10-1 must first be determined. The design seismic force is then divided by the diaphragm area to determine the horizontal loading in pounds per square foot. These values are used for determining diaphragm shears (and also collector forces). The design seismic force shall not be less than $0.2S_{DS}Iw_{px}$ nor greater than $0.4S_{DS}Iw_{px}$.

7a. Roof diaphragm check

The roof diaphragm will be checked in two steps. First, the shear in the diaphragm will be determined and compared to allowables. Next, the diaphragm deflection will be calculated. In Part 7b, the diaphragm deflection is used to determine whether the diaphragm is flexible or rigid.

Check diaphragm shear

The roof diaphragm consists of $^{15}/_{32}$-inch-thick sheathing with 10d @ 6 inches o/c and panel edges are unblocked. Loading on the segment between C and E, where

$$v = \frac{(8.41)48.0\text{ ft}(32.0\text{ ft})0.7}{(480.0\text{ ft})2} = 94\text{ plf}$$

Diaphragm span = 32.0 ft

Diaphragm depth = 48.0 ft

Diaphragm shears are converted to allowable stress design by multiplying by 0.7.

From IBC Table 2306.3.1, the allowable shear of 190 plf is based on $^{15}/_{32}$-inch DOC PS1 or PS2 (APA or TECO performance) rated wood structural panels with unblocked edges and 10d nails spaced at 6 inches o/c at boundaries and supported panel edges. APA or TECO performance-rated wood structural panels may be either plywood or oriented strand board (OSB).

Check diaphragm deflection

The code specifies that the deflection is calculated on an equivalent tributary lateral load basis. In other words, the diaphragm deflection should be based on the same load as the load used for the lateral resisting elements, not F_{px} total force at the level considered. Since the code requires building drifts to be determined by the strength level forces specified in §12.8, strength loads on the building diaphragm must be determined.

The basic equation to determine seismic forces on a diaphragm is

$$F_{px} = \frac{\sum_{i=x}^{n} F_i}{\sum_{i=x}^{n} w_i} w_{px} \qquad \text{Eq 12.10-1}$$

$$F_{p\ roof} = \frac{(44.5 \times 134.25)}{134.25} = 44.5 \text{ k}$$

For simplicity, the wall weights parallel to the direction of loading are included in w_{px}.

For the uppermost level, the above calculation will always produce the same force as computed in Equation 12.8-11. Then divide by the area of the diaphragm to find the equivalent uniform force.

$$f_{p\ roof} = \frac{44.5 \times 1000}{5288} = 8.41 \text{ psf}$$

In this example, the roof and floor diaphragms spanning between C and E will be used to illustrate the method. The basic 4-term equation to determine the deflection of a diaphragm is

$$\Delta = \frac{5vL^3}{8EAb} + \frac{vL}{4Gt} + 0.188Le_n + \frac{\Sigma(\Delta_c X)}{2b} \qquad \text{(Eq 23-1)}$$

The equation above is based on a uniformly nailed, simple span diaphragm with panel edges blocked and is based on monotonic tests conducted by the American Plywood Association (APA). The equation has four parts. The first part accounts for beam bending, the second accounts for shear deformation, the third accounts for nail slippage/bending, and the fourth part accounts for chord slippage.

The SDPWS now has a simplified 3-term expression for determining diaphragm deflections.

$$\Delta = \frac{5vL^3}{8EAW} + \frac{0.25vL}{1000\,G_a} + \frac{\Sigma(X\Delta_c)}{2W} \qquad \text{SDPWS Eq 4.2-1}$$

For the purpose of this design example, the diaphragm is assumed to be a simple span supported at C and E (refer to Figure 2-4). In reality, with continuity, the actual deflection will be less.

$$v = \frac{(8.41)48.0 \text{ ft}(32.0 \text{ ft})}{(48.0 \text{ ft})2} = 135 \text{ plf}$$

With nails at 6 inches o/c the strength load per nail is 135 × 6⁄12 = 67 lb/nail = V_n
Other terms in the deflection equation are

$L = 32.0$ ft

$b = 48.0$ ft

$G = 50{,}000$ psi — PDS T 3

$E = 1{,}700{,}000$ psi

$A_{2x4\ chords} = 5.25$ sq in × 2 = 10.50 sq in

Fastener slip/nail deformation values (e_n) are obtained as follows:

APA Research Report 138 requires the nail slip values, e_n, be decreased 50 percent for seasoned lumber. This means that the table is based on nails being driven into *green* lumber and the engineer must use half of these values for nails driven in *dry* (seasoned) lumber. The values are based on tests conducted by the APA. The 50-percent nail-slip reduction for dry lumber is a conservative factor. The actual tested slips with dry lumber were less than 50 percent of the green lumber slips.

For 10d common nails, there are 2 basic equations

When the nails are driven into *green* lumber: $e_n = (V_n/977)^{1.894}$ — APA T B-4

When the nails are driven into *dry* lumber: $e_n = (V_n/769)^{3.276}$ — APA T B-4

where

V_n is the fastener load in pounds per fastener.

These values are based on Structural-I sheathing and must be increased by 20 percent when the sheathing is not Structural-I. Footnote a in APA Research Report 138, Table B-4 states "Fabricated green/tested dry (seasoned) . . ." is very misleading. The values in the table are actually *green values*, because the lumber is fabricated when green. Again, don't be misled by the word "seasoned."

$e_n = 1.20(67/769)^{3.276} = 0.0004$

$t = 0.298$ in (for CDX or Standard Grade) — PDS T 2

Assume chord-splice at the mid-span of the diaphragm that will be nailed. The allowable loads for fasteners are based on limit state design. In other words, the deformation is set at a limit rather than the strength of the fastener. The deformation limit is 0.05 diameter of the fastener. For a 16d nail, a conservative slippage of 0.01 inch will be used.

Using strength level diaphragm shear

$$\Sigma(\Delta_c X) = (0.01)\ 16.0\text{ ft}\ (2) = 0.32\text{ in-ft}$$

$$\Delta = \frac{5(135)32.0^3}{8(1.7E6)10.50(48.0)} + \frac{135(32.0)}{4(50{,}000)0.298} + 0.188(32.0)0.0004 + \frac{0.32}{2(48.0)} = 0.08\text{ in}$$

This deflection is based on a *blocked* diaphragm. The IBC does not have a formula for an *unblocked* diaphragm. The roof diaphragm is also sloped at 6:12, which is believed to increase the deflection (but this has not been confirmed by tests). This design example has unblocked panel edges for the floor and roof diaphragms. The floors will similarly neglect the stiffening effects of lightweight concrete fill and gluing of sheathing.

As a comparison, determine diaphragm deflection for the unblocked roof diaphragm using the SDPWS 3-term expression.

$$\Delta = \frac{5vL^3}{8EAW} + \frac{0.25vL}{1000\,G_a} + \frac{\Sigma(X\Delta_c)}{2W}$$ SDPWS Eq 4.2-1

v = 135 plf

L = 32.0 ft

E = 1.7E6

A = 10.5 sq in

W = 48.0 ft

G_a = 14.0 kip/in SDPWS T 4.2B

$\Sigma(x\Delta_c)$ = 0.32 in-ft

$$\Delta = \frac{5(135)32^3}{8(1.7E6)10.5(48.0)} + \frac{0.25(135)32.0}{1000(14.0)} + \frac{0.32}{2(48.0)} = 0.08 \text{ in}$$

7b. Flexible versus rigid diaphragms §12.3.1.3

In this example, the maximum diaphragm deflection was estimated as 0.08 inch. This assumes a simple span for the diaphragm, and the actual deflection would probably be less. The average story drift is on the order of 0.10 inch at the roof (see Step 3c for the computed deflections of the shear walls). For the diaphragms to be considered flexible, the maximum diaphragm deflection will have to be more than 2 times the average story drift. This is right at the limit of a definition of a flexible diaphragm. The other diaphragm spans would easily qualify as "rigid" diaphragms. As defined by the code, the diaphragms in this design example are considered rigid.

In reality, some amount of diaphragm deformation will occur, and the true analysis is highly complex and beyond the scope of what is normally done for this type of construction. Diaphragm deflection analysis and testing has been performed on level/flat diaphragms. There has not been any testing of sloped and complicated diaphragms, as found in the typical wood-framed structure. Therefore, some engineers perform their design based on the roof diaphragm as flexible and the floor diaphragms as rigid.

In using this procedure, the engineer should exercise careful judgment in determining if the higher load of the two methodologies is actually required. For example, if the load to two walls by rigidity analysis is found to be 5 percent to line A and 95 percent to line B, but by flexible analysis it is found to be 50 percent to line A and 50 percent to line B, the engineer should probably design for the larger of the two loads for the individual walls. Note that though the same definition of a flexible diaphragm has been in the UBC since the 1988 edition and in the IBC since the first edition in 2000 it is usually not enforced by building officials for Type V construction.

8. Tiedown forces for the shear wall on line C

Tiedowns are required to resist the uplift tendency on shear walls caused by overturning moments. In this step, tiedown forces for the three-story shear wall on line C are determined. The design chosen uses continuous tiedowns below the third floor. At the third floor, conventional premanufactured straps are used.

Not included in this design example, but it should be noted: the IBC has two provisions for 1-hour wall assemblies—Footnotes l and m of Table 720.1(2) in the code. Footnote 1 requires longer fasteners for gypsum sheathing when the sheathing is applied over wood structural panels. Footnote m requires values for F'_c to be reduced to 78 percent of the allowable in 1-hour walls.

8a. Discussion on continuous tiedown systems

The continuous tiedown system is a relatively new method for resisting shear wall overturning. Similar to the many metal connectors used for wood framing connections, most are proprietary and have National Evaluation Service or ICC approval. All the systems have some type of rod and hardware connector system that goes from the foundation to the top of the structure. A common misconception among engineers regarding these types of systems is that elongation of the rod will produce large displacements in the shear walls. Contrary to that perception, these systems are in many instances superior to the one-sided bolted tiedowns.

Investigations after the Northridge earthquake as well as independent testing of the conventional one-sided bolted tiedowns, have concluded that there can be large displacements associated with this type of connection. The large displacements are a result of eccentricity with the boundary element, deflection of the tiedown, wood shrinkage, wood crushing, and oversized holes for the through-bolts.

Some of the proprietary systems compensate for shrinkage either by pre-tensioning of the rod or by a self-ratcheting connector device. Shrinkage-compensating devices are desirable in multi-level wood frame construction. These devices will also compensate for other slack in the tiedown system caused by crushing of plates, seating of posts, studs, etc.

8b. Determine strength shear wall forces

The shear wall on line C is shown on Figure 2-7. Forces at each story are determined as follows (from Table 2-20)

$$F_{roof} = 13{,}099/2 = 6550 \text{ lb}$$

$$F_{third} = (24{,}574 - 13{,}099)/2 = 5738 \text{ lb}$$

$$F_{second} = (29{,}690 - 24{,}574)/2 = 2558 \text{ lb}$$

The distance between the centroid of the boundary forces that represent the overturning moment at each level must be estimated. This is shown below.

e = the distance to the center of tiedown rod and boundary studs or collector studs (Figure 2-12)

$$e = 2 \times 2.5 \text{ in} + (13/2) = 11.5 \text{ in} = 0.958 \text{ ft}$$

Use $e = 1.0$ ft

d = the distance between centroids of the tiedown and the boundary studs, in feet. (Note that it is also acceptable to use the distance from the end of the shear wall to the centroid of the tiedown.)

$d = 21.5 \text{ ft} - 2(1.0 \text{ ft}) = 19.5$ ft at second floor for third level (Figure 2-12)

$d = 21.5 \text{ ft} - (2 \times 0.125 \text{ ft}) = 21.25$ ft at third floor for roof level (Figure 2-11)

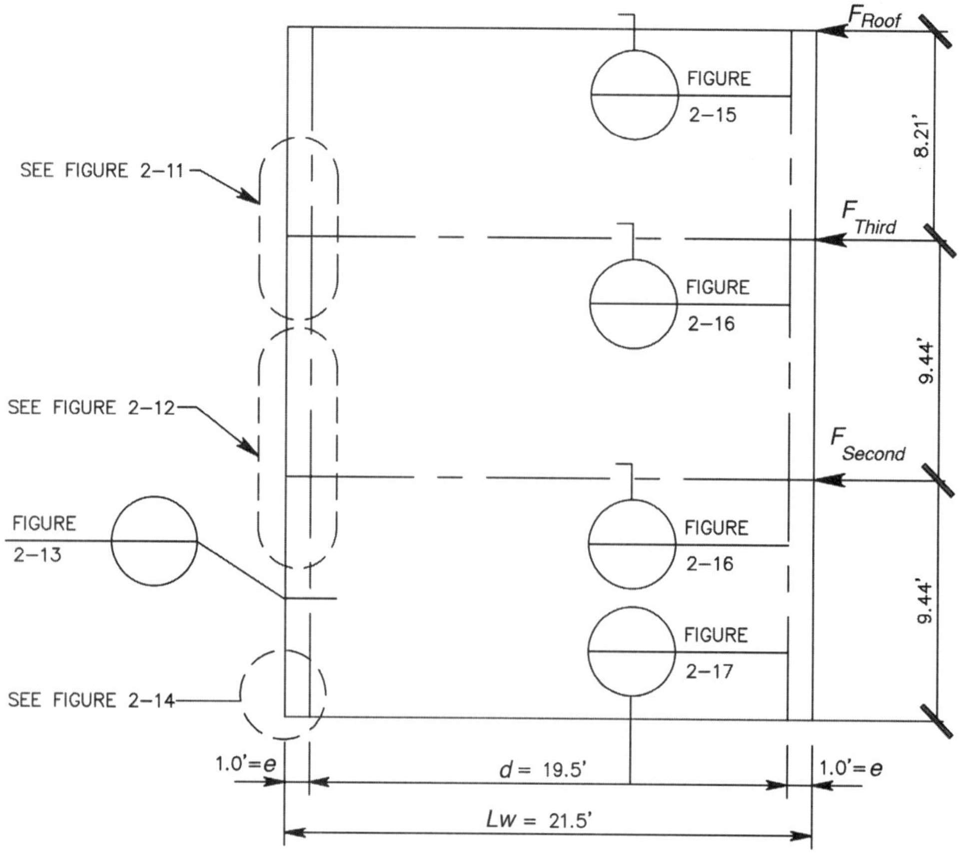

Figure 2-7. Shear wall C elevation

The resisting moment M_R is determined from the following loads

$W_{roof} = 13.5 \text{ psf } (2.0 \text{ ft}) = 27.0 \text{ plf}$

$W_{floor} = 25.0 \text{ psf } (2.0 \text{ ft}) = 50.0 \text{ plf}$

$W_{wall} = 10.0 \text{ psf}$

Table 2-21. Tiedown forces for shear wall C

Level	M_{ot} (ft-lb)	M_R (ft-lb)	$M_R \times (0.6 - 0.14\ S_{DS})^1$ (ft-lb)	Uplift $\frac{(M_{ot} \times 0.7) - (0.6 - 0.14\ S_{DS})M_R}{d}$ (lb)	Differential Load[2] (lb)
Roof	53,775	25,216	10,843	1,374	1,374
Third	169,774	58,590	25,194	4,802	3,428
Second	309,920	91,965	39,545	9,097	4,295

Notes:

1. Where $(0.6 - 0.14\ S_{DS}) = 0.43$
2. The differential is the load difference between the uplift force at level x and the level above.

9. Design tiedown connection at the third floor for the shear wall on line C

Figure 2-11 illustrates the typical tiedown connection for the shear wall on line C at the third floor. This is the conventional pre-manufactured strap and is fastened to the framing with nails.

The total uplift force at this level is 1374 lb

$$P_1 = 1374 \text{ lb}$$

The tiedowns will be planned using allowable stress design. §12.4.2.3

With a 16-gage by 1.25-in strap and 10d common nails.

Allowable load per nail is $ZC_D = 116(1.6) = 185$ lb/nail NDS T 11P

Number of nails required = 1374/185 = 7.4 $\therefore$ use 8

With nails at 1.5 inches o/c the length of strap required is

$$2(0.75 \text{ in} + 7 \times 1.5 \text{ in}) + 6 \text{ in} = 28.5 \text{ in}$$

$\therefore$ use 30-inch-strap

10. Design tiedown connection at the second floor for the shear wall on line C

As previously mentioned, the second floor tiedown will be part of the continuous tiedown system used below the third level. Refer to Figure 2-12 for illustration of this system and the location of forces P_1, P_2, and P_3.

The total uplift force at the second floor is 4802 lb (Table 2-21).

$P_1 = P_2$ = total uplift force from above = 1374

P_3 = uplift force for the collector studs = differential load/2 = 3428 lb/2 = 1714 lb

Since the strap from above is only connected to one pair of collector studs, the total uplift force for the outside set of collectors is equal to the uplift force plus the uplift force on the second floor shear wall from the third floor.

Taking a free-body diagram of the system, the tension in the tiedown rod is increased because of cantilever action between the centroids of the forces. A downward component is actually applied to the interior-most support stud (Figure 2-8).

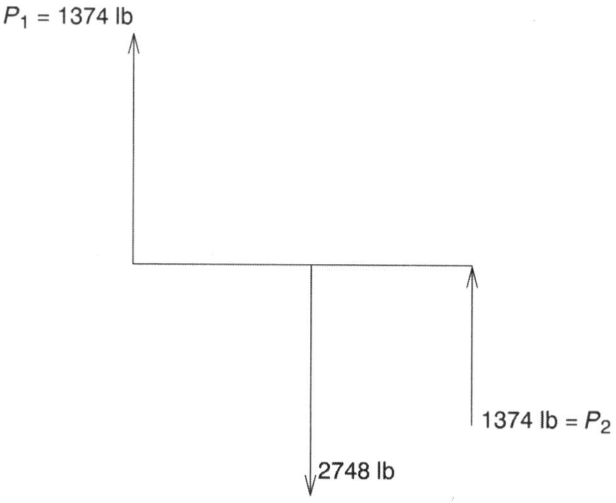

Figure 2-8. Free-body force diagram of compression bridge

Next, the tension in the tiedown rod between the second floor and the compression bridge is the differential load plus the tension load, as computed above. This will produce the total force, P_2, on support stud (Figure 2-9).

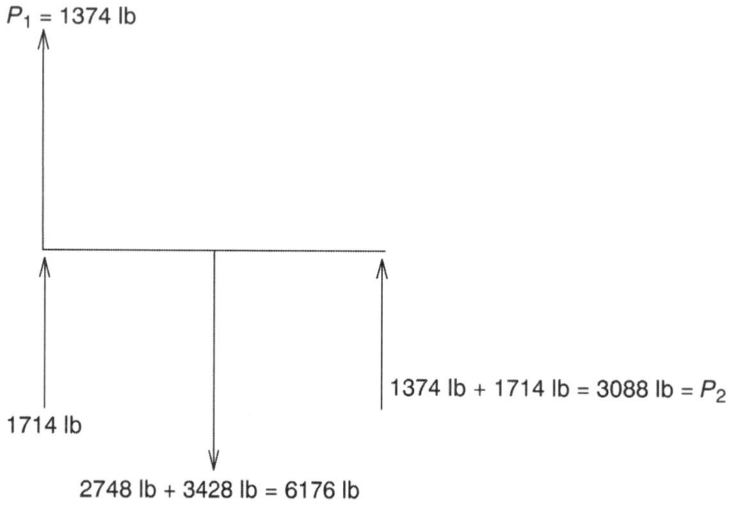

Figure 2-9. Free-body force diagram of compression bridge

Determine spacing for the flat nailing

$P_{max} = 3088$ lb

The allowable lateral load for a 16d common nail in a 1½-inch side member is

$ZC_D = 141(1.6) = 225$ lb NDS T 11N

With 2 rows of 16d nails, the number of nails per row is 3088 lb/2 × 225 = 6.8 nails

∴ use 7 nails

Maximum spacing = 48 in/(6 + 1) = 6.9 in

∴ Use 6-inch o/c for the flat nailing

Check compression perpendicular to grain for the bridge support studs to compression bridge

Critical at P_2.

$f_{c\,max} = 3088 \text{ lb}/(1.5 \times 3.5) = 588 \text{ psi} < F_{c\perp} = 625$ psi . . . *o.k.* NDS Supp. T 4A

Check the bearing perpendicular to grain on bearing plate

$F = T_1 = 6{,}176$ lb

$f_{c\perp} = 6176 \text{ lb}/3.25 \times 5.0 = 380 \text{ psi} < F_{c\perp} = 625$ psi . . . *o.k.*

Check bearing perpendicular to grain on the top plate from the collector studs from below: First floor is framed with 3x4 studs

Force at $P_3 = 1714$ lb

$f_{c\perp} = P/A = 1714 \text{ lb}/(2.5 \times 3.5) = 195 \text{ psi} < F_{c\perp} = 625$ psi . . . *o.k.*

Check shear on 4x8 compression bridge (assume tiedown is at center of wall and not at party wall, see Figure 2-12)

$T_1 = 6176$ lb

Assuming compression bridge to take all shear

$$V = \frac{T_1}{2} = \frac{6176}{2} = 3{,}088 \text{ lb}$$

$$f_V = \frac{3088 \times 1.5}{3.5 \times 7.25} = 183 \text{ psi}$$

For Douglas Fir-Larch No. 1

$F'_V = F_V C_D = 180 \times 1.6 = 288$ psi . . . *o.k.* NDS Supp. T 4A

Check bending on 4x8 compression bridge

$$T_1 = 6176 \text{ lb}$$

$$M = \frac{T_1 \times L}{4} = \frac{6176 \times (10 + 1.5)}{4} = 17{,}756 \text{ in-lb}$$

S_x for 4x8 with hole for ⅝-in rod = (3.5 – 0.69) $7.25^2/6$ = 24.6 in^3

$$f_b = \frac{M}{S} = \frac{17{,}756}{24.6} = 722 \text{ psi}$$

For Douglas Fir-Larch No. 1 — NDS Supp. T 4A

$$F_b' = F_b C_D C_F = 1000(1.6)(1.3) = 2080 \text{ psi} \ldots \textit{o.k.}$$

Check shear on plates at floor

Tiedown connector reaction is the *differential load,* which is 3595 lb. Recognizing the fact that 4802 lb of the 9097-lb total force is already in the tiedown rod

$$T = 4{,}295 \text{ lb}$$

Assuming 2 sill plates and 2 top plates to take all shear

$$V = \frac{T}{2} = \frac{4295}{2} = 2150 \text{ lb}$$

$$f_V = \frac{2150 \times 1.5}{4(1.5 \times 3.5)} = 153 \text{ psi}$$

Because plates have no spits, $C_H = 2.0$ (plates rarely check on the edges)

$$F_V' = F_V C_H C_D = 180(2.0)(1.6) = 576 \text{ psi} \ldots \textit{o.k.}$$

Therefore, the tiedown connection shown on Figure 2-12 meets the requirements of code.

11. Design tiedown connection and anchor bolt spacing for shear wall on line C

11a. Design anchor bolt spacing of sill plate on line C

See discussion about fasteners for pressure-preservative treated wood in Step 19.

From Table 2-20

$$V = 29{,}690 \text{ lb}$$

$$v = \frac{V}{L} = \frac{29{,}690 \text{ lb}}{43 \text{ ft}} = 690 \text{ lb/ft}$$

For a side member, thickness = 2.5 inches in Hem-Fir wood (note that designing for Hem-Fir will require a tighter nail and bolt spacing)

Z_{11} = 1400 lb/bolt NDS T 11E

$$\text{Required spacing} = \frac{Z_{11}C_D}{v} = \frac{(1400)(1.6)}{690(0.7)} = 4.6 = 55 \text{ in}$$

where

0.7 is the strength conversion factor

∴ Use ¾-in-diameter bolts at 48 inches o/c

Sill bolts in concrete:

ASCE 7-05 §14.2.2.17 amends ACI 318-05 Appendix D, sections D.3.3.4 and D.3.3.5 to require that anchors be designed to be governed by tensile or shear strength of a ductile steel element or that they shall be designed such that the attachment undergoes ductile yielding. It is questionable if wood sill plates transferring shear to the concrete anchor bolts can attain the requirements of ASCE 7-05 §14.2.2.17. SEAOC is currently in the process of testing wood sill plates with anchor bolts (performed at the Simpson Strong-Tie Stockton facility) in the hopes of developing a publication regarding the design of these anchors.

11b. Determine tiedown anchor embedment

In this calculation, the tiedown anchor will be assumed to occur at the center of the exterior wall. This will produce a lower capacity than if the rod were located at the double-framed wall shown in Figure 2-13.

From Table 2-21

$$T = 9097 \text{ lb}$$

$$T_u = \frac{9097}{0.7} = 12{,}995 \text{ lb}$$

Determine factored design loads

$$N_u = T_u = 12{,}995 \text{ lb}$$

$$V_u = 0 \text{ lb}$$

Provide an oversized hole for the tiedown rod in the foundation sill plate. The rod has no nut or washer to the sill plate, therefore, assume $V = 0$ lb in the rod. Tiedown bolts resist vertical loads only, anchor bolts are designed to resist the lateral loads.

Determine the design tensile strength, ϕN_n, where ϕN_n is the smallest of the design tensile strengths as controlled by steel, ϕN_{sa}, concrete breakout, ϕN_{cb}, pullout, ϕN_{pn}, and side-face blowout, ϕN_{sb}

$\phi N_{sa} = \phi\, n\, A_{se} f_{uta}$ ACI Eq D-3

where

$\phi = 0.75$ for the anchor rod ACI §D.4.4

Per Table 1 of "Strength Design of Anchorage to Concrete," the ASTM A307 Grade-A bolt meets the *Ductile Steel Element* definition.

A_{se} = 0.462 for a ⅞-inch ϕ bolt Strength Design of Anchorage to Concrete, T 2

f_{ut} = 60,000 psi Strength Design of Anchorage to Concrete, T 1

n = 1 fastener

$\phi N_{sa} = 0.75(1)(0.462)(60{,}000) = 20{,}800$ lb

Concrete breakout strength, ϕN_{cb}

Since there is no supplemental reinforcement provided, $\phi = 0.75$ ACI D.4.5

$$N_{cb} = \frac{A_{Nc}}{A_{Nco}} = \psi_{ec,N} \psi_{ed,N} \psi_{c,N} \psi_{cp,N} N_b$$ ACI Eq D-5

A_{NC} is the projected area of the tensile failure surface as approximated by a rectangle with edges bounded by 1.5 h_{ef}

$Tr_y\ h_{ef} = 15$ in

$1.5\ h_{ef} = 1.5 \times 15 = 22.5$ in

$A_{Nc} = (12.0 + 22.5)(22.5 + 22.5) = 1550\ \text{in}^2$

$A_{Nco} = 9\ h_{ef}^{\ 2} = 9(15)^2 = 2025\ \text{in}^2$ ACI Eq D-6

See figures 2-7 and 2-13 for tiedown location.

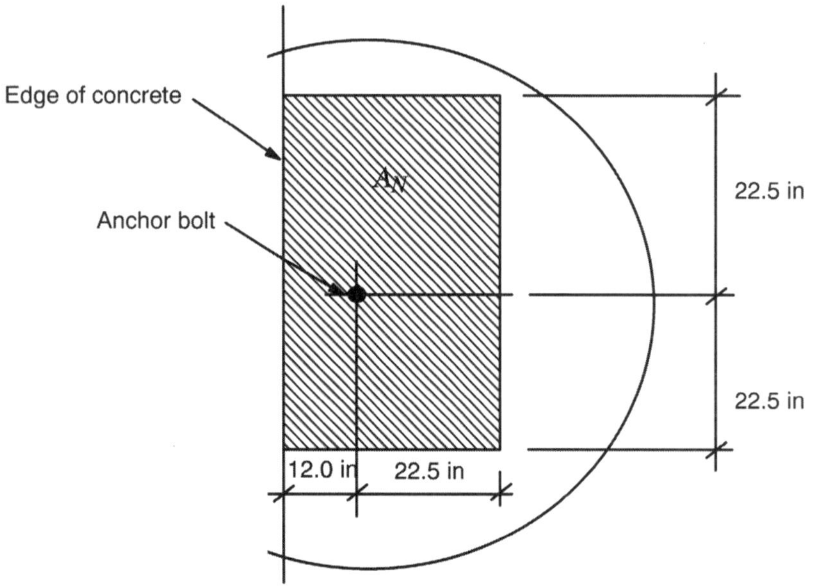

Figure 2-10. Tiedown bolt showing projected area A_N at surface

$\psi_{ec,N} = 1.0$ because there is no eccentricity

Determine $\psi_{ed,N}$ for the fastening

$$\psi_{ed,N} = 0.7 + 0.3\,\frac{C_{a,min}}{1.5\,h_{ef}}$$ ACI Eq D-11

$$\psi_{ed,N} = 0.7 + 0.3\left(\frac{12}{(1.5 \times 15)}\right) = 0.86$$

$\psi_{c,N}$ = 1.25 for no cracking at service loads

$\psi_{cp,N}$ = 1.0 for cast-in anchors

Determine N_b for the fastening

$$N_b = k\sqrt{f'_c}\,h_{ef}^{1.5}$$ ACI Eq D-7

where

$k = 24$

$$N_b = 24\sqrt{3000}(15)^{1.5} = 76{,}000 \text{ lb}$$

Substituting into Equation D-5

$$N_{cb} = \left[\frac{1550}{2025}\right](0.86)(1.25)(76{,}000) = 62{,}500 \text{ lb}$$

$$\phi N_{cb} = 0.70(62{,}500) = 44{,}000 \text{ lb}$$

ϕ = 0.70 for supplementing reinforcement not provided

Determine pullout strength, ϕN_{pn}

$$\phi N_{pn} = \phi\psi_{c,p}N_p$$ ACI Eq D-14

where

ϕ = 0.70 because there is no supplemental reinforcement provided

$\psi_{c,p}$ = 1.4 with no cracking at service loads

$N_p = 8\,A_{brg}f'_c$ ACI Eq D-15

A_{brg} = 0.891 for ⅞-in hex head bolt — Strength Design of Anchorage to Concrete T 2

Pullout strength, ϕN_{pn}

$$\phi N_{pn} = 0.70(1.4)(0.891)8(3000) = 20{,}950 \text{ lb}$$

Determine side-face blowout strength, ϕN_{sb}.

The side-face blowout failure mode must be investigated when the edge distance, c_a, is less than 0.4 h_{ef}

$$0.4\,h_{ef} = 0.4 \times 15 = 6.0 \text{ in} < 12.0 \text{ in}$$ ACI §D.5.4.1

Therefore, the side-face blowout strength need not be determined — ACI Eq D-17

Summary of steel strength, concrete breakout strength, pullout strength, and side-face blowout strength for tension

Steel strength	ϕN_{sa} : 20,800 lb
Embedment strength–concrete breakout	ϕN_{cb} : 44,000 lb
Embedment strength–pullout	ϕN_{pn} : 20,950 lb

Hence, steel strength governs as required by ACI §D.3.3.4 for a structure in Seismic Design Category D. Also, from ACI §D.3.3.3 an additional reduction factor of 0.75 is required. Therefore

$$0.75\ \phi N_{sa} = 0.75 \times 20{,}800 = 15{,}600 \text{ lb} > T_u = 12{,}955 \text{ lb} \ldots \textit{o.k.}$$

The limiting value is embedment strength—pullout.

11c. Check the bearing perpendicular to grain on sill plates

Assuming all compressive force for overturning will be resisted by end boundary elements, the critical load combination is

$$(1.0 + 0.105 S_{DS})\, D + 0.525\ \rho E + 0.75L + 0.75L_r \qquad \text{§12.4.2.3}$$

where

$$0.105\ S_{DS} = 0.125$$
$$\rho = 1.0$$

From Table 2-21, the strength level overturning moment is

$$M_{ot} = 309{,}920 \text{ ft-lb}$$

The seismic compressive force is obtained by dividing by the distance, *d*.

$$P_{seismic} = \frac{M_{ot}}{d} = \frac{309{,}920}{19.5} = 15{,}900 \text{ lb}$$

$$P_{DL} = [W_{roof} + (W_{floor}) + W_{wall}(27 \text{ ft})]$$

$$P_{DL} = \left[15.1 \times \frac{16}{12} + 2\left(25.0 \times \frac{24}{12}\right) + 10(27 \text{ ft})\right]\left(\frac{16 \text{ in} + 8 \text{ in}}{12 \text{ in}}\right) = 780 \text{ lb}$$

$$P_{LL} = \left(40 \text{ psf} \times \frac{24}{12} \times 2\right)\left(\frac{16 \text{ in} + 8 \text{ in}}{12 \text{ in}}\right) = 320 \text{ lb}$$

$$PL_r = \left[16 \text{ psf} \times \frac{16}{12}\right]\left(\frac{16 \text{ in} + 8 \text{ in}}{12 \text{ in}}\right) = 45 \text{ lb}$$

$$\Sigma P = 1.12\ (780) + 0.525\ (15{,}900) + 0.75\ (320) + 0.75\ (45) = 9500 \text{ lb}$$

with full-width bearing studs bearing on both sill plates (Figure 2-13), the bearing area is equal to six 3x4 studs.

$$f_{c\,max} = \frac{9500}{6(8.75)} = 180 \text{ psi} < F'_{c\perp} = 625 \text{ psi} \ldots \textit{o.k.} \qquad \text{NDS Supp T 4A}$$

where the area of a 3x4 is 8.75 square inches. Note that if a Hem-Fir sill plate is used, the allowable compression perpendicular to grain $F'_{c\perp} = 405$ psi.

$$f_c < 0.73\, F'_{c\perp} = 0.73(405) = 295 \text{ psi} \qquad \text{NDS Supp. T 4A}$$

Therefore, the assumed crushing effect of 0.02 inch (Table 2-13) is correct.

As discussed in the notes for Table 2-4, the ratcheting effect of the continuous tiedown system will compensate for the crushing.

12. Detail of tiedown connection at the third floor for shear wall on line C

Note that since the boundary element is a double stud and the wall panel edge nailing is nailed to the end stud, the 16d at 12 inches o/c internailing of the two tiedown studs should have the capacity to transfer one-half the force to the interior stud (Figure 2-11). These nails may be installed from either side (normally nailed from the outside). See Figure 2-16 for the location of the top plates and commentary about plate locations.

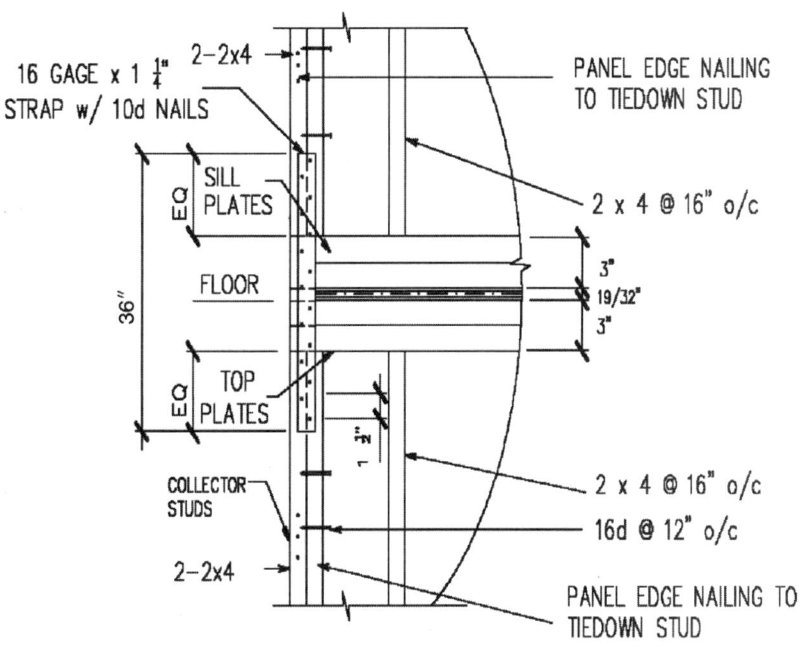

Figure 2-11. Tiedown connection at the third floor for shear wall C.

13. Detail of tiedown connection at the second floor for shear wall on line C

This tiedown rod system (Figure 2-12) may also be extended to the third floor instead of using the conventional metal strap shown in Figure 2-11. See Figure 2-16 for the location of the top plates and commentary about plate locations.

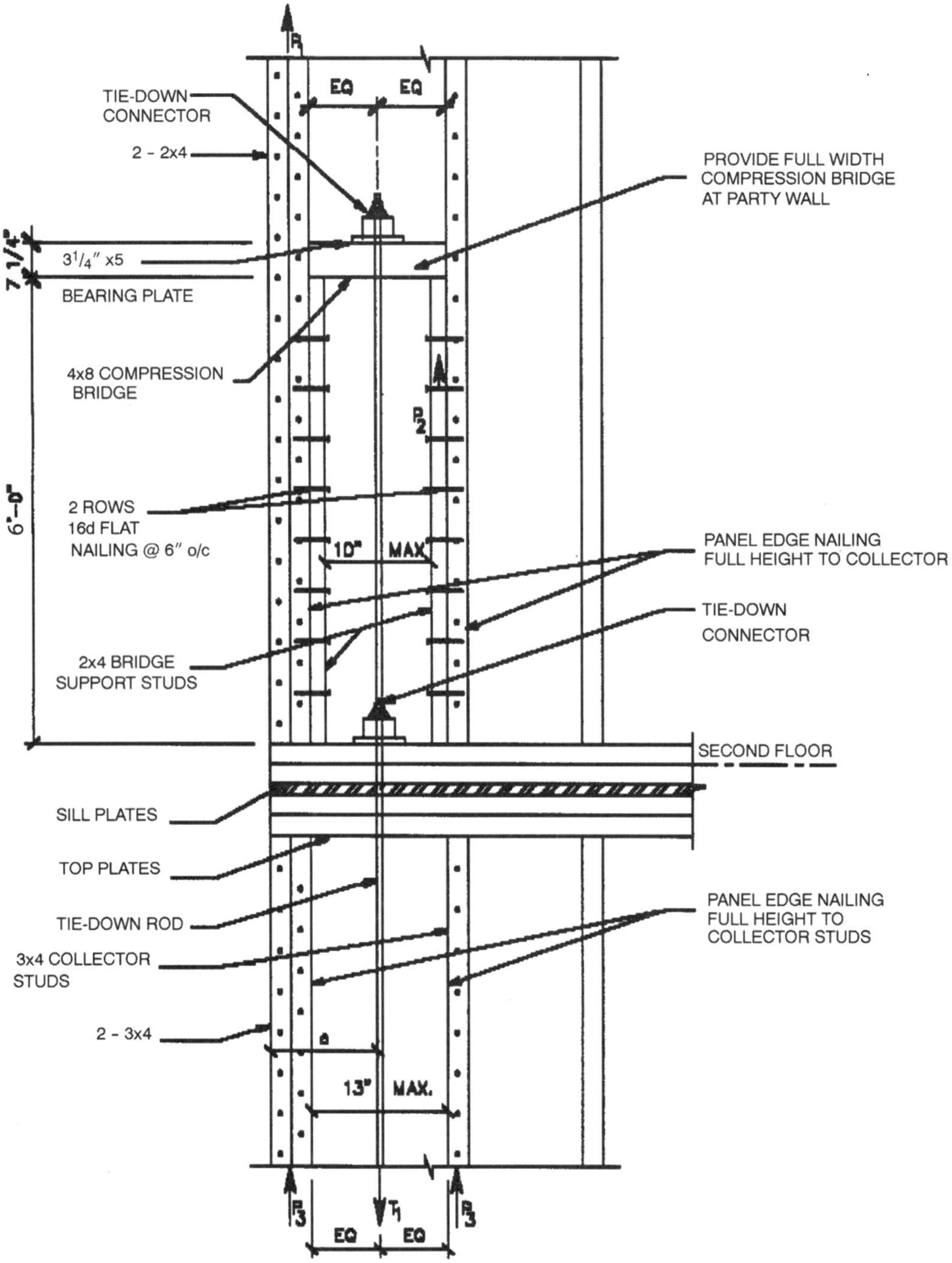

Figure 2-12. Tiedown connection at second floor for shear wall C

14. Detail of wall intersection at exterior walls

The detail shows full-width studs at tiedown (Figure 2-13). This is desirable when sheathing is applied to both stud walls. It is also desirable for bearing perpendicular to grain because the bearing area is doubled. When full-width studs are used for bearing, both sill plates will need to be 3x thickness (*not* as shown in Figure 2-17). Tiedowns may be located at the center of the stud wall that is also sheathed. It is good practice to tie the wall together. In this case, there is no design requirement or minimum shear wall-to-shear wall connection requirement other than that outlined in the IBC standard fastening schedule (IBC Table 2304.9.1).

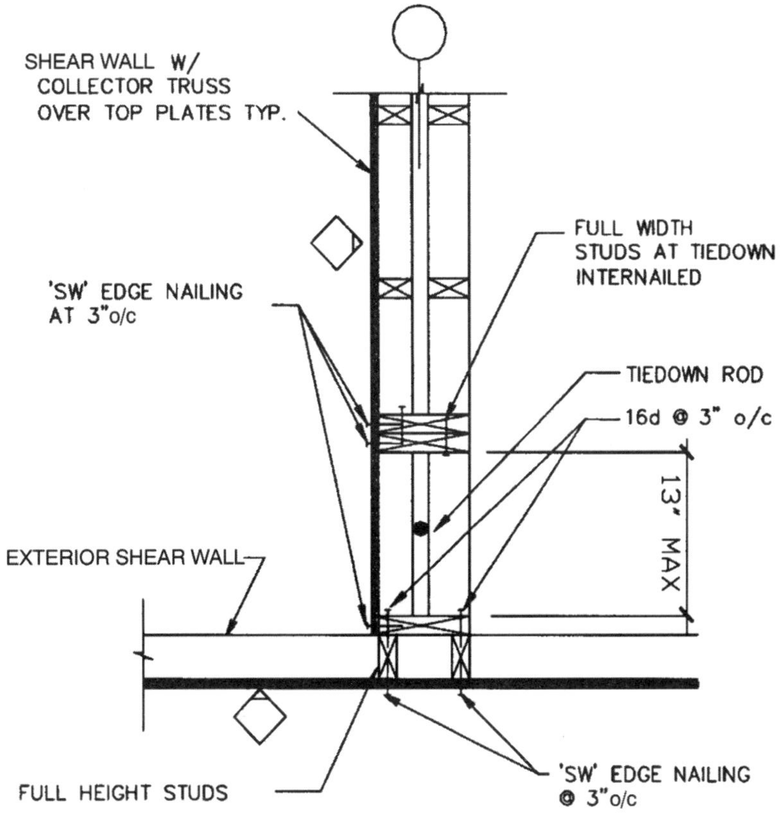

Figure 2-13. Wall intersection at shear wall (plan view)

15. Detail of tiedown connection at foundation

The manufacturer of the tiedown system usually requires the engineer of record to specify the tiedown forces at each level of the structure. This can easily be done in a schedule (Figure 2-14).

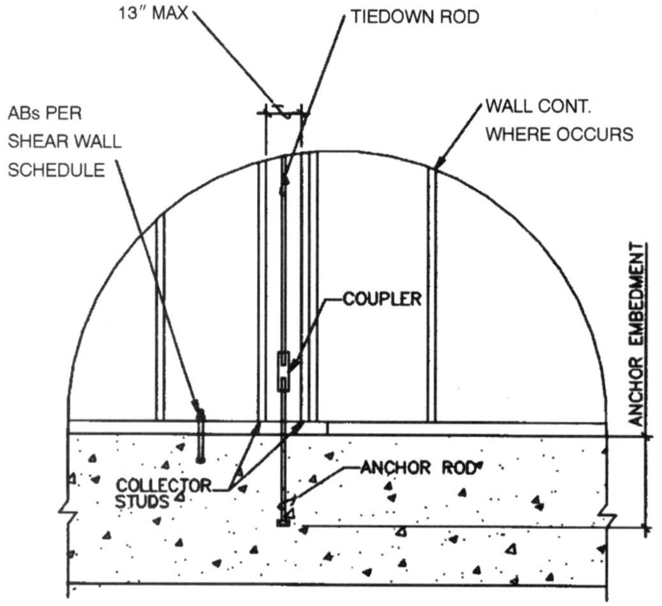

Figure 2-14. Tiedown connection at foundation

16. Detail of shear transfer at interior shear wall at roof

Note: Edge nailing from roof sheathing to collector truss may need to be closer than the roof sheathing edge nailing because of shears being collected from each side of the truss. It is also common to use a double collector truss at these locations. The 2x4 braces at the top of the shear wall need to be designed for compression or to provide tension bracing on each side of the wall (Figure 2-15).

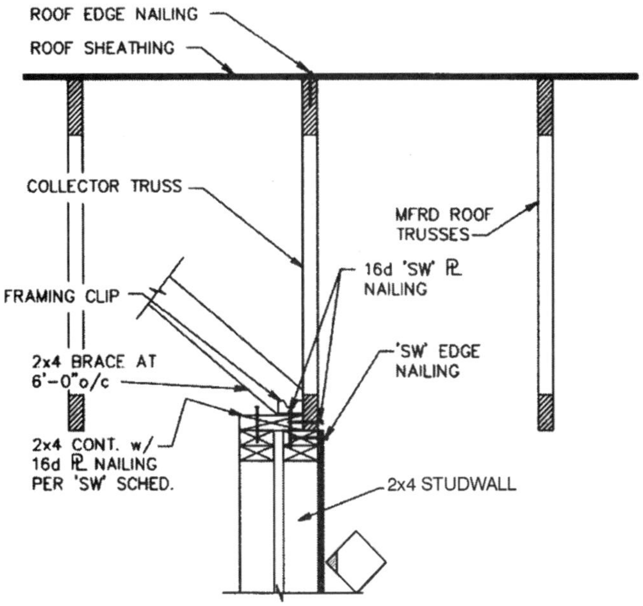

Figure 2-15. Shear transfer at interior shear wall at roof

17. Detail of shear transfer at interior shear wall at floors

This detail uses the double top plates at the underside of the floor sheathing (Figure 2-16). This is advantageous for shear transfer. Another often-used detail is to bear the floor joists directly on the top plates. However, when the floor joist is on top of the top plates, shear transfer is required through the glue joint in the webs and heavy nailing from the joist chord to the top plate.

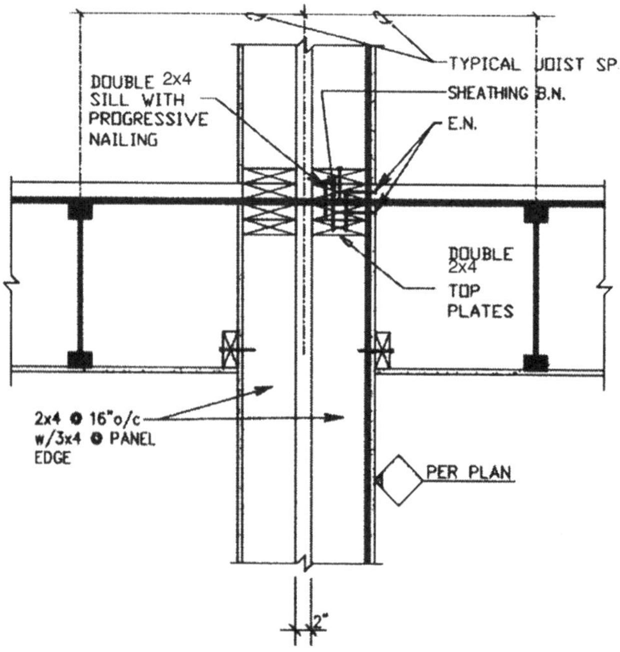

Figure 2-16. Shear transfer at interior shear wall at floor

Note: The nailers for the drywall ceiling need to be installed after the wall sheathing and wall drywall have been installed.

18. Detail of shear transfer at interior shear wall at foundation

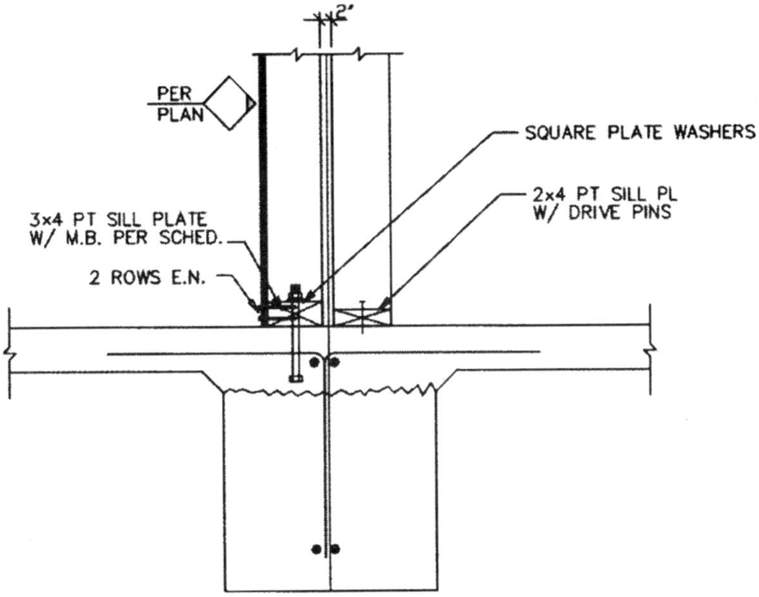

Figure 2-17. Shear transfer at foundation

19. Detail of sill plate at foundation edge

Fasteners for pressure- or preservative-treated wood

Section 2304.9.5 of the 2006 IBC requires corrosion-resistant fasteners in treated sill plates. This requires hot-dipped zinc-coated galvanized nails and anchor bolts. The exception that has existed (§2311.1 in the 1994 UBC) exempted this requirement when "not below grade or exposed to weather." The language in the code was submitted by the wood industry, and §2304.9.5 is from a report in the Wood Handbook by the Forest Products Lab where fasteners were found to react with the preservative treatment when ". . . in the presence of moisture . . ." This can create a construction problem because hot-dipped zinc-coated nails have to be hand driven, requiring the framer to put down his nail gun and change nailing procedures.

An additional caution for sill plates is the type of wood used. The most common species used on the west coast for pressure treatment is Hem-Fir, which has lower fastener values for nails and bolts than Douglas-Fir-Larch. A tighter nail spacing to the sill plate is necessary, or a double stagger row can be used. Figure 2-18 shows two rows of edge nailing to the sill plate as a method of compensating for a Hem-Fir sill plate.

Gap at bottom of sheathing.

Investigations into wood-framed construction have found that plywood or oriented strand board (OSB) sheathing that bears on concrete at perimeter exterior edges can "wick" moisture up from the concrete and cause corrosion of the fasteners and rotting in the sheathing. To help prevent this problem, the sheathing can be placed with a gap above the concrete surface. A ¼-inch gap is recommended for a 3x sill plate and a ⅛-inch gap is recommended for a 2x sill plate (Figure 2-18).

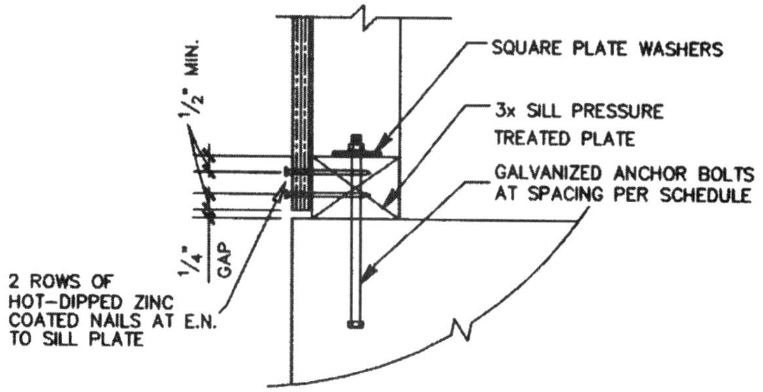

Figure 2-18. Sill plate at foundation edge

Note: The SDPWS only requires a minimum edge distance of ⅜ inch for nails in sheathing. Tests have shown that sheathing with greater edge distances has performed better.

In addition, the ¼-inch gap shown at the bottom of the sheathing should be a minimum. Recent cyclic testing has shown that when this gap is ½ inch the shear wall has performed better.

20. Detail of shear transfer at exterior wall at roof

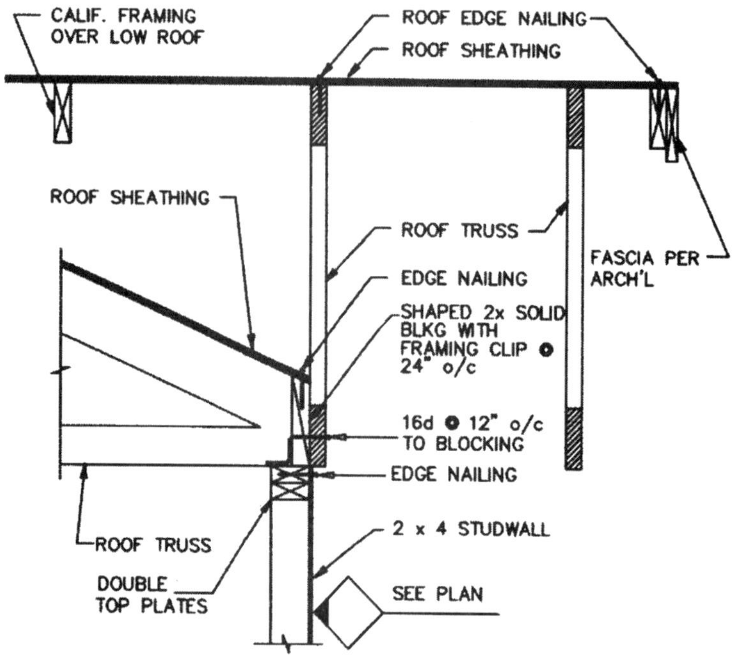

Figure 2-19. Shear transfer at exterior wall at roof

Note: The roof truss directly above the exterior wall is also a "collector" truss. Roof edge nailing to this truss and the 16d nails to the blocking need to be checked for the "collector" load. Double top plates are also a chord and collector.

21. Detail of shear transfer at exterior wall at floor

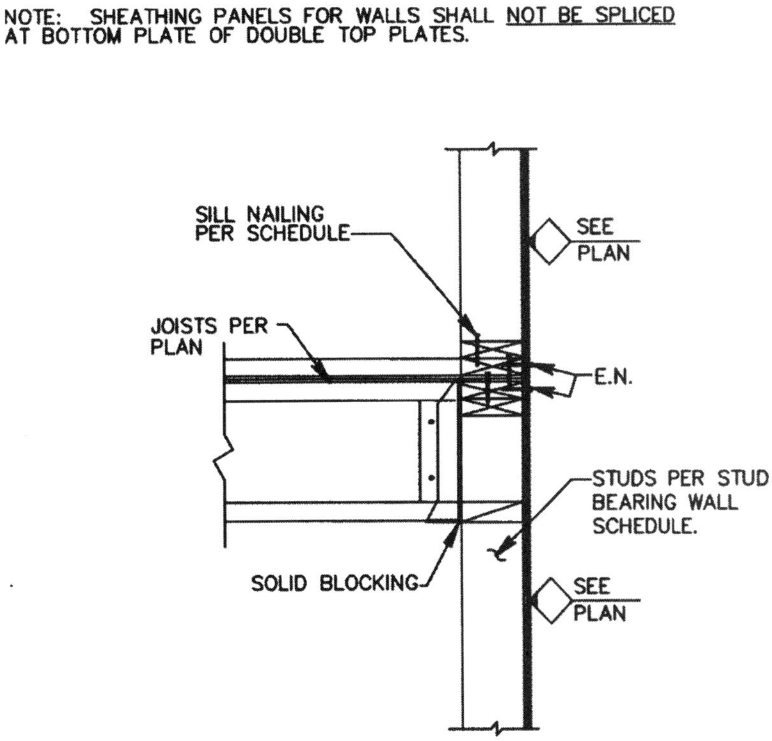

Figure 2-20. Shear transfer at exterior wall at floor

Note: This detail uses double top plates at the underside of the floor sheathing. Another often-used detail is bearing the floor joists on the double top plates. See Figure 2-16 for additional commentary.

References

American Forest and Paper Association, 2005, AF&PA Supplement Special Design Provisions for Wind and Seismic, American Forest and Paper Association, Washington, D.C.

American Forest and Paper Association, 2005, *National Design Specification for Wood Construction Including Supplements, NDS-05*. American Forest and Paper Association, Washington D.C.

American Forest and Paper Association, 1996, *Wood Construction Manual*. American Forest and Paper Association, Washington D.C.

American Plywood Association, 1997, *Design/ Construction Guide—Diaphragms and Shear Walls*. From L350, Engineered Wood Association, Tacoma, Washington.

American Plywood Association, 1997, *Diaphragms and Shear Walls*. Engineered Wood Association, Tacoma, Washington.

American Plywood Association, 1993, revised, *Wood Structural Panel Shear Walls*. Report 154, Engineered Wood Association, Tacoma, Washington.

American Plywood Association, 1994, *Northridge, California Earthquake*. Report T-94-5. Engineered Wood Association, Tacoma, Washington.

American Plywood Association, *Performance Standards and Policies for Structural-Use Panels [Sheathing Standard, Sec. 2.3.3]*. Standard PRP-108. Engineered Wood Association, Tacoma, Washington.

American Plywood Association, 1997, *Plywood Design Specifications*, From Y510, Engineered Wood Association, Tacoma, Washington.

American Plywood Association, 1988, *Plywood Diaphragms*, Research Report 138. American Plywood Association, Tacoma, Washington.

American Plywood Association, 2002. Effect of Green Lumber Framing on Wood Structural Panel Shear Wall Performance. APA Report T2002-53. American Plywood Association, Tacoma, Washington.

American Plywood Association, 2005, Using Narrow Pieces of Wood Structural Panel Sheathing in Wood Shear Walls, APA T 2005-08, The Engineered Wood Association, Tacoma, Washington.

Applied Technology Council, 1995, *Cyclic Testing of Narrow Plywood Shear Walls ATC R-1*. Applied Technology Council, Redwood City, California.

Applied Technology Council, 1981, *Guidelines for Design of Horizontal Wood Diaphragms*, ATC-7. Applied Technology Council, Redwood City, California.

Applied Technology Council, 1980, *Proceedings of a Workshop on Design of Horizontal Wood Diaphragms*, ATC-7-1. Applied Technology Council, Redwood City, California.

Building Seismic Safety Council, 1997, *National Earthquake Hazard Reduction Program, Recommended Provisions for Seismic Regulations for New Buildings*. Building Seismic Safety Council, Washington D.C.

Breyer, Donald E., Kenneth J. Fridley, David G. Pollock, Jr. and Kelly E. Cobeen, 2007. *Design of Wood Structures ASD*. McGraw-Hill Book Co., New York, New York

Bugni, David A., 1999, "A Linear Elastic Dynamic Analysis of a Timber Framed Structure." *Building Standards*, International Conference of Building Officials, Whittier, California

Cobeen, K. E., 1996, "Performance Based Design of Wood Structures." *Proceeding: Annual SEAOC Convention*. Structural Engineers Association of California, Sacramento, California.

Coil, J., 1999, "Seismic Retrofit of an Existing Multi-Story Wood Frame Structure," *Proceedings: Annual SEAOC Convention*. Structural Engineers Association of California, Sacramento, California.

Commins, A. and Gregg, R., 1996, *Effect of Hold Downs and Stud-Frame Systems on the Cyclic Behavior of Wood Shear Walls*, Simpson Strong-Tie Co., Pleasanton, California.

Cook, R. A., 1999, "Strength Design of Anchorage to Concrete." Portland Cement Association, Skokie, Illinois.

Countryman, D., and Col Benson, 1954, *1954 Horizontal Plywood Diaphragm Tests*. Laboratory Report 63, Douglas Fir Plywood Association, Tacoma Washington.

CUREe, 1999, *Proceedings of the Workshop on Seismic Testing, Analysis, and Design of Wood Frame Construction*. California University for Research in Earthquake Engineering.

Dolan, J. D., 1996, *Experimental Results from Cyclic Racking Tests of Wood Shear Walls with Openings*. Timber Engineering Report No. TE- 1996-001. Virginia Polytechnic Institute and State University, Blacksburg, Virginia.

Dolan, J. D. and Heine, C. P., 1997a, *Monotonic Tests of Wood Frame Shear Walls with Various Openings and Base Restraint Configurations*. Timber Engineering Report No. TE-1997-001, Virginia Polytechnic Institute and State University, Blacksburg, Virginia.

Dolan, J. D. and Heine, C. P., 1997b, *Sequential Phased Displacement Cyclic Tests of Wood Frame Shear Walls with Various Openings and Base Restrain Configurations*. Timber Engineering Report No. TE-1997-002, Virginia Polytechnic Institute and State University, Blacksburg, Virginia.

Dolan, J. D., and Heine, C. P., 1997c, *Sequential Phased Displacement Test of Wood Frame Shear Walls with Corners*. Timber Engineering Report No. TE-1997-003, Virginia Polytechnic Institute and State University, Blacksburg, Virginia.

Earthquake Engineering Research Institute, 1996, "Northridge Earthquake of January 17, 1994," *Reconnaissance Report, Earthquake Spectra*. Vol. 11, Supplement C. Earthquake Engineering Research Institute, Oakland, California.

Faherty, Keith F., and Williamson, Thomas G., 1995, *Wood Engineering Construction Handbook*. McGraw Hill, Washington D.C.

Federal Emergency Management Agency, 2003, *National Earthquake Hazard Reduction Program, Recommended Provisions for Seismic Regulations for New Buildings and Other Structures and Commentary*. Federal Emergency Management Agency, Washington D.C.

Ficcadenti, S. K., T. A. Castle, D. A. Sandercock, and R. K. Kazanjy, 1996, "Laboratory Testing to Investigate Pneumatically Driven Box Nails for the Edge Nailing of 3/8" Plywood Shear Walls," *Proceedings: Annual SEAOC Convention*. Structural Engineers Association of California, Sacramento, California.

Foliente, Greg C., 1994, *Analysis, Design and Testing of Timber Structures Under Seismic Loads*. University of California Forest Products Laboratory, Richmond, California.

Foliente, Greg C., 1997, *Earthquake Performance and Safety of Timber Structures*. Forest Products Society, Madison Wisconsin.

Forest Products Laboratory, 1999, *Wood HandbookPublication FPL—GTR—113*. Madison, Wisconsin.

Goers R. and Associates, 1976, *A Methodology for Seismic Design and Construction of Single-Family Dwellings*. Applied Technology Council, Redwood City, California.

Ju, S. and Lin, M., 1999, "Comparison of Building Analysis Assuming Rigid or Flexible Floors," *Journal of Structural Engineering*. American Society of Civil Engineers, Washington, D.C.

Matteson, Thor, 2004 Wood-Framed Shear Wall Construction. International Code Council, Country Club Hills, Illinois.

Mendes, S., 1987, "Rigid versus Flexible: Inappropriate Assumptions Can Cause Shear Wall Failures!" *Proceedings: Annual SEAOC Convention*. Structural Engineers Association of California, Sacramento, California.

Mendes, S., 1995, "Lessons Learned From Four Earthquake Damaged Multi-Story Type V Structures," *Proceedings: Annual SEAOC Convention*. Structural Engineers Association of California, Sacramento, California.

Rose, J. D., 1998, *Preliminary Testing of Wood Structural Panel Shear Walls Under Cyclic (Reversed) Loading*. Research Report 158, APA—Engineered Wood Association, Tacoma, Washington.

Rose, J. D., and E. L. Keith, P. E., 1996, *Wood Structural Panel Shear Walls with Gypsum Wallboard and Window [Sheathing Standard, Sec. 2.3.3]*. Research Report 158. APA—The Engineered Wood Association, Tacoma Washington.

SEAOC, 1999, *Acceptable Diaphragm-Rigidity Assumptions for Distribution of Horizontal Forces in Light-Frame Construction.* Structural Engineers Association of California, Sacramento, California.

SEAOC, 1997, *Seismic Detailing Examples for Engineered Light Frame Timber Construction.* Structural Engineers Association of California, Sacramento, California.

SEAOC, 1999, *Guidelines for Diaphragms and Shear Walls*. Structural Engineers Association of California, Sacramento, California.

SEAOC, 1999, *Plan Review—Codes and Practice*. Structural Engineers Association of California, Sacramento, California.

Shipp, John and D. Thompson, 2001 Timber Design I. II, and III, Volumes VIII, IX, and X. Professional Engineering Development Publications, Inc., Irvine, California.

Steinbrugge, J., 1994, "Standard of Care in Structural Engineering Wood Frame Multiple Housing," *Proceedings: Annual SEAOC Convention*. Structural Engineers Association of California, Sacramento, California.

Design Example 3
Cold-formed Steel Light-frame Three-story Structure

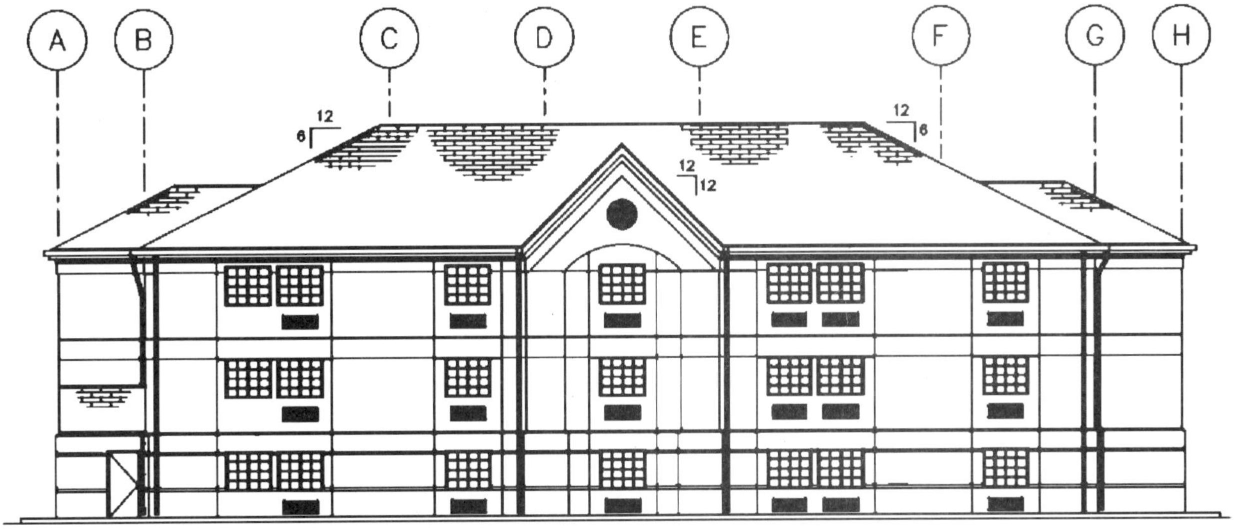

Figure 3-1. Cold-formed light-frame three-story structure elevation

Foreword

The building in this example has cold-formed light-gage steel framing, and shear walls and diaphragms that are sheathed with wood structural panels. This example presents a new approach to the seismic design of this type of building. This is because past and present design practice in seismic design of light-framed structures has almost exclusively considered flexible diaphragm assumptions when determining shear distribution to shear walls. However, since the 1988 UBC, there has been a *definition* in the code (§12.3.1.1 of ASCE/SEI 7-05) that defines diaphragm flexibility. The application of this definition often requires the use of the rigid diaphragm assumption, and calculation of shear wall rigidities for distribution to shear walls. While the latter is rigorous and complies with the letter of the code, it does not reflect present-day practice. In actual practice, for reasons of simplicity and precedence, many structural engineers routinely use the flexible diaphragm assumption.

ASCE/SEI 7-05 exempts one- and two-family residential buildings of light-frame construction from a rigid or semi-rigid structural analysis (§12.3.1.1), while the 2006 IBC exempts nearly all diaphragms of light-frame construction from a rigid or semi-rigid structural analysis (IBC §1613.6.1). This design example will follow the ASCE/SEI 7-05 requirements.

A rigid diaphragm analysis is recommended where the shear walls can be judged by observation to be flexible compared to the diaphragm, and particularly where one or more lines of either shear walls, moment frames, or cantilever columns are more flexible than the rest of the shear walls.

This design example has floor diaphragms with lightweight concrete fill over the floor sheathing (for sound insulation), making the diaphragms significantly stiffer than those determined by using the standard IBC diaphragm deflection equations.

Before beginning design, users of this manual should check with the local jurisdiction regarding the level of analysis required for cold-formed light-framed structures.

Overview

This design example illustrates the seismic design of a three-story, cold-formed (i.e., light-gage) steel structure. The structure is shown in Figures 3-2, 3-3, and 3-4. The building in this example is the same as in Design Example 2, with the exception that light-gage metal framing is used in lieu of wood. The structure has wood structural panel shear walls, and roof and floor diaphragms. The roofs have composite shingles over the wood panel sheathing that is supported by light-gage metal trusses. The floors have 1½ inches of lightweight concrete fill and are framed with metal joists.

The following steps illustrate a detailed analysis of some of the important seismic requirements of the ASCE/SEI 7-05. As stated in the introduction of the manual, this example is not a complete building design. Many aspects have not been included, and only selected steps of the seismic design have been illustrated. As is common for Type V construction (see ASCE/SEI 7-05, Chapter 6), a complete wind design is also necessary, but is not given here.

Although code requirements recognize only two diaphragm categories, flexible and rigid, the diaphragms in this example are judged to be semi-rigid because the diaphragms do deflect. The code also requires only one type of analysis, flexible or rigid. The analysis in this design example will use the *envelope* method. The envelope method considers the worst loading condition from both flexible and rigid diaphragm analyses to determine the design load on each shear-resisting element. It should be noted that the envelope method is not a code requirement, but is deemed appropriate for this design example, because neither flexible nor rigid diaphragm analysis may accurately model the structure.

Outline

This example will illustrate the following parts of the design process

1. **Design base shear and vertical distributions of seismic forces**
2. **Rigidities of shear walls**
3. **Distribution of lateral forces to the shear walls**
4. **Reliability/redundancy factor ρ**
5. **Tiedown forces for shear wall on line C**
6. **Allowable shear and nominal strength of No. 10 screws**

7. **Diaphragm deflections to determine if diaphragm is flexible or rigid**

8. **Tiedown connection at third floor for wall on line C**

9. **Tiedown connection at the second floor for shear wall on line C**

10. **Boundary studs for first floor wall on line C**

11. **Shear transfer at second floor on line C**

12. **Shear transfer at foundation for walls on line C**

13. **Shear transfer at roof at line C**

Given Information

Roof weights (slope 6:12):

Roofing	3.5 psf
½-inch sheathing	1.5
Trusses	3.5
Insulation	1.5
Miscellaneous	0.7
Gyp ceiling	2.8
DL (along slope)	13.5 psf

Floor weights:

Flooring	1.0 psf
Lt. wt. concrete	14.0
⅝-inch sheathing	1.8
Floor framing	5.0
Miscellaneous	0.4
Gyp ceiling	2.8
	25.0 psf

DL (horiz. proj.) = 13.5 (13.41/12) = 15.1 psf

Stair landings do not have lightweight concrete fill

Area of floor plan is 5288 sq ft

Weights of respective diaphragm levels, including tributary exterior and interior walls

$W_{roof} = 134{,}250$ lb

$W_{3rd\,floor} = 228{,}750$ lb

$W_{2nd\,floor} = 228{,}750$ lb

$W = 591{,}750$ lb

The same roof, floor, and wall weights used in Design Example 2 are used in this example to better illustrate a side-by-side comparison of cold-formed light-gage steel construction with the more traditional wood frame construction used in Design Example 2. This side-by-side comparison gives the engineer a better "feel" for the similarities and differences between structures with wood studs and structures with cold-formed metal studs. It should be noted that roof, floor, and wall weights for light-gage steel-framed structures are typically lighter than similar structures having wood framing. Because light-gage steel-framed structures are lighter, a more accurate estimate of building weight for this structure would be about 560 kips instead of the 591.75 kips used in this example. Consequently, wall shears and overturning forces would be reduced accordingly.

Weights of diaphragms are typically determined by taking one-half the height of walls at the third floor to the roof and full height of walls for the third and second floor diaphragms.

Wall framing is ASTM A653, grade 33, 4-inch by 18-gage metal studs at 16 inches on center (o/c). These have a 1⅝-inch flange with a ⅜-inch return lip. The ratio of tensile strength to yield point is at least 1.08. Studs are painted with primer. ASTM A653 steel is one of three ASTM steel specifications used in light-frame steel construction. The others are A792 and A875. The differences between the specifications are primarily the coatings that are galvanized, 55 percent aluminum-zinc (A792), and zinc-5 percent aluminum (A875), respectively. The recommended minimum coating classifications are G60, AZ50, and GF60, respectively. It should be noted that the studs do not require painting with primer.

It should also be noted that changing stud sizes or thicknesses of studs at various story heights is common (as in wood construction). The thickness of studs and tracks should be identified by visible means such as coloring or metal stamping of gages/sizes on studs and tracks.

DOC PS-1 or PS-2-rated wood structural panels with a trademark of an approved testing and grading agency (APA or TECO performance) for shear walls will be 15/32-inch-thick Structural-I, 32/16 span rating, 5-ply with Exposure I glue is specified.

Framing screws are No. 8 by ⅝-inch wafer head self-drilling with a minimum head diameter of 0.292 inch, as required by AISI-Lateral.

The roof is 15/32-inch-thick DOC PS-1 or PS-2-rated sheathing, 32/16 span rating with Exposure I glue.

The floor is 19/32-inch-thick DOC PS-1 or PS-2-rated Sturd-I-Floor 24-inch o/c rating (or APA-rated sheathing, 48/24 span rating) with Exposure I glue.

Seismic and site data:

$S_s = 1.78$	F22-1
$S_1 = 0.55$	F22-2
$S_{DS} = 1.19$	Eq 11.4-3
Seismic Design Category D	T 11.6-1
$I = 1.0$	T 11.5-1

Site Class C has been determined by geotechnical investigation. Without such investigation, Site Class D shall be used as a default value.

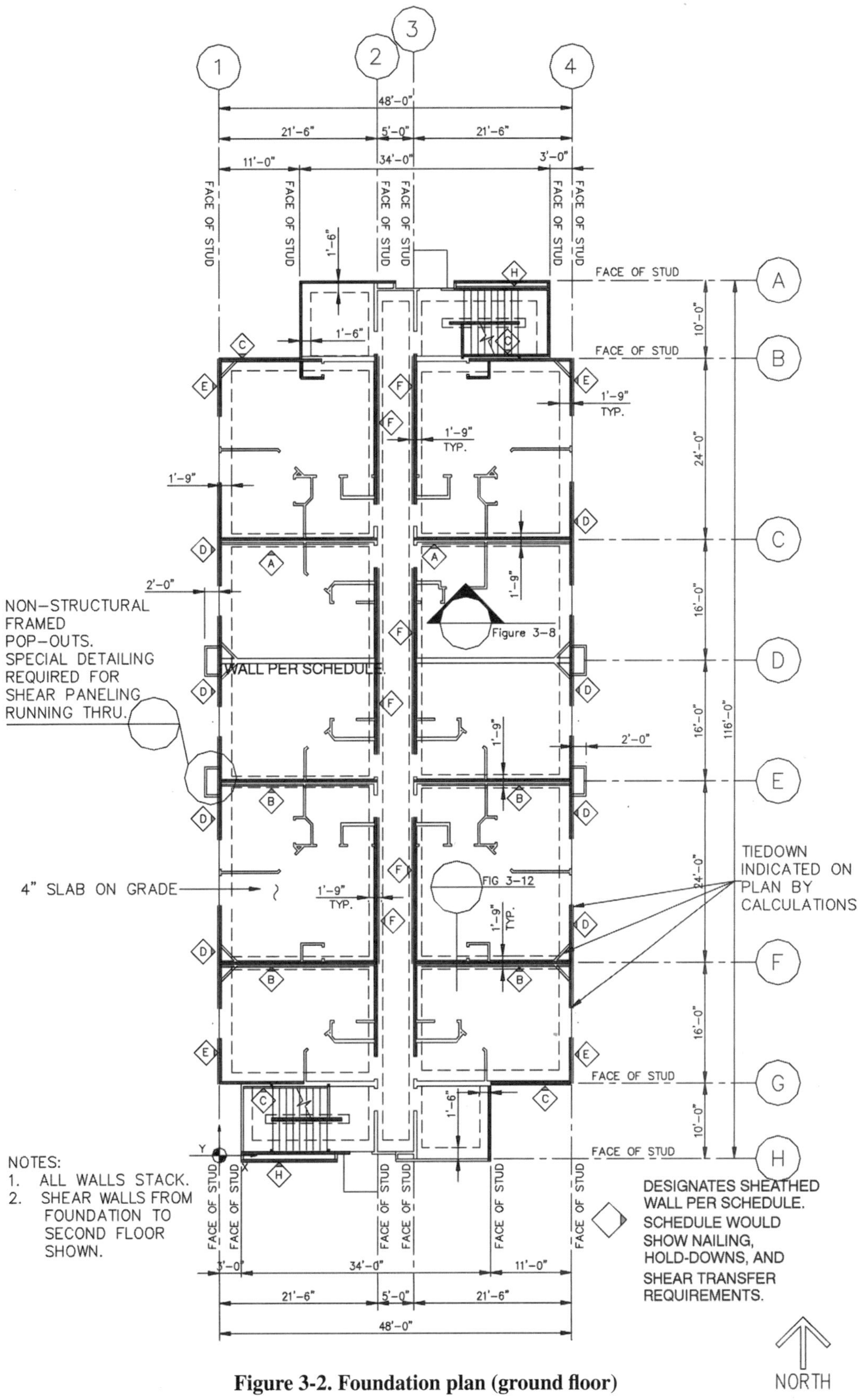

Figure 3-2. Foundation plan (ground floor)

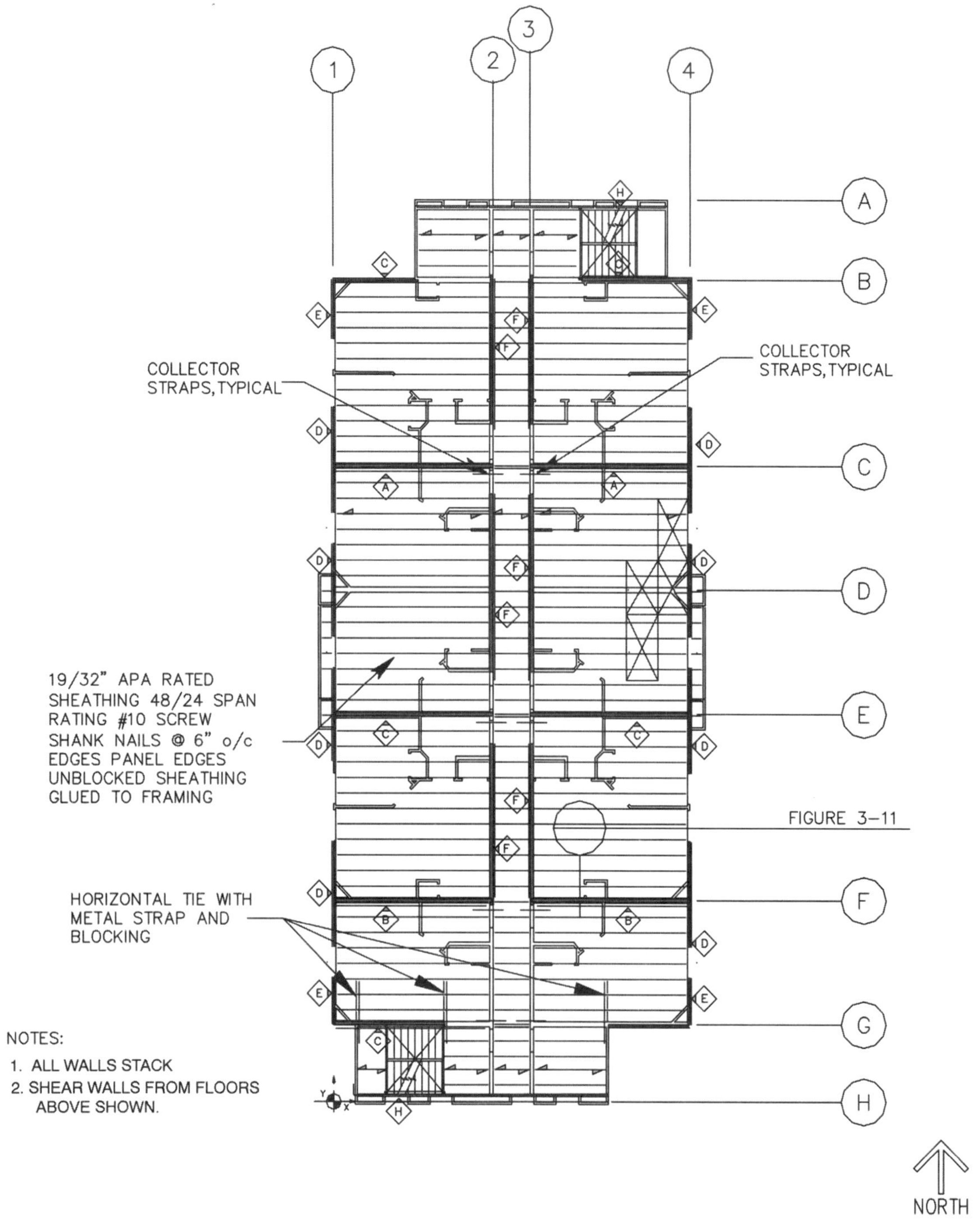

Figure 3-3. Floor framing plan (second and third floors)

Note: Shear walls on lines 2 and 3 do not extend from the third floor to the roof.

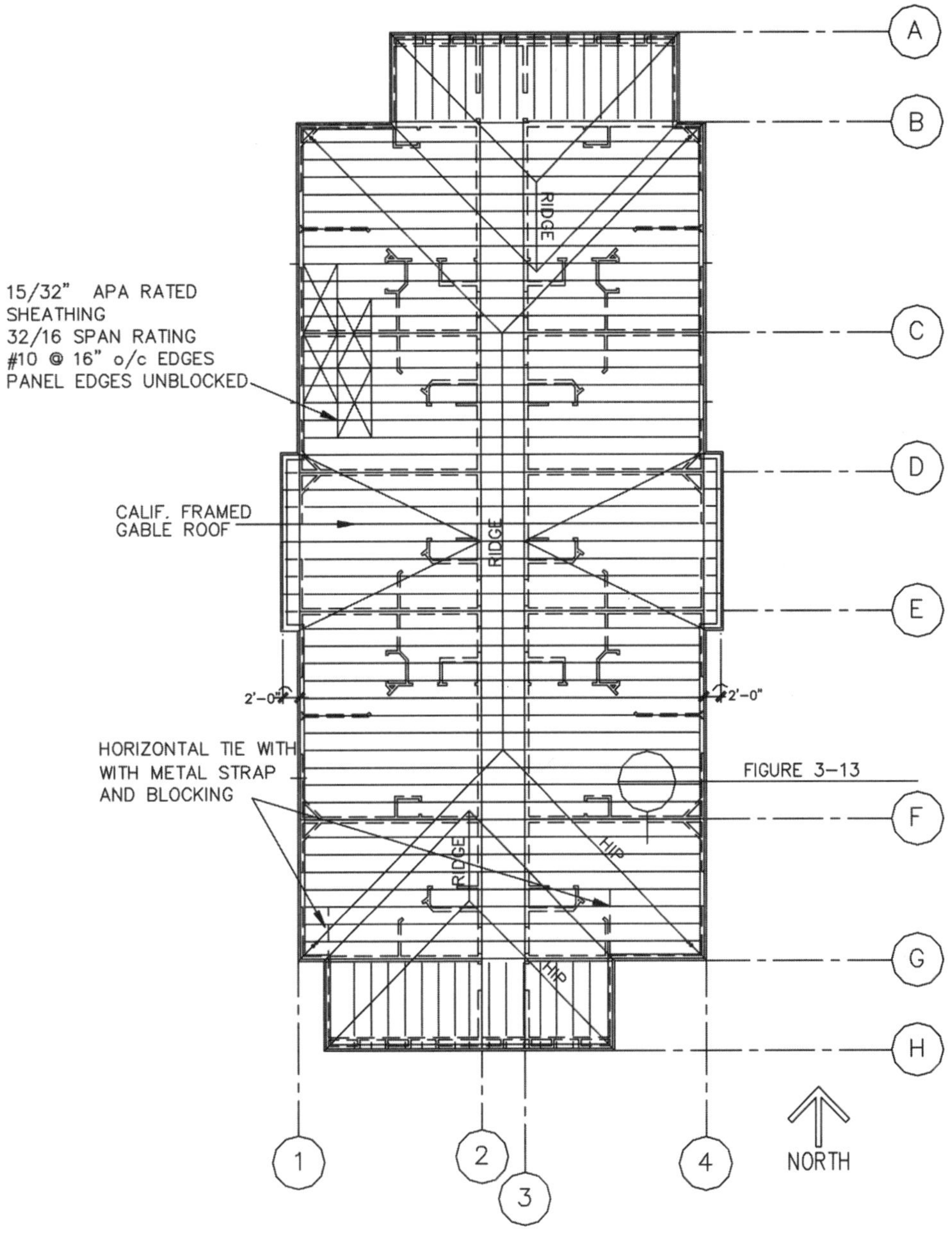

Figure 3-4. Roof framing plan

Factors That Influence Design

AISI Standard for Cold-Formed Steel Framing-Lateral Design provisions for seismic forces for shear walls with wood structural panels framed with cold-formed steel studs. The table for shear walls and diaphragms are primarily based on static and cyclic tests conducted by the Light Gauge Steel Research Group at the Santa Clara University Engineering Center for the American Iron and Steel Institute (AISI).

Before starting the example, several important aspects of cold-formed construction will be discussed. These are:

Stud thickness

Screw type

Material strength

Use of pre-manufactured roof trusses to transfer lateral forces

Proper detailing of shear walls at building "pop-outs"

Stud thickness.

ASI Lateral Design §C2.2 states that the uncoated base metal thickness for the studs used with wood structural panels shall be a minimum thickness of 0.033 inch, and shall not have a thickness greater than 0.054 inch.

The industry has departed from the use of the gage designation and is, for the purposes of framing applications, switching to a mil (thousandths of an inch) designation. In the future, thicknesses of studs, joists, and track will be expressed in mils.

Table 3-1. Stud thicknesses

Mils	Min. Delivered Thickness	Min. Design Thickness	Gage Reference
33	0.033 inch	0.0346 inch	20
43	0.043 inch	0.0451 inch	18
54	0.054 inch	0.0566 inch	16

The failure mode of the tests with 33-mil studs for screw spacings of 3 inches and 2 inches o/c was end stud *compression* failure. Assemblies using 43-mil end studs, have higher capacities, and those using 43-mil studs throughout have even higher capacities.

The values in Table 14.1-1 are for seismic forces and are *nominal* shear values. Values are to be modified using the strength level forces of ASCE/SEI 7-05. The design shear values are determined by multiplying the nominal shear values by a resistance factor (ϕ) of 0.55. The IBC no longer includes this table, which showed a conversion factor (Ω) of 2.5 for allowable stress design. Comparing the difference in the two designs: 2.5(0.55) = 1.375. In other words, design shears for LRFD (or strength design) are 1.375 times higher than shears for ASD or working stress design. This is consistent with the ASD conversion factor of (1/0.7) in §12.4.2.3.

The values in Table 14.1-1 for $^{15}/_{32}$-inch Structural-I sheathing using #8 screws are almost identical to the values for the same sheathing applied to Douglas Fir with 8d common nails at the same spacing.

Screw type.

Footnote b of Table 14.1-1 requires the framing screws to be *self-tapping*. The reason for the self-tapping screws (or drill point screws) is to be able to penetrate 43-mil (and thicker) steel. Self-piercing screws can also be used in 33-mil steel, but with some difficulty. Both self-tapping and self-piercing screws have performed equally well in the shear tests. The screws are to be of sufficient length to penetrate through the framing member by at least three exposed threads.

There is a significant concern in screw installation when there is a gap between the stud flange and the sheathing after installation (e.g., jacking). When jacking occurs, the stiffness of the shear wall is *significantly* reduced. The drill point alone will not prevent jacking, which occurs when the drill point spins for a rotation or two before piercing the metal. Only a blank shaft (i.e., smooth with no threads) for the depth of the sheathing will remove the jacking created by the drill point spin prior to piercing. A detailed drawing or explicit specifications should be included in the design drawings and should specify that the distance from the screw head to the beginning of the thread portion be equal to or less than the thickness of the plywood or oriented strand board (OSB). The "unused portion" of the screw protruding from the connection of sheathing and metal stud can be used as a simple inspection gauge to see if jacking has occurred.

Material strength

Common practice is for material 16-gage and heavier to have a yield strength of 50,000 psi; for 18-gage and lighter, 33,000 psi. This practice holds true for studs and track, but not for manufactured hardware (straps, clips, and tiedown devices).

Use of pre-manufactured roof trusses to transfer lateral forces

The structural design in this design example utilizes pre-manufactured roof trusses to transfer the lateral forces from the roof diaphragm to the tops of the interior shear walls. Special considerations need to be included in the design and detailed on the plans, including:

1. Provision that any trusses used as collectors (i.e., drag struts) should be clearly indicated on the structural framing plan.
2. The magnitude of the forces, the means by which the forces are applied to the trusses, and how the forces are transferred from the trusses to the shear walls should be shown.
3. If the roof sheathing at the hip ends breaks above the joint between the end jack trusses and the supporting girder truss, the lateral forces to be resisted by the end jacks should be specified so that an appropriate connection can be provided to resist these forces.
4. The drawings should also specify the load combinations and whether or not a stress increase is permitted.
5. If ridge vents are being used, special detailing for shear transfers needs to be indicated in the details.

Proper detailing of shear walls at building "pop-outs"

The structure for this design example has double-framed walls for party walls, exterior "planted-on" box columns (pop-outs). The designer should not consider these walls as shear walls unless special detailing and analysis are provided to substantiate that there is a viable lateral force path to that wall and the wall is adequately braced.

Calculations and Discussion | **Code Reference**

1. Design base shear and vertical distributions of seismic forces §12.8.1

1a. Design base shear

Period using Method A (See Figure 3-5 for section through structure):

$$T_a = C_t(h_n)^x = 0.020(33.63)^{3/4} = 0.28 \text{ sec}$$ Eq 12.8-7

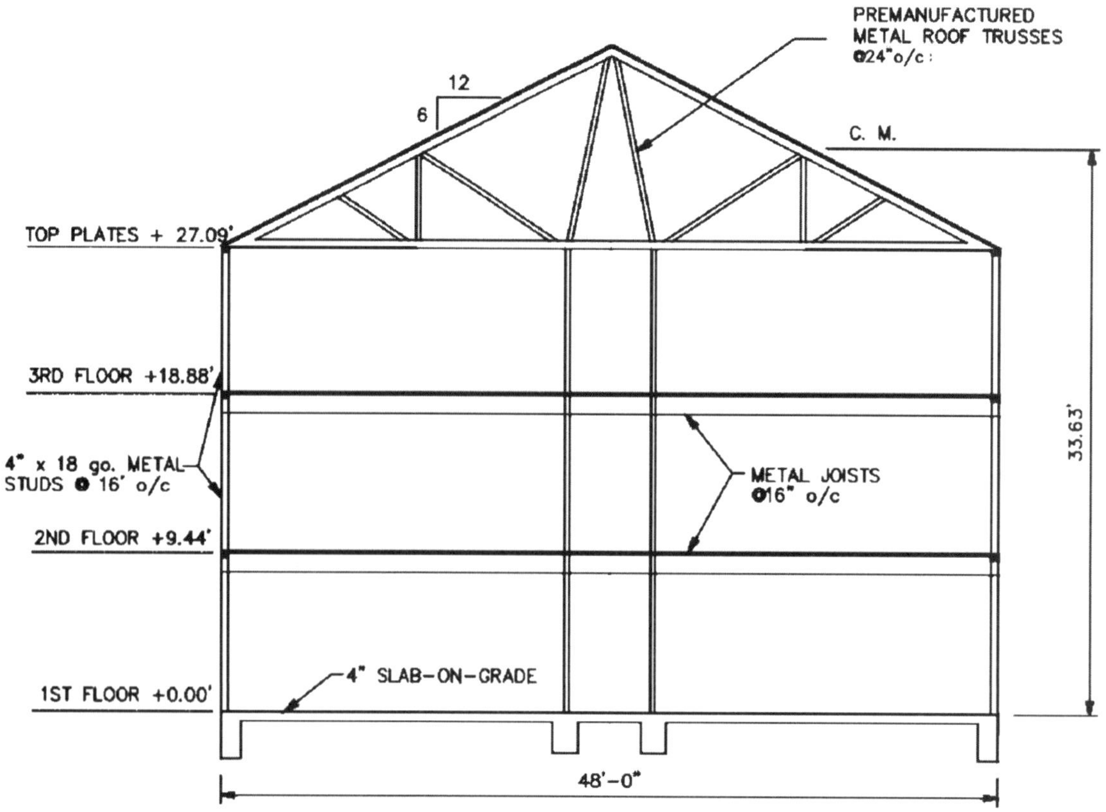

Figure 3-5. Typical cross section through building

Since the stud walls are both wood structural panel shear walls and bearing walls

$$R = 6.5$$ T 12.2-1

Design base shear is

$$V = C_s W$$ Eq 12.8-1

where

$$C_s = \frac{S_{DS}}{\left(\frac{R}{I}\right)}$$ Eq 12.8-2

Note that design base shear is on a strength design basis

$$I = 1.0$$
$$R = 6.5$$
$$S_{DS} = \tfrac{2}{3}\, S_{MS} = \tfrac{2}{3}\,(1.78) = 1.19$$
$$S_{MS} = F_a S_S = 1.0 \times 1.78 = 1.78$$

where:

$$F_a = 1.0$$
$$S_S = 1.78$$

Site Class C

$$C_s = \frac{1.19}{\left(\frac{6.5}{1.0}\right)} = 0.183$$

but need not exceed

$$C_s = \frac{S_{D1}}{T\left(\frac{R}{I}\right)} \qquad \text{(Eq 16-36)}$$

$$S_1 = 0.55$$
$$S_{M1} = F_v S_1 = 1.3 \times 0.55 = 0.72$$
$$S_{D1} = \tfrac{2}{3}\, S_{M1} = \tfrac{2}{3}\,(0.72) = 0.48$$
$$F_v = 1.3$$
$$C_s = \frac{0.48}{\left(\frac{6.5}{1.0}\right)0.28} = 0.264 > 0.183 \quad \therefore \text{ does not control}$$
$$C_s = 0.183$$

but shall not be less than

$$C_s = 0.01$$
$$V = 0.183W$$

Comparison of the above result with the simplified static method permitted under §12.14 shows that it is more advantageous to use the simplified standard method of determining the design base shear.

$$V = \frac{FS_{DS}}{R} W = \frac{1.2(1.0)}{6.5} W = 0.185W > 0.183W \qquad \text{Eq 12.14-11}$$

where

$$F = 1.2 \text{ for three-story buildings}$$
$$S_{DS} = (\tfrac{2}{3}) \times 1.0 \times 1.5 = 1.0 \qquad \text{§12.14.8.1}$$
$$S_a = 1.5 \text{ max}$$
$$F_a = 1.0 \text{ for } S_S \geq 1.25$$

Check for maximum S_a value: §12.8.1.3
Since the structure does not have any Horizontal or Vertical Irregularities listed in Tables 12.3-1 and 12.3-2, the structure is classified as "regular." Also, since the structure is five stories or fewer in height and the period, T, is less than 0.5 seconds, C_S is permitted to be calculated using a value of 1.5 for S_S.

$$S_{MS} = F_a\, S_S = 1.0 \times 1.5 = 1.5$$

$$S_{DS} = 2/3\, S_{MS} = 2/3(1.5) = 1.0$$

$$C_S = \frac{1.0}{\left(\frac{6.5}{1.0}\right)} = 0.154$$

$$V = 0.154W < 0.183W$$

The engineer can choose which method to use. This design example illustrates the Equivalent Lateral Force Procedure using the full value of $S_S = 1.78$.

$$V = 0.183W$$

$$\therefore\ V = 0.183\ (591{,}750) = 108{,}290 \text{ lb}$$

In addition, equation (12.8-6) requires an additional check for C_S minimum for structures that are located where S_1 is equal to or greater than 0.6g:

$$C_S = \frac{0.5S_1}{\left(\frac{R}{I}\right)} = \frac{0.5(1.0)}{\left(\frac{6.5}{1.0}\right)} = 0.07$$

However, this condition does not control.

In this design example, the designer may choose either allowable strength design or strength design. In Design Example 2, however, allowable strength design must be used.

It is desirable to use the strength level forces throughout the design of the structure for two reasons:

1. Errors in calculations can occur and may cause confusion as to which load, (strength or allowable stress design) is being used. This design example uses the following format

   $V_{base\ shear}$ = strength

   F_{px} = strength

   F_x = force-to-wall (strength)

   v = wall shear at element level (strength)

2. This design example is not intended to pave the way for the future when the code will be all strength design.

Seismic load effect E:

Where the effects of gravity and the seisimic ground motion are additive, the seismic load E is defined as

$$E = \rho Q_E + 0.2\, S_{DS}\, D \qquad \text{Eq 12.4-1}$$

Where the effects of gravity and the seisimic ground motion counteract, the seismic load E is defined as

$$E = \rho Q_E - 0.2\, S_{DS}\, D \qquad \text{Eq 12.4-2}$$

The redundancy ρ will be assumed to be 1.0, and in most cases $\rho = 1.0$ for Type V construction with interior shear walls. The assumed value for ρ will have to be verified. This is done later in Part 4.

The basic load combination for allowable stress design for horizontal forces is

$$(1.0 + 0.14\, S_{DS})\, D + 0.7\, \rho Q_E$$

and §12.4.2.3

$$(1.0 + 0.105\, S_{DS})\, D + 0.525\, \rho Q_E + 0.75\, L + 0.75\, L_r$$

For vertical downward

$$(1.0 + 0.105\, S_{DS})\, D + 0.75\, L + 0.75\, S + 0.525\, \rho Q_E \qquad §12.4.2.3$$

where

$$(1.0 + 0.105\, S_{DS})\, D = (1.0 + 0.105 \times 1.19)\, D = 1.125\, D \qquad §12.4.2.3$$

For vertical uplift

$$(0.6 - 0.14\, S_{DS})\, D + 0.7\, \rho Q_E \qquad §12.4.2.3$$

1b. Vertical distributions of forces

The design base shear must be distributed to each level, as follows:

$$F_x = C_{vx} V \qquad \text{Eq 12.8-11}$$

$$C_{vx} = \frac{\omega_x h_x^k}{\sum_{i=1}^{n} \omega_i h_i^k} \qquad \text{Eq 12.8-12}$$

where h_x is the height from the base to level x in feet.

k is a distribution exponent related to the building period

Since $T = 0.28$ seconds < 0.5 seconds, $k = 1$

Determination of F_x is shown in Table 3-2.

Table 3-2. Vertical distribution of seismic forces

Level	w_x (k)	h_x (ft)	$w_x h_x$ (k-ft)	$\frac{w_x h_x}{\Sigma w_i h_i}$	F_x (k)	$\frac{F_x}{w_x}$	F_{tot} (k)
Roof	134.25	33.6	4,511	41.1	44.5	0.330	44.5
3rd Floor	228.75	18.9	4,323	39.4	42.7	0.186	87.2
2nd Floor	228.75	9.4	2,150	19.5	21.1	0.092	108.3
Σ	591.75		10,984		108.3		

Note: Although not shown here, designers must also check wind loading. In this example, wind load may control the design in the east-west direction.

2. Rigidities of shear walls

2a. Vertical distributions of forces

The American Iron and Steel Institute has a new publication "AISI Standard for Cold-Formed Steel Framing-Lateral Design," 2004 Edition with commentary. In this publication is a formula for a blocked wood structural panel or steel sheet shear wall. The formula is as follows:

$$\frac{8vh^3}{E_s A_c b} + \omega_1, \omega_2 \frac{vh}{\rho G t_{\text{sheathing}}} + \omega_1^{5/4} \omega_2 \omega_3 \omega_4 \left(\frac{v}{\beta}\right)^2 + d_a$$

where:

A_c = gross cross-sectional area of the boundary member in square inches

b = the shear wall length in feet

E_s = modulus of elasticity of steel = 29,500,000 psi

G = shear modulus of sheathing from Table 3, Plywood Design Specifications, in pounds per square inch (psi)

h = wall height, in feet

s = maximum fastener spacing at panel edges, in inches

$t_{\text{sheathing}}$ = nominal panel thickness, in inches

t_{stud} = framing designation thickness, in inches

v = shear in the wall in pounds per linear foot (plf)

β = 810 for plywood and 660 for OSB

d_a = displacement of the tie down due to anchorage details in inches

ρ = 1.85 for plywood and 1.05 for OSB

ω_1 = $s/6$

ω_2 = $0.033 / t_{\text{stud}}$

ω_3 = $\dfrac{\sqrt{(h/b)}}{2}$

ω_4 = 1.0 for wood structural panels

2b. Calculation of shear wall rigidities

In this design example, shear wall rigidities (k) are computed using the basic stiffness equation.

$$F = k\Delta$$

or

$$k = \frac{F}{\Delta}$$

To simplify the calculations (compared to the more rigorous approach used in Design Example 2), this example uses wall rigidities based on the chart in Figure 3-6. This chart is based on the shear wall deflection AISI Equation C2.1-1. It should be noted that Design Example 2 considered wood shrinkage and tiedown displacements. With metal framing, shrinkage is zero. This design example also assumes a fixed base and pinned top for all shear walls. The chart in Figure 3-6 uses a tiedown displacement (e.g., elongation) of 1/16 inch, which is based on judgment and considered appropriate for this structure.

Actual determinations of shear wall rigidities at the roof, third floor, and second floor are shown in Tables 3-3, 3-4, and 3-5, respectively.

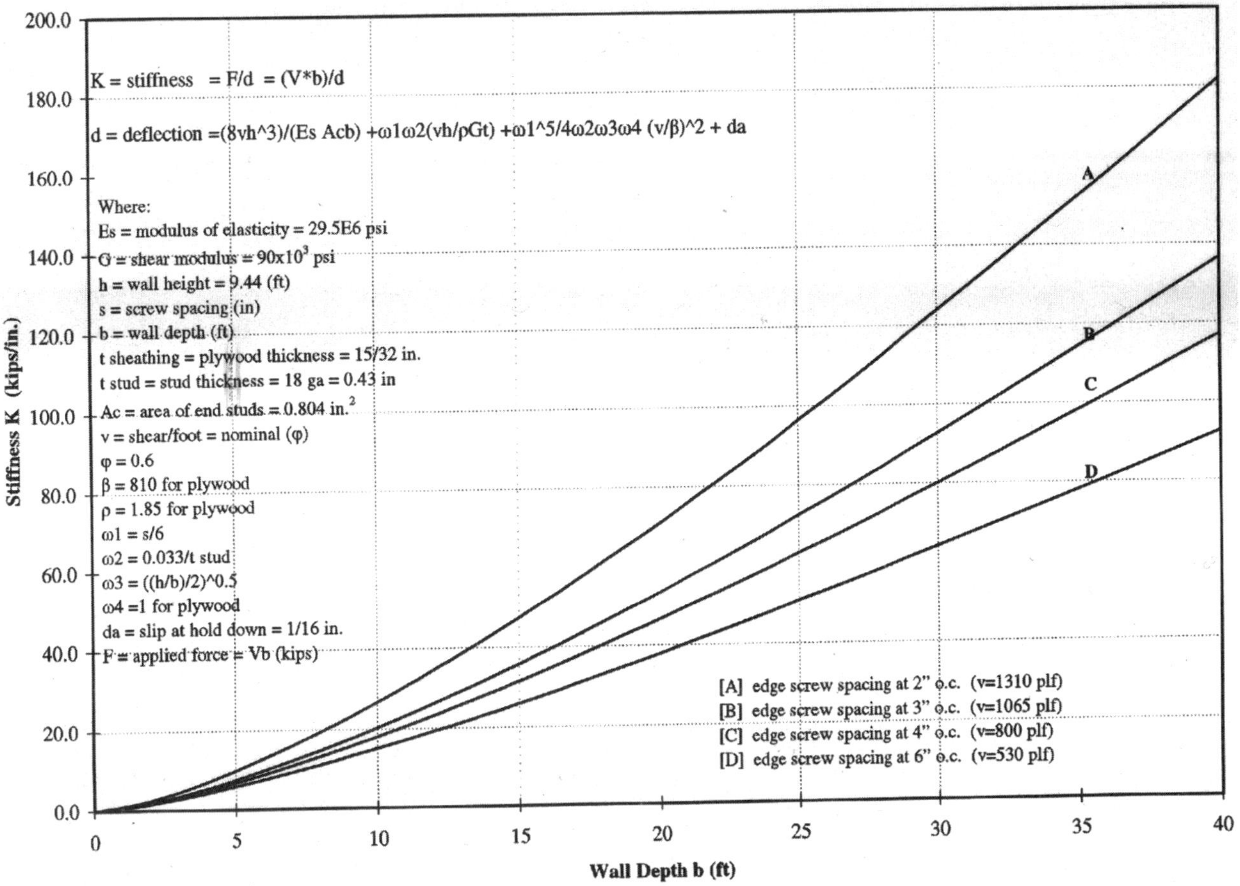

Figure 3-6. Stiffness of one-story Structural-I 15/32-inch plywood shear walls

Table 3-3. Shear wall rigidities at roof level

Wall	Wall Depth b (ft)	Edge Fastener Spacing (in)	k (From Fig 3-6) (k/in)	k_{total} (k/in)
A	12.5	6	20.0	20.0
B1	11.0	6	17.0	—
B2	11.0	6	17.0	—
B	—	—	34.0	34.0
C1	21.5	6	42.0	—
C2	21.5	6	42.0	—
C	—	—	84.0	84.0
E1	21.5	6	42.0	—
E2	21.5	6	42.0	—
E	—	—	84.0	84.0
F1	21.5	6	42.0	—
F2	21.5	6	42.0	—
F	—	—	84.0	84.0
G1	11.0	6	17.0	—
G2	11.0	6	17.0	—
G	—	—	34.0	34.0
H	12.5	6	20.0	20.0
1a, 4a	8.0	6	11.0	—
1b, 4b	14.0	6	24.0	—
1c, 4c	11.5	6	18.0	—
1d, 4d	11.5	6	18.0	—
1e, 4e	11.5	6	18.0	—
1f, 4f	8.0	6	11.0	—
1, 4	—	—	100.0	100.0

Table 3-4. Shear wall rigidities at third floor

Wall	Wall Depth b (ft)	Edge Fastener Spacing (in)	k (From Fig 3-6) (k/in)	k_{total} (k/in)
A	12.5	6	20.0	20.0
B1	11.0	4	20.5	—
B2	11.0	4	20.5	—
B	—	—	41.0	41.0
C1	21.5	4	51.0	—
C2	21.5	4	51.0	—
C	—	—	102.0	102.0
E1	21.5	4	51.0	—
E2	21.5	4	51.0	—
E	—	—	102.0	102.0
F1	21.5	4	51.0	—
F2	21.5	4	51.0	—
F	—	—	102.0	102.0
G1	11.0	4	20.5	—
G2	11.0	4	20.5	—
G	—	—	41.0	41.0
H	12.5	6	20.0	20.0
1a, 4a	8.0	4	13.0	—
1b, 4b	14.0	4	29.0	—
1c, 4c	11.5	4	22.0	—
1d, 4d	11.5	4	22.0	—
1e, 4e	11.5	4	22.0	—
1f, 4f	8.0	4	13.0	—
1, 4	—	—	121.0	121.0
2a, 3a	18.0	6	33.0	—
2b, 3b	24.0	6	48.0	—
2c, 3c	18.0	6	33.0	—
2, 3	—	—	114.0	114.0

Table 3-5. Shear wall rigidities at second floor

Wall	Wall Depth b (ft)	Edge Fastener Spacing (in)	k (From Fig 3-6) (k/in)	k_{total} (k/in)
A	12.5	6	20.0	20.0
B1	11.0	3	23.0	—
B2	11.0	3	23.0	—
B	—	—	46.0	46.0
C1	21.5	3	59.0	—
C2	21.5	3	59.0	—
C	—	—	118.0	118.0
E1	21.5	3	59.0	—
E2	21.5	3	59.0	—
E	—	—	118.0	118.0
F1	21.5	3	59.0	—
F2	21.5	3	59.0	—
F	—	—	118.0	118.0
G1	11.0	3	23.0	—
G2	11.0	3	23.0	—
G	—	—	46.0	46.0
H	12.5	6	20.0	20.0
1a, 4a	8.0	4	13.0	—
1b, 4b	14.0	4	29.0	—
1c, 4c	11.5	4	22.0	—
1d, 4d	11.5	4	22.0	—
1e, 4e	11.5	4	22.0	—
1f, 4f	8.0	4	13.0	—
1, 4	—	—	121.0	121.0
2a, 3a	18.0	6	33.0	—
2b, 3b	24.0	6	48.0	—
2c, 3c	18.0	6	33.0	—
2, 3	—	—	114.0	114.0

2c. Determination of the design story drift Δ §12.12

For both strength and allowable stress design, the code requires that building drifts be determined by using strength level forces.

Shear wall displacements for structures of this type (generally) are well below the maximum allowed by code and the computation of these displacements is considered unnecessary. Refer to Design Example 2 for an illustration of this procedure.

3. Distribution of lateral forces to the shear walls. §12.8.4

In this part, story shears are distributed to shear walls with the diaphragms assumed to be rigid. (Refer to Design Example 2 for a code confirmation of the accuracy of this assumption.)

It has been common practice for engineers to *assume* flexible diaphragms and distribute loads to shear walls based upon tributary areas. The procedures used in this design example are not intended to imply that seismic design of light-frame construction in the past should have been performed in this manner. Recent earthquakes and testing of wood panel shear walls have indicated that drifts can be considerably higher than what was known or assumed in the past. Knowledge of the increased drifts of short wood panel shear walls has increased the need for the engineer to consider relative rigidities of shear walls.

Section 12.8.4.2 requires the center of mass (CM) to be displaced from the calculated CM a distance of 5 percent of the building dimension at that level perpendicular to the direction of force. The code requires the most severe load combination to be considered and also permits the negative torsional shear to be subtracted from the direct load shear. The net effect of this is to add 5 percent accidental eccentricity to the actual eccentricity.

The direct shear force F_{vi} in wall i is determined from

$$F_{vi} = F\frac{R}{\Sigma R}$$

and the torsional shear force F_{ti} in wall i is determined from

$$F_{ti} = T\frac{R_i d_i}{J}$$

where

i = wall number

$J = \Sigma R d_x^2 + \Sigma R d_y^2$

R = shear wall rigidity

d = distance from the lateral resisting element (e.g., shear wall) to the center of rigidity (CR)

$T = Fe$

F = story shear

e = eccentricity

3a. Determine center of rigidity, center of mass, eccentricities for roof diaphragm

Forces in the east-west (x) direction

$$\bar{y}_r = \frac{\Sigma k_{xx} y}{\Sigma k_{xx}} \quad or \quad \bar{y}_r = \Sigma k_{xx} + \Sigma k_{xx} y$$

Using the rigidity values k from Table 3-3 and the distance y from line H to the shear wall

$$\bar{y}\,(20.0 + 34.0 + 84.0 + 84.0 + 84.0 + 34.0 + 20.0) = 20.0\,(116) + 34.0(106) + 84.0(82.0) + 84.0(50.0) + 84.0(26.0) + 34.0(10.0) + 20.0(0)$$

$$\therefore\ \bar{y}_r = \frac{19{,}536}{360.0} = 54.5 \text{ ft}$$

The building is symmetrical about the x-axis and the CM is determined as

$$\bar{y}_m \quad = \frac{116.0}{2} = 58.0 \text{ ft}$$

The minimum 5-percent accidental eccentricity for east-west forces, e'_y is computed from the length of the structure perpendicular to the applied story force

$$e'_y \quad = (0.05)(116 \text{ ft}) = \pm\, 5.8 \text{ ft}$$

The $\bar{y}_m$ to the displaced CM = 58.0 ft ± 5.8 ft = 63.8 ft *or* 52.2 ft

The total eccentricity is the distance between the displaced CM and the CR $y_r = 54.5$ ft

$$\therefore\ e'_y = 63.8 - 54.5 = 9.3 \text{ ft} \quad or \quad 52.2 - 54.5 = -2.3\text{ft}$$

Note that the distance is slightly different from that shown in Design Example 2.

Note that in this design example, displacing the CM by 5 percent can result in the CM being on either side of the CR and can produce added torsional shears to all walls.

Note that the 5 percent may not be conservative. The contents-to-structure weight ratio can be higher in light-framed structures than in heavier types of construction. Also, the location of the calculated CR is less reliable for light-framed structures than for other structural systems. Be judicious when selecting the eccentricity e.

Forces in the north-south (y) direction

The building is symmetrical about the y-axis. Therefore, the distance to the CM and CR is

$$\bar{x}_m \quad = \frac{48.0}{2} = 24.0 \text{ ft}$$

$$\min e'_x = (0.05)(48 \text{ ft}) = \pm\, 2.4 \text{ ft}$$

Because the CM and CR locations coincide

$$e_x \quad = e'_x$$

$$\therefore\ e_x = 2.4 \quad or \quad -2.4 \text{ ft}$$

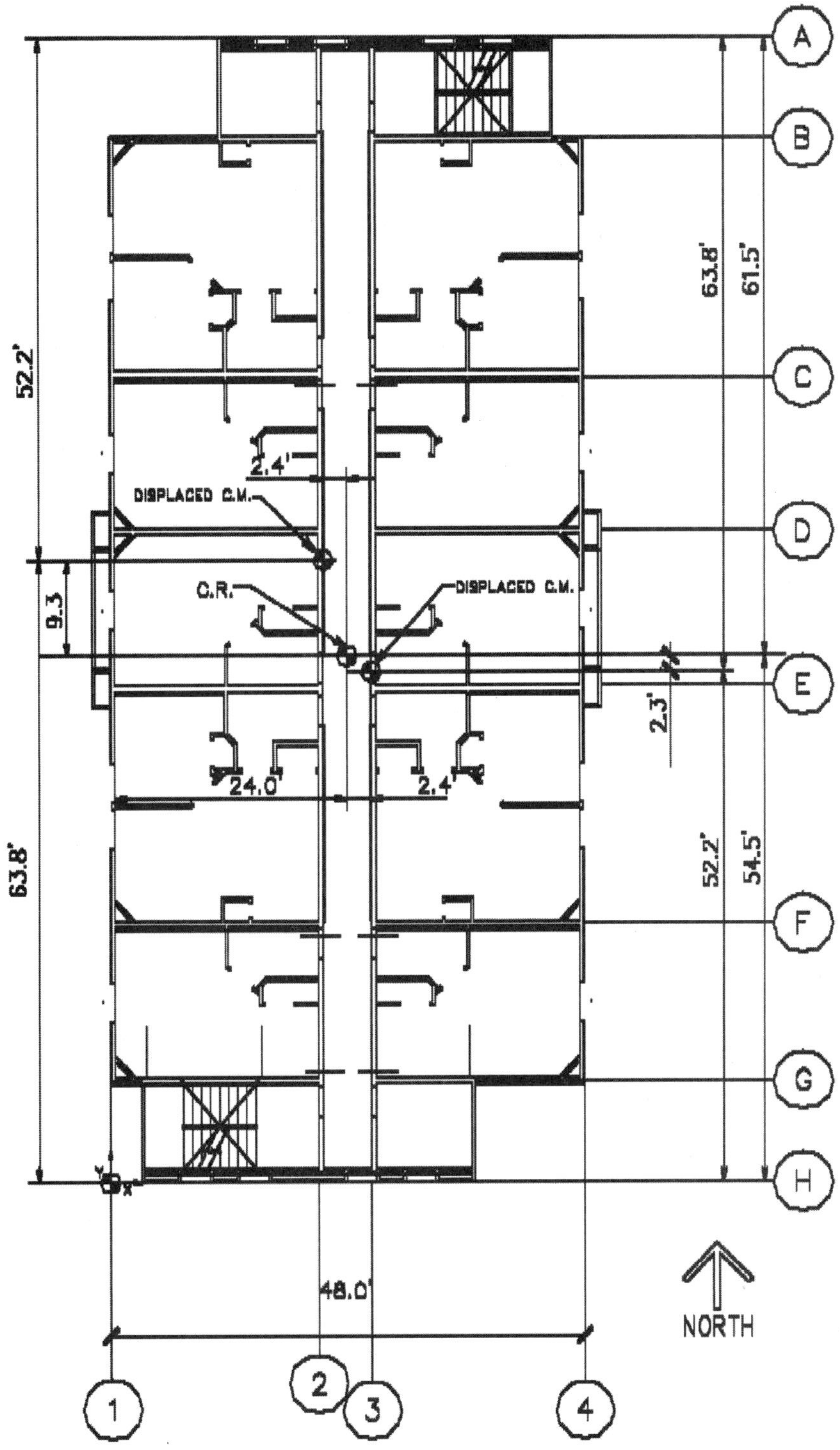

Figure 3-7. Center of rigidity and location of displaced centers of mass for diaphragms

3b. Determine total shears on walls at roof level

The total shears on the walls at the roof level are the direct shears, F_v, and the shears due to torsion (combined actual and accidental torsion), F_{ti}.

Torsion on the roof diaphragm is computed as

$$T_x = Fe_y = 44{,}500 \text{ lb } (9.3 \text{ ft}) = 413{,}850 \text{ ft-lb for walls A, B, and C}$$

or $$T_x = 44{,}500 \text{ lb } (2.3 \text{ ft}) = 102{,}350 \text{ ft-lb for walls E, F, G, and H}$$

$$T_y = Fe_x = 44{,}500 \text{ lb } (2.4 \text{ ft}) = 106{,}800 \text{ ft-lb for walls 1 and 4}$$

Because the building is symmetrical for forces in the north-south direction, the torsional forces can be subtracted for those walls located on the opposite side from the displaced CM. However, when the forces are reversed, the torsional forces will be additive. As required by the ASCE/SEI 7-05, the larger values are used in this design example. The walls are then designed using the critical force. Table 3-6 summarizes the spreadsheet for determining combined forces on the roof level walls.

Table 3-6. Distribution of forces to shear walls below the roof level

	Wall	R_x	R_y	d_x	d_y	Rd	Rd^2	Direct Force F_v	Torsional Force F_t	Total Force $F_v + F_t$
East-West	A	20.0			61.5	1,230	75,645	2,475	940	3,415
	B	34.0			51.5	1,751	90,177	4,200	1,340	5,540
	C	84.0			27.5	2,310	63,525	10,385	1,765	12,150
	E	84.0			4.5	378	1,701	10,385	70	10,455
	F	84.0			28.5	2,394	68,229	10,385	450	10,835
	G	34.0			44.5	1,513	67,329	4,200	285	4,485
	H	20.0			54.5	1,090	59,405	2,475	205	2,680
	Σ	360.0					426,011	44,505		
North-South	1		100	24.0		2,400	57,600	22,250	475	22,725
	4		100	–24.0		–2,400	57,600	22,250	–475	21,775
	Σ		200				115,200	44,500		
	Σ						541,211			

3c. Determine center of rigidity, center of mass, and eccentricities for the third floor diaphragm

Because the walls stack with uniform fasteners, it can be assumed that the CR for the third floor and the second floor diaphragms will coincide with the CR of the roof diaphragm.

Torsion on the third floor diaphragm is

$$F = (44{,}500 + 42{,}700) = 87{,}200 \text{ lb}$$

$$T_x = Fe_y = 87{,}200 \text{ lb } (9.3 \text{ ft}) = 810{,}960 \text{ ft-lb for walls A, B, and C}$$

or $$87{,}200 \text{ lb } (2.3 \text{ ft}) = 200{,}560 \text{ ft-lb for walls E, F, G, and H}$$

$$T_y = Fe_x = 87{,}200 \text{ lb } (2.4 \text{ ft}) = 209{,}280 \text{ ft-lb for walls 1 and 4}$$

Results for the third floor are summarized in Table 3-7.

Table 3-7. Distribution of forces to shear walls below the third floor level

	Wall	R_x	R_y	d_x	d_y	Rd	Rd^2	Direct Force F_v	Torsional Force F_t	Total Force $F_v + F_t$
East-West	A	20.0			61.5	1,230	75,645	4,075	1,590	5,665
	B	41.0			51.5	2,112	108,745	8,355	2,730	11,085
	C	102.0			27.5	2,805	77,138	20,780	3,625	24,405
	E	102.0			4.5	459	2,065	20,780	145	20,925
	F	102.0			28.5	2,907	82,850	20,780	930	21,710
	G	41.0			44.5	1,825	81,190	8,355	585	8,940
	H	20.0			54.5	1,090	59,405	4,075	350	4,425
	Σ	428.0					487,038	87,200		
North-South	1		121.0	24.0		2,904	69,696	22,450	970	23,420
	2		114.0	2.5		285	713	21,150	95	21,245
	3		114.0	–2.5		–285	713	21,150	–95	21,055
	4		121.0	–24.0		–2,904	69,696	22,450	–970	21,480
	Σ		470.0				140,818	87,200		
	Σ						627,856			

3d. Determine center of rigidity, center of mass, and eccentricities for the second floor diaphragm

Torsion on the second floor diaphragm is

$$F = (44{,}500 + 42{,}700 + 21{,}100) = 108{,}300 \text{ lb}$$

$$T_x = Fe_y = 108{,}300 \text{ lb } (9.3 \text{ ft}) = 1{,}007{,}190 \text{ ft-lb for walls A, B, and C}$$

or

$$108{,}300 \text{ lb } (2.3 \text{ ft}) = 249{,}090 \text{ ft-lb for walls E, F, G, and H}$$

$$T_y = Fe_x = 108{,}300 \text{ lb } (2.4 \text{ ft}) = 259{,}920 \text{ ft-lb for walls 1 and 4}$$

Results for the second floor are summarized in Table 3-8.

Table 3-8. Distribution of forces to shear walls below second floor level

	Wall	R_x	R_y	d_x	d_y	Rd	Rd^2	Direct Force F_v	Torsional Force F_t	Total Force $F_v + F_t$
East-West	A	20.0			61.5	1,230	75,645	4,455	1,830	6,285
	B	46.0			51.5	2,369	122,003	10,250	3,525	13,775
	C	118			27.5	3,245	89,238	26,295	4,830	31,125
	E	118			4.5	531	2,390	26,295	195	26,490
	F	118			28.5	3,363	95,845	26,295	1,240	27,535
	G	46.0			44.5	2,047	91,092	10,250	755	11,005
	H	20.0			54.5	1,090	59,405	4,455	400	4,855
	Σ	486					535,618	108,300		
North-South	1		121.0	+24.0		2,904	69,696	27,880	1,115	28,995
	2		114.0	+2.5		285	713	26,270	110	26,380
	3		114.0	−2.5		−285	713	26,270	−110	26,160
	4		121.0	−24.0		−2,904	69,696	27,880	−1,115	26,765
	Σ		470.0				140,818	108,300		
	Σ						676,436			

3e. Comparison of flexible vs. rigid diaphragm results

Table 3-9 summarizes wall forces determined under the separate flexible and rigid diaphragm analysis. Fastener requirements were established in Part 2 of Design Example 2. These determinations should be checked for results of the rigid diaphragm analysis and adjusted if necessary (also shown in Table 3-9).

Table 3-9. Comparison of loads on shear walls using flexible versus rigid diaphragm results and recheck of wall fastening

Wall	$F_{flexible}$[2] (lb)	F_{rigid} (lb)	Rigid/ Flexible ratio %	b (ft)	$v = \frac{F_{max}}{(b)}$ (plf)	Plywood 1 or 2 sides	Design Strength[1] (plf)	Edge Fastener Spacing (in)
Roof Level								
A	1,430	3,415	239	12.5	275	1	530	6
B	6,280	5,540	88	22.0	285	1	530	6
C	11,310	12,150	107	43.0	285	1	530	6
E	11,310	10,455	92	43.0	265	1	530	6
F	8,080	10,835	134	43.0	255	1	530	6
G	4,660	4,485	96	22.0	215	1	530	6
H	1,430	2,680	187	12.5	215	1	530	6
1	22,250	22,725	102	64.5	355	1	530	6
4	22,250	22,725[3]	102	64.5	355	1	530	6
Third Floor								
A	2,805	5,665	201	12.5	455	1	530	6
B	12,305	11,085	90	22.0	560	1	800	4
C	22,160	24,405	110	43.0	570	1	800	4
E	22,160	20,925	94	43.0	515	1	800	4
F	15,830	21,710	137	43.0	505	1	800	4
G	9,135	8,940	98	22.0	415	1	800	4
H	2,805	4,425	158	12.5	355	1	530	6
1	31,955	23,420	73	64.5	495	1	800	4
2	11,645	21,245	182	60.0	355	1	530	6
3	11,645	21,295[3]	182	60.0	355	1	530	6
4	31,955	23,420[3]	73	64.5	495	1	800	4
Second Floor								
A	3,485	6,285	180	12.5	505	1	530	6
B	15,280	13,775	90	22.0	695	1	1,065	3
C	27,525	31,125	113	43.0	725	1	1,065	3
E	27,525	26,490	96	43.0	640	1	1,065	3
F	19,660	27,535	140	43.0	640	1	1,065	3
G	11,345	11,005	97	22.0	515	1	1,065	3
H	3,485	4,855	139	12.5	390	1	530	6
1	36,750	28,995	79	64.5	570	1	800	4
2	17,400	26,380	152	60.0	440	1	530	6
3	17,400	26,380[3]	152	60.0	440	1	530	6
4	36,750	28,995[3]	79	64.5	570	1	800	4

Notes:

1. Design strength values are determined from AISI-Lateral Design Table C2.1-3 for 15⁄32-inch Structural-I sheathing using nominal shear values multiplied by the factor (ϕ) of 0.60. Sheathing may be either plywood or OSB.
2. Forces taken from Design Example 2.
3. Designates the force used was the higher force for the same wall at the opposite side of the structure.

Comment: Wall rigidities used in this analysis are approximate. The initial rigidity *R* can be significantly higher than estimated because of the stiffening effects of stucco, drywall walls not considered, and areas over doors and windows. During an earthquake, some low stressed walls may maintain their stiffness and others may degrade in stiffness. Some walls and their collectors may attract significantly more lateral load than anticipated in either a flexible or rigid diaphragm analysis. It must be understood that the method of analyzing a structure using rigid diaphragms takes significantly more engineering effort. This rigid diaphragm analysis method indicates that some lateral resisting elements can attract significantly higher seismic demands than those determined under tributary area analysis methods.

4. Redundancy coefficient ρ §12.3.4.2

The redundancy coefficent penalizes lateral-force-resisting systems without adequate redundancy. In this design example, Part 1, the reliability/redundancy factor was previously assumed to be $\rho = 1.0$.

The method for determining the redundancy factor ρ is different in the ASCE/SEI 7-05. The code now requires structures in Seismic Design Categories D, E, or F to use a $\rho = 1.3$, unless one of two exceptions is met, in which case $\rho = 1.0$.

Step 1: Determine if one of the exceptions is met.

a) Each story resisting more than 35 percent of the base shear in the direction under consideration complies with Table 12.3-3. For shear walls with a height-to-length ratio of more than 1.0, the removal of that wall would not result in more than a 33-percent reduction in the overall story strength. From Table 3-2 of this design example, all three levels resist more than 35 percent of the base shear. However, all of the shear walls for this structure have a height-to-length ratio less than 1.0. Therefore, this exception is met.

b) Structure is regular in plan and there are at least two bays of lateral-load-resisting perimeter framing on each side of the structure. Therefore, this exception is also met.

Therefore, for both directions and all levels, no increase in base shear is required due to lack of redundancy.

5. Diaphragm deflections to determine if the diaphragm is flexible or rigid

This step is shown only as a reference for how to calculate horizontal diaphragm deflections. Since the shear wall forces were determined using both flexible and rigid diaphragm assumptions, there is no requirement to verify that the diaphragm is actually rigid or flexible.

The roof diaphragm has been selected to illustrate the methodology. The design seismic force in the roof diaphragm using Equation 12.10-1 must first be determined. The design seismic force is then divided by the diaphragm area to determine the horizontal loading in pounds per square foot. These values are used for determining diaphragm shears (and also collector forces). The design seismic force shall not be less than $0.2S_{DS}I_w\omega_{px}$ nor greater than $0.4S_{DS}I_w\omega_{px}$.

5a. Roof diaphragm check

The roof diaphragm will be checked in two steps. First, the shear in the diaphragm will be determined and compared to allowables. Next, the diaphragm deflection will be calculated. In Part 5b, the diaphragm deflection is used to determine whether the diaphragm is flexible or rigid.

Check diaphragm shear

The roof diaphragm consists of 15⁄32-inch-thick sheathing with #8 screws @ 6 inches o/c and panel edges are unblocked. Loading on the segment between C and E, where

Roof area = 5288 sq ft

$$f_{p\ roof} = \frac{44.5 \times 1{,}000}{5288} = 8.41 \text{ psf}$$

$$v = \frac{(8.41)48.0 \text{ ft}(32.0 \text{ ft})}{(48.0 \text{ ft})2} = 135 \text{ plf}$$

Diaphragm span = 32.0 ft

Diaphragm depth = 48.0 ft

From ASI-Lateral Design Table D2.1, the nominal shear of 555 plf is based on 15⁄32-inch DOC PS1 or PS2 (APA) rated wood structural panels with unblocked edges and #8 screws spaced at 6 inches o/c at boundaries and supported panel edges. APA or TECO performance-rated wood structural panels may be either plywood or oriented strand board (OSB).

Design shear strength = 555ϕ = 555 (0.6) = 335 plf > 135 . . . *o.k.*

Check diaphragm deflection

The code specifies that the deflection is calculated on an equivalent tributary lateral load basis. In other words, the diaphragm deflection should be based on the same load as the load used for the lateral resisting elements, not F_{px} total force at the level considered. Since the code requires building drifts to be determined by the strength level forces specified in §12.8, strength loads on the building diaphragm must be determined.

The basic equation to determine seismic forces on a diaphragm is shown as

$$F_{px} = \frac{\sum_{i=x}^{n} F_i}{\sum_{i=x}^{n} \omega_i}\,\omega_{px} \qquad \text{Eq 12.10-1}$$

In this example, the roof and floor diaphragms spanning between C and E will be used to illustrate the method. The basic 4-term equation to determine the deflection of a diaphragm is

$$\Delta = \frac{5vL^3}{8E_sA_cb} + \omega_1\omega_2\frac{vL}{\rho Gt_{\text{sheathing}}} + \omega_1^{5/4}\omega_2(a)\left(\frac{v}{2\beta}\right)^2 + \frac{\sum_{j=1}^{n}(\Delta_{ci}X_i)}{2b} \qquad \text{AISI-Lateral Eq D2.1-1}$$

The equation above is based on a uniformly nailed, simple span diaphragm with panel edges blocked. The equation has four parts. The first part accounts for beam bending, the second accounts for shear deformation, the third accounts for fastener slippage/bending, and the fourth part accounts for chord slippage.

For the purpose of this design example, the diaphragm is assumed to be a simple span supported at C and E (refer to Figure 3-4). In reality, with continuity, the actual deflection will be less.

L = 32.0 ft

b = 48.0 ft

G = 50,000 psi PDS T 3

E = 29,500,000 psi

t = 0.298 in (for C_{Dx} or Standard Grade)

v = 135 plf PDST2

ρ = 1.05 for OSB and 1.85 for plywood

β = 810 for plywood and 660 for OSB

A_c = gross cross-sectional area of chord member

A_c = 0.293 in^2 for a 4-inch × 18 GA 400 T125-43 stud track

ω_1 = S/6 = screw spacing in inches/6

ω_1 = 6/6 = 1.0

$\omega_2 = 0.033/t_{stud} = 0.033/0.043 = 0.767$

α = ratio of the average load per screw based on a non-uniform nail pattern to the average load per screw based on a uniform screw pattern. (= 1.0 for a uniformly screwed pattern)

Assume chord splice at the mid-span of the diaphragm setting the deformation at the strength for the fastener, a slippage of 0.01 inch will be used.

$$\Sigma\Delta_c X = 0.01\ (16.0)\ 2 = 0.32 \text{ in-ft}$$

$$\Delta = \frac{5(135)32.0^3}{8(29.5E6)0.293(48.0)} + 1.0\,(0.767)\frac{135(32.0)}{1.85(50{,}000)0.298}$$

$$+ (1.0)^{5/4}(0.767)1.0\left(\frac{135}{2(810)}\right)^2 + \frac{0.32}{2(48.0)} = 0.14 \text{ in}$$

This deflection is based on a blocked diaphragm. For unblocked diaphragms, the deflection is to be multiplied by 2.50 (from AISI-Lateral Design).

$$\Delta = 0.14 \times 2.5 = 0.34 \text{ in}$$

Check shear wall deflection.

The shear wall deflections at lines C and E between the third floor and roof are determined by using the basic stiffness equation

$$F = k\Delta$$

or

$$\Delta = \frac{F}{k}$$

where

$$F_c = 12{,}150 \text{ lb} \qquad k_c = 42.0 \text{ k/in} \qquad \Delta_c = \frac{12.15}{42.0} = 0.29 \text{ in}$$

$$F_E = 10{,}455 \text{ lb} \qquad k_e = 42.0 \text{ k/in} \qquad \Delta_E = \frac{10.45}{42.0} = 0.25 \text{ in}$$

$$\Delta_{\text{ave}} = \frac{0.29 + 0.25}{2} = 0.27 \text{ in}$$

5b. Flexible versus rigid diaphragms §12.3.1.3

In this example, the maximum diaphragm deflection was estimated as 0.34 inch. This assumes a simple span for the diaphragm, and the actual deflection would probably be less. The average story drift is on the order of 0.27 inch at the roof. For the diaphragms to be considered flexible, the maximum diaphragm deflection will have to be more than 2 times the average story drift. The diaphragm spans would easily qualify as "rigid" diaphragms. As defined by the code, the diaphragms in this design example are considered rigid.

In reality, some amount of diaphragm deformation will occur, and the true analysis is highly complex and beyond the scope of what is normally done for this type of construction. Diaphragm deflection analysis and testing has been performed on level/flat diaphragms. There has not been any testing of sloped and complicated diaphragms, as found in the typical wood-framed structure. Therefore, some engineers perform their design based on the roof diaphragm as flexible and the floor diaphragms as rigid.

In using this procedure, the engineer should exercise careful judgment in determining if the higher load of the two methodologies is actually required. For example, if the load to two walls by rigidity analysis is found to be 5 percent to line A and 95 percent to line B, but by flexible analysis it is found to be 50 percent to line A and 50 percent to line B, the engineer should probably design for the larger of the two loads for the individual walls. Note that though the same definition of a flexible diaphragm has been in the code since the 1988 UBC edition, it has not been enforced by building officials for Type V construction.

6. Tiedown forces for the shear wall at line C AISI-Lateral §C5.3

6a. Determination of tiedown forces

Tiedowns are required to resist the uplift tendency of shear walls caused by overturning moments. In this step, tiedown forces for the three-story shear wall on line C (Figure 3-8) are determined. AISI-Lateral requires vertical boundary members and the anchorage of the boundary members to have the nominal strength to resist amplified (overstrength) seismic loads.

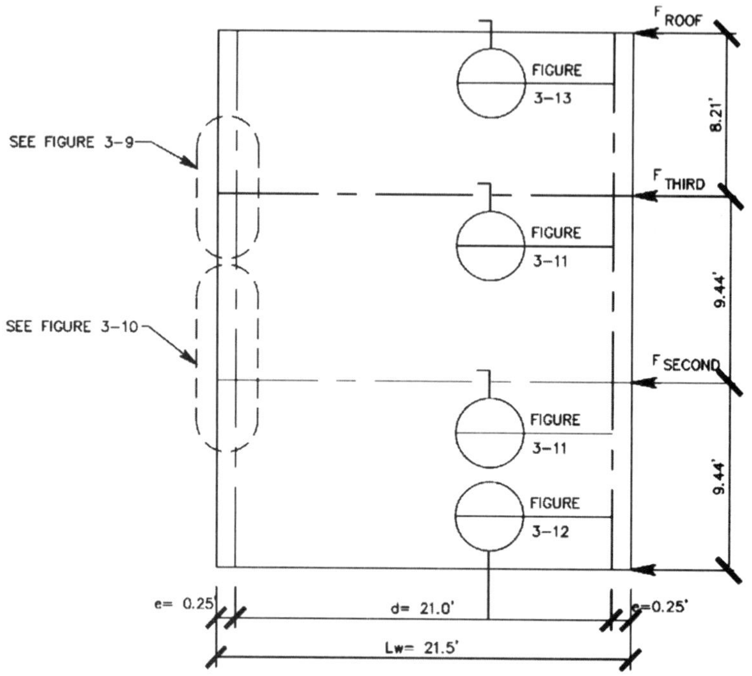

Figure 3-8. Typical shear wall C elevation

Because there are two identical shear walls on line C, forces from Table 3-6 must be divided by 2. Computation of story forces for one of the two walls is shown below. Note that forces are on *strength design basis*.

F_{roof} = 12,150/2 = 6075 lb/wall (two walls on line C)

F_{third} = (24,405 – 12,150)/2 = 6130 lb

F_{second} = (31,125 – 24,405)/2 = 3360 lb

Ω_O = 3.0 bearing wall system **T 12.2-1**

The distance between the centroid of the boundary forces that represent the overturning moment at each level must be estimated. This is shown below.

e = the distance to the center of tiedown and boundary studs or collectors studs (Figure 3-10)

e = 3 in = 0.25 ft

d = the distance between centroids of the tiedowns and the boundary studs. Note that it is also considered acceptable to use the distance from the end of the shear wall to the centroid of the tiedown.

d = 21.5 ft – 2(0.25) = 21.0 ft

The resisting moment M_R is determined from the following dead loads

w_{roof} = 13.5 psf(1.33 ft) = 18.0 plf

w_{floor} = 25.0 psf(1.33 ft) = 33.0 plf

w_{wall} = 10 plf

Overturning resisting moments are determined from simple statics. Calculations are facilitated by use of a spreadsheet. Table 3-10 summarizes the tiedown (i.e., uplift) forces for the shear walls on line C.

Table 3-10. Tiedown forces for shear wall C

Level	M_{ot} (ft-lb)	$\Omega_O M_{ot}$ (ft-lb)	M_R (ft-lb)	Strength Uplift $\dfrac{\Omega_o M_{ot} - (0.9 - 0.2S_{DS} M_R)}{d}$ (lb)	ASD Uplift $\dfrac{\Omega_o M_{ot} - M_R(0.6 - 0.14S_{DS})0.7}{d}$ (lb)
Roof	49,875	149,625	23,135	6,320	4,655
Third	165,090	495,270	52,580	21,750	15,755
Second	312,025	936,075	82,025	41,725	30,025

where

$$(0.6 - 0.14\,S_{DS}) = 0.43$$

$$(0.9 - 0.14\,S_{DS}) = 0.73$$

6b. Load combinations using allowable stress design

AISI-Lateral Section C.5.3 specifies requirements for steel stud wall boundary members. Section C.5.3 specifies requirements for boundary element connections and requires Ω_O to be applied to the connections. SDPWS deals with wood stud walls and does not have any such special requirements. In the case of identical building types (as in Design Examples 2 and 3 of this manual) this gives an apparent advantage to wood framing. As a comparison, refer to Table 2-21 in Design Example 2.

7. Allowable shear and nominal shear strength of #10 screws

Tiedown connections for the line C shear wall will use 12-gage straps at the third floor. This part shows determination of the shear strength of the #10 screws that will be used to connect the tiedown straps to the 18-gage boundary studs.

There are two basic ways of determining the shear strength of the screws. The first is to use the values established in an ICC Evaluation Report with appropriate conversion to strength design. The second is to compute the shear strength of a screw using the '01 AISI specification. Both methods are shown below.

7a. Nominal shear strength determined from ICC Evaluation Report

The Metal Stud Manufacturers' Association provides ICC Evaluation Report ICC-ES ER-4943. Shear values on an ASD basis are provided for various gage studs having a minimum yield strength of 33 ksi and a minimum ultimate strength of 45 ksi.

For #10 screws in an 18-gage stud, the allowable shear is given as 258 pounds per screw. This must be increased as shown below to convert to the strength design basis used in this example.

$$P_{ns} = \Omega P_{as}$$

where

P_{ns} = nominal shear strength per screw

P_{as} = allowable shear strength per screw

Ω = 3.0 — 01AISI(E4)

$P_{ns} = 3.0(258 \text{ lb}) = 774$ lb per screw

Note that ER-4943 also specifies a minimum edge distance and a minimum on-center spacing of 9⁄16 inch for #10 screws.

7b. Calculation of nominal shear strength using strength design

The nominal shear strength is the screw capacity without the appropriate reduction factors for allowable stress design (Ω) or load and resistance factor design (ϕ).

d = 0.190 in

F_{u1} = 45,000 psi

Note: some connector straps and hardware have an F_u = 65,000 psi, which will give higher screw capacities.

F_{u2} = 45,000 psi

Case I: Strap applied to stud flange (Figure 3-9)

Assume 12-gage galvanized strap

t_1 = 0.1017 in

With 18-gage studs

t_2 = 0.0451 in

$t_2/t_1 = 0.0451/0.1017 = 0.44 < 1.0$

$P_{ns} = 4.2(t_2^3 d)^{1/2} F_{u2} = 789$ lb — 01 AISI (E4.3.1-1)

$P_{ns} = 2.7 t_1 d F_{u1} = 2348$ lb — 01 AISI (E4.3.1-2)

$P_{ns} = 2.7 t_2 d F_{u2} = 1041$ lb — 01 AISI (E4.3.1-3)

Using the smallest value of P_{ns}

$$P_{ns} = 789 \text{ lb per screw}$$

Note how this value is almost equal to the 774 lb determined from Part 6a, above.

Case 2: Strap applied to double stud webs (Figure 3-10)

Assume 10-gage galvanized strap

$$t_1 = 0.138 \text{ in}$$

With 18-gage studs.

Because there are two stud webs, thickness t_2 is doubled

$$t_2 = 0.0451 \times 2 = 0.0902 \text{ in}$$

$$t_2/t_1 = 0.0902/0.138 = 0.65 < 1.0$$

$$P_{ns} = 4.2(t_2^3 d)^{1/2} F_{u2} = 2232 \text{ lb}$$ 01 AISI (E4.3.1-1)

$$P_{ns} = 2.7 t_1 d F_{u1} = 3186 \text{ lb}$$ 01 AISI (E4.3.1-2)

$$P_{ns} = 2.7 t_2 d F_{u2} = 2082 \text{ lb}$$ 01 AISI (E4.3.1-3)

Using the smallest value of P_{ns}

$$P_{ns} = 2{,}082 \text{ lb}$$

7c. Calculation of allowable shear using ASD

Case I: Strap applied to stud flange (Figure 3-9)

From Part 6b, above

$$P_{ns} = 789 \text{ lb}$$

$$P_{as} = P_{ns}/\Omega = 789/3.0 = 236 \text{ lb per screw}$$

Case II: Strap applied to double stud webs (Figure 3-10) From Part 7b, above

$$P_{ns} = 2082 \text{ lb}$$

$$P_{as} = P_{ns}/\Omega = 2082/3.0 = 694 \text{ lb per screw}$$

8. Tiedown connection at third floor for wall on line C

Shown below is the strength design of the tiedown strap to be used for the shear walls on line C at the third floor. The configuration at the tiedown is shown on Figure 3-9.

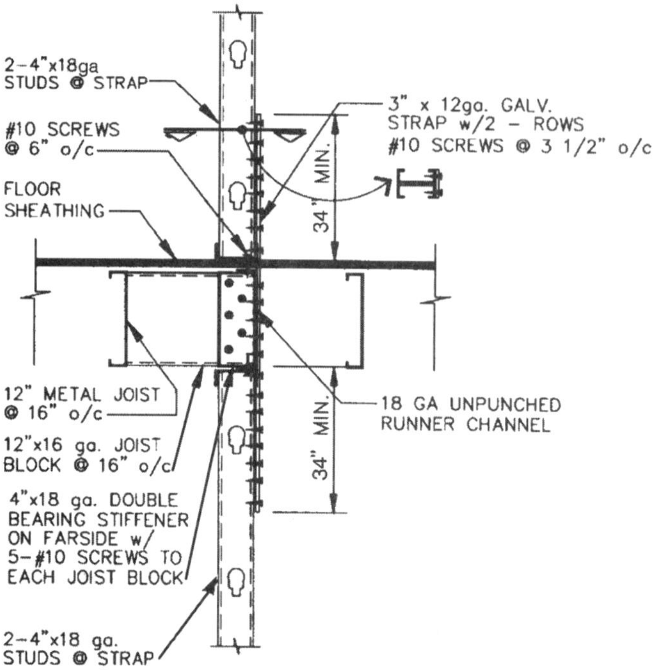

Figure 3-9. Typical tiedown connection at the third floor on line C.

Uplift = 6320 lb

Try a 12-gage × 3-in strap and #10 screws

P_{ns} = 789 lb per screw

LRFD design strength = ϕP_{ns} 01 AISI (E4.3)

where

$\phi = 0.50$

$\phi P_{ns} = 0.5(789) = 395$ lb

Number of screws required

6320/395 = 16

∴ Use 16 minimum

With 2 rows of #10 screws @ 3½ inches o/c, determine the length of strap required.

Strap is pre-manufactured. Use half spacing for end distance or 1¾ inches. Net spacing of screws is 1¾ inches o/c. Need to add in thickness of 1½-inch lightweight concrete and ¾-inch sheathing, plus the 12-inch depth for the floor joist

$(1.75 + (1 + 16)1.75 + 1.75)2 + (1.5 + 0.75 + 12) = 79$

∴ Use 80-inch-long strap

Check capacity of strap for tension.

Strap to be used will be a pre-manufactured strap for which there is an ICC Evaluation Report. The rated capacity, including 33-percent increase for wind or seismic loading, is given as 9640 lb.

9640 lb > 6320 lb . . . *o.k.*

If the strap does not have an ICC rated capacity, the manufacturer should be contacted to determine the strength of the steel used. It is probable that the steel used in the strap will have strengths that differ from the steel used in the studs. Generally, strengths differ from one manufacturer to another.

Checking capacity of strap

$$T_n = A_n F_y$$ 01 AISI (C2-1)

b = 3.0 in (strap width)

t = 0.1046 in (strap thickness)

d = 0.171 in (diameter of holes)

$A_n = 0.2987\ \text{in}^2$ (net area of strap)

F_y = 45,000 psi (yield strength of particular manufacturer)

T_n = 0.2987(45,000) = 13,443 lb (nominal strength of strap)

For ASD

Tie force = 4655 lb (Table 3-10)

$$\frac{T_n}{\Omega_t} = \text{allowable tension} = \frac{13{,}443}{1.67} = 8050\ \text{lb} > 4655\ \ldots\ \text{o.k.}$$

For LRFD

Tie force = 6320 lb (Table 3-10)

ϕT_n = tension strength = 0.95(13,443) = 12,770 lb => 6320 lb . . . *o.k.*

Use 12-gage × 3-in × 72-in strap with 12 #10 screws @ 3½ inches o/c each end.

9. Tiedown connection at second floor for wall on line C

Design of the pre-manufactured tiedowns for the second floor shear walls on line C is shown below. Figure 3-10 shows the configuration of the tiedown.

Uplift = 15,755 lb from Table 3-10

The connector is an ICC-approved, pre-manufactured hold-down device. The rated capacity, including the 33-percent increase for wind or seismic loading, is 9900 lb.

Using two hold-downs, one on each boundary stud, the capacity is

$$2 \times 9900 = 19{,}800 > 15{,}755 \text{ lb} \ldots o.k.$$

In general, when using pre-manufactured tiedowns, consult with ICC Evaluation Service or the manufacturer for the necessary approvals for hardware selection.

10. Boundary studs for first floor wall on line C

The studs at each end of the shear walls on line C must be designed to resist overturning forces. In this example, double studs as shown in Figure 3-10 will be used at each end. The critical aspect of design is checking the studs for axial compression. This is shown below.

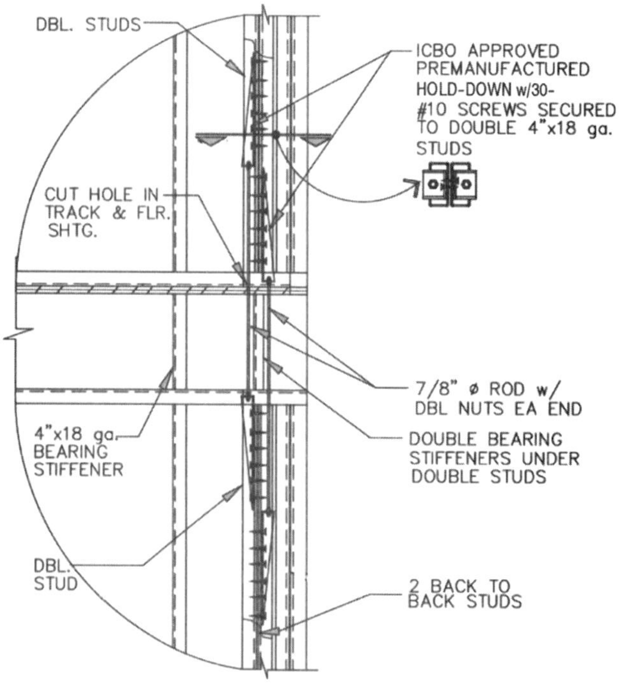

Figure 3-10. Typical tiedown connection at the second floor on line C

For axial compression, the load combination to be used is

$$(1.0 + 0.10\, S_{DS})\, D + 0.525\, \rho E + 0.75L \qquad \text{§12.4.2.3}$$

where

$S_{DS} = 1.19$

$\rho = 1.0$

$P_{LL} = [40 \text{ psf} + (16/12)(16/12)]\, 2 = 145 \text{ lb}$

$P_{DL} = [13.5 \text{ psf} + (25.0)2]\,(16/12) + [10 \text{ psf}\,(8 \text{ ft})(16/12)(3)] = 405 \text{ lb}$

From Table 3-10

$M_{ot} = 312{,}025$ lb-ft

$P_E = 312{,}025$ ft-lb/(21.0 ft) = 14,850 lb

Thus, the design load to boundary studs using Equation 16-17 is

$1.12(405) + 0.75(145) + 0.525\,(14{,}850) = 8400$ lb

With a computer program using 2001 AISI Specifications, the allowable axial load for a 4-inch by 18-gage stud with 2-inch flanges is 4760 lb with the flanges braced at mid-height.

$$\text{Number of studs required} = \frac{8400}{4760} = 1.8$$

Note that the allowable stress increase is no longer permitted by the code.

Therefore, use two studs at ends of wall (Figure 3-10).

11. Shear transfer at second floor on line C

Shear forces in the second floor diaphragm are transferred to the shear walls below as shown in Figure 3-11.

From Table 3-9, the strength shear in the wall is

$v = 570$ lb/ft

The allowable shear in the wall is $570 \times 0.7 = 400$ lb/ft

Try using #8 screws, 18-gage metal side plates, and Douglas Fir plywood

$Z = 90$ lb/screw NDS-05 T 11M

$C_D = 1.6$ NDS-05 2.3.2

$$\text{Maximum spacing} = \frac{ZC_D}{v} = \frac{90(1.6)12}{400} = 4.3 \text{ in}$$

∴ Use #8 screws at 3 inches o/c.

Capacity of the #8 screws in the 18-gage tracks and runner channels are O.K. by inspection.

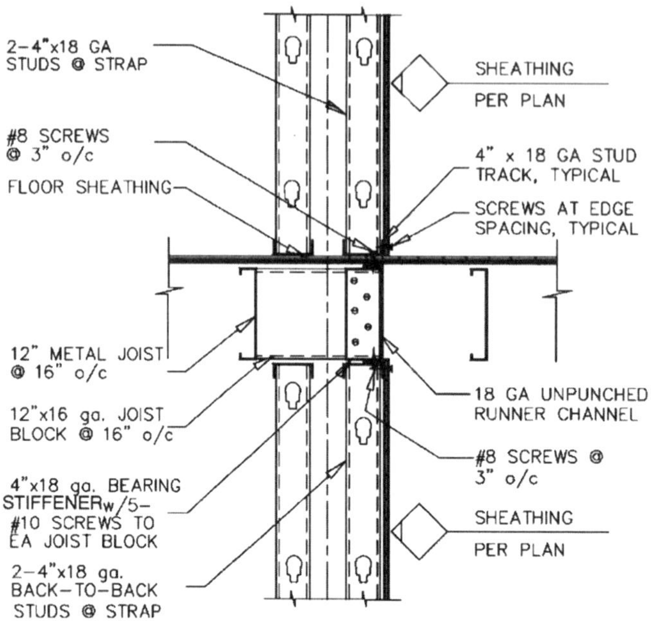

Figure 3-11. Typical detail for shear transfer through floor on line C

12. Shear transfer at foundation for walls on line C

Shown below is the design of the connection to transfer the shear force in the walls on line C to the foundation. This detail is shown in Figure 3-12.

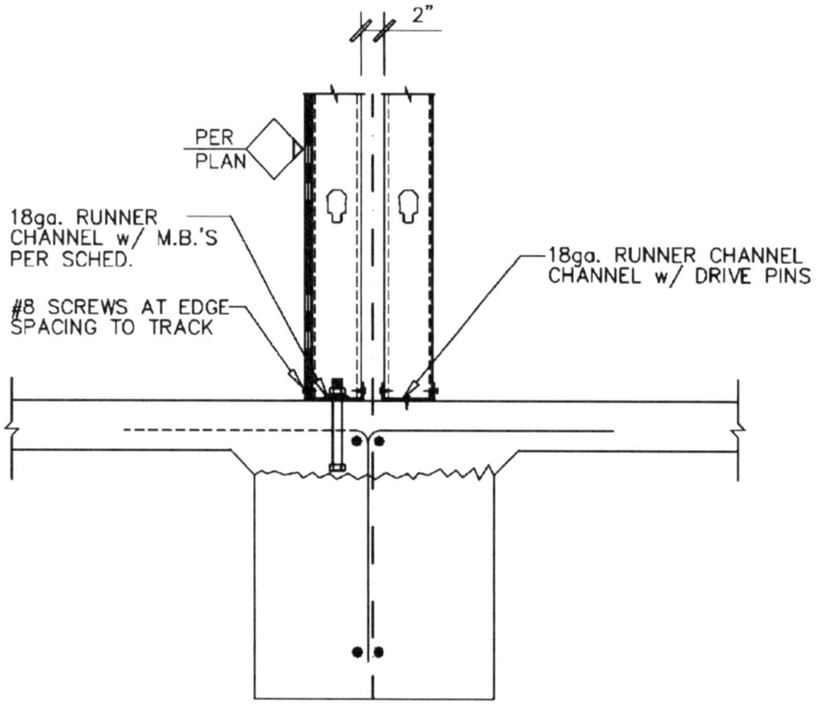

Figure 3-12. Detail for shear transfer at foundation on line C.

From Table 3-9

$v = 400$ lb/ft

Allowable load based on bolt bearing on track 01 AISI (E3.3)

For ⅝-inch bolts and 18-gage track

$P_n = 2.22 F_u d$ 01 AISI, T (E3.3-2)

where

P_n = nominal resistance

F_u = 45 ksi (minimum value)

d = 0.625 in

t = 0.0451 in

$$P_n = 2.22(45)(0.625)(0.0451) = 2.82 \frac{k}{\text{bolt}}$$

Allowable service load on embedded bolts in concrete is determined as follows:

For ⅝-inch bolts and 3000 psi concrete

$$\text{Allowable shear} = 2750 \frac{\text{lb}}{\text{bolt}}$$

Therefore the bolt in concrete governs the required spacing

$$\text{Maximum spacing } \frac{2750}{400} = 6.8 \text{ ft o/c}$$

∴ Use ⅝-inch-diameter bolts at 4 feet o/c spacing

For comments on sill bolts in concrete see Section 11a in Design Example 2.

13. Shear transfer at roof at line C

Shear forces in the roof diaphragm are transferred to the shear walls below as shown in Figure 3-13. From Table 3-9, find the strength shears in the wall.

$v = 285$ lb/ft

The allowable shear in the wall

$v = 285 \times 0.7 = 200$

From manufacturer's catalog, allowable load for the 6⅜-inch-long framing clip is 915 lb.

With framing clips at 4-ft centers, the design ASD force is

$(200)(4) = 800 < 915$ lb . . . *o.k.*

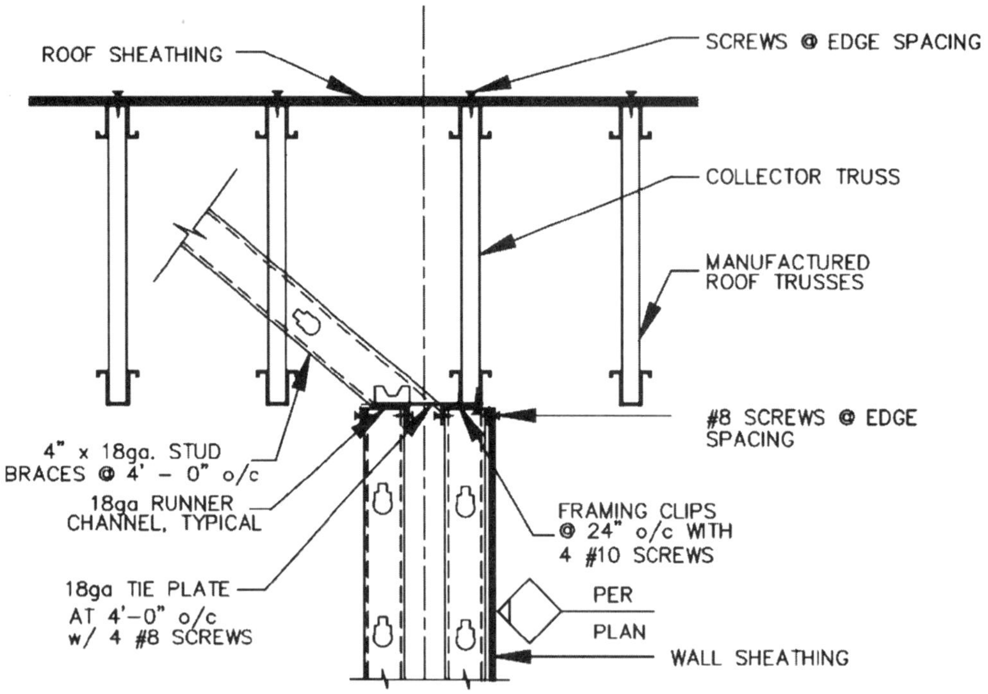

Figure 3-13. Shear transfer at roof at line C

Note that double studs are used for sound control, but that only one stud is considered in shear wall calculations.

Commentary

The AISI does not have conventional construction provisions for cold-formed steel similar to the conventional light-frame construction provisions for wood. The 2006 *International Residential Code*® (IRC) has included prescriptive provisions for cold-formed steel for one- and two-family dwellings. It should be noted that the structure shown in this example could not use the IRC prescriptive provisions. Inasmuch as there is no one standard for the manufacturing of studs, the process to design gravity load members is tedious and should not be done by prescriptive means.

The AISI Specification for Design of Cold-Formed Steel Structural Members has complex equations and is considered by most engineers to be too difficult to be readily used in design.

Because of the complex nature of the equations, the AISI code recommends that engineers designing in cold-formed steel use computer software for design.

References

American Forest and Paper Association, 2005, AF&PA Supplement Special Design Provisions for Wind & Seismic, American Forest and Paper Association, Washington, D.C.

American Forest and Paper Association, 2005. *National Design Specification for Wood Construction* including Supplements NDS-05. American Forest and Paper Association, Washington D.C.

American Iron and Steel Institute, 1996. *Cold-Formed Steel Design Manual, 1996 Edition.* American Iron and Steel Institute, Washington, D.C.

American Iron and Steel Institute, 1986. *Cold-Formed Steel Design Manual, 1986 Edition.* American Iron and Steel Institute, Washington, D.C.

American Iron and Steel Institute, 2004. Standard for Cold-Formed Steel Framing—General Provisions. American Iron and Steel Institute, Washington, D.C.

American Iron and Steel Institute, 2004. Standard for Cold-Formed Steel Framing—Header Design. American Iron and Steel Institute, Washington, D.C.

American Iron and Steel Institute, 2004. Standard for Cold-Formed Steel Framing—Lateral Design. American Iron and Steel Institute, Washington, D.C.

American Iron and Steel Institute, 2004. Standard for Cold-Formed Steel Framing—Wall Stud Design. American Iron and Steel Institute, Washington, D.C.

American Plywood Association, 1997. *Design/Construction Guide—Diaphragms and Shear Walls.* Report 105, Engineered Wood Association, Tacoma, Washington.

American Plywood Association, 1997. Plywood Design Specification. Engineered Wood Association.

American Plywood Association, 1993. *Revised. Wood Structural Panel Shear Walls.* Report 154, Engineered Wood Association, Tacoma, Washington.

Applied Technology Council, 1995. *Cyclic Testing of Narrow Plywood Shear Walls, ATC R-1.* Applied Technology Council, Redwood City, California.

BSSC, 1997. *National Earthquake Hazard Reduction Program, Recommended Provisions for Seismic Regulations for New Buildings, 1997.* Building Seismic Safety Council, Washington, D.C.

Building Seismic Safety Council, 1997. *National Earthquake Hazard Reduction Program, Recommended Provisions for Seismic Regulations for New Buildings.* Building Seismic Safety Council, Washington D.C.

Cobeen, K. E., 1996. "Performance Based Design of Wood Structures," *Proceedings, Annual SEAOC Convention.* Structural Engineers Association of California, Sacramento, California.

Countryman, D., and Col Benson, 1954. *1954 Horizontal Plywood Diaphragm Tests.* Laboratory Report 63, Douglas Fir Plywood Association, Tacoma, Washington.

Dolan, J. D., 1996. *Experimental Results from Cyclic Racking Tests of Wood Shear Walls with Openings.* Timber Engineering Report No. TE-1996-001. Virginia Polytechnic Institute and State University, Blacksburg, Virginia.

Dolan, J. D. and Heine, C. P., 1997a. *Monotonic Tests of Wood-frame Shear Walls with Various Openings and Base Restraint Configurations.* Timber Engineering Report No. TE-1997-001. Virginia Polytechnic Institute and State University, Blacksburg, Virginia.

Dolan, J. D. and Heine, C. P., 1997b. *Sequential Phased Displacement Cyclic Tests of Wood frame Shear Walls with Various Openings and Base Restrain Configurations.* Timber Engineering Report No. TE-1997-001. Virginia Polytechnic Institute and State University, Blacksburg, Virginia.

Dolan, J. D., and Heine, C. P., 1997c, *Sequential Phased Displacement Test of Wood-frame Shear Walls with Corners.* Timber Engineering Report No. TE-1997-003. Virginia Polytechnic Institute and State University, Blacksburg, Virginia.

Earthquake Engineering Research Institute, 1996. "Northridge Earthquake of January 17,1994, Reconnaissance Report," *Earthquake Spectra.* Vol. 11, Supplement C. Earthquake Engineering Research Institute, Oakland, California.

Foliente, Greg C., 1994. *Analysis, Design and Testing of Timber Structures Under Seismic Loads.* University of California Forest Products Laboratory, Richmond, California.

Foliente, Greg C., 1997. *Earthquake Performance and Safety of Timber Structures.* Forest Products Society, Madison Wisconsin.

Forest Products Lab, 1999. *Wood Handbook Publication FPL-GTR-113.* Madison, Wisconsin.

Goers R. and Associates, 1976. *A Methodology for Seismic Design and Construction of Single-Family Dwellings.* Applied Technology Council, Redwood City, California.

Ju, S., and Lin M., 1999. "Comparison of Building Analysis Assuming Rigid or Flexible Floors," *Journal of Structural Engineering.* American Society of Civil Engineers, Washington D.C.

Light Gauge Steel Engineers Association, Tech Note 558b-1. *Lateral Load Resisting Elements: Diaphragm Design Values.* Light Gauge Steel Engineers Association., 2400 Crestmoor Road, Nashville, Tennessee 37215.

Light Gauge Steel Engineers Association, Tech Note 556a-6, *Vertical Lateral Force Resisting System Boundary Elements.* Light Gauge Steel Engineers Association, 2400 Crestmoor Road, Nashville, Tennessee 37215.

Light Gauge Steel Engineers Association, Tech Note 556a-4. *Shear Transfer at Top Plate: Drag Strut Design.* Light Gauge Steel Engineers Association, 2400 Crestmoor Road, Nashville, Tennessee 37215.

Light Gauge Steel Engineers Association, Tech Note 565c. *Screw Fastener Selection for Light-Gauge Steel Frame Construction.* Light Gauge Steel Engineers Association. 2400 Crestmoor Road, Nashville, Tennessee 37215, February 1997.

Metal Stud Manufacturer's Association, 1993. *ICBO Evaluation Report No. 4943.* Metal Stud Manufacturer's Association, P.O. Box 1211, Corvallis, Oregon 97339, revised December 1993.

Performance Standards and Policies for Structural-Use Panels [Sheathing Standard, Sec. 2.3.31. APA Standard PRP-108. APA-The Engineered Wood Association. Tacoma, Washington 98411.

Rose, J. D., 1998, *Preliminary Testing of Wood Structural Panel Shear Walls Under Cyclic (Reversed) Loading.* Research Report 158. APA-The Engineered Wood Association, Tacoma, Washington.

Rose, J. D., and E. L. Keith, 1996. *Wood Structural Panel Shear Walls with Gypsum Wallboard and Window [Sheathing Standard, Sec. 2.3.31.* Research Report 158. APA—The Engineered Association, Tacoma Washington.

SEAOC, 1999. *Acceptable Diaphragm-Rigidity Assumptions for Distribution of Horizontal Forces in Light-Frame Construction.* Structural Engineers Association of California, Sacramento, California.

SEAOC, 1999. *"Recommended Lateral Force Requirements and Commentary,"* Structural Engineers Association of California, Sacramento, California.

Serrette, R. 1996. *Final Report: Shear Wall Values for Lightweight Steel Framing.* Santa Clara University Engineering Center, Santa Clara, California 95053.

Yu, Wei-wen, 1991. *Cold-Formed Steel Design*, John Wiley & Sons, Inc., New York, New York.

Design Example 4
Masonry Shear Wall Building

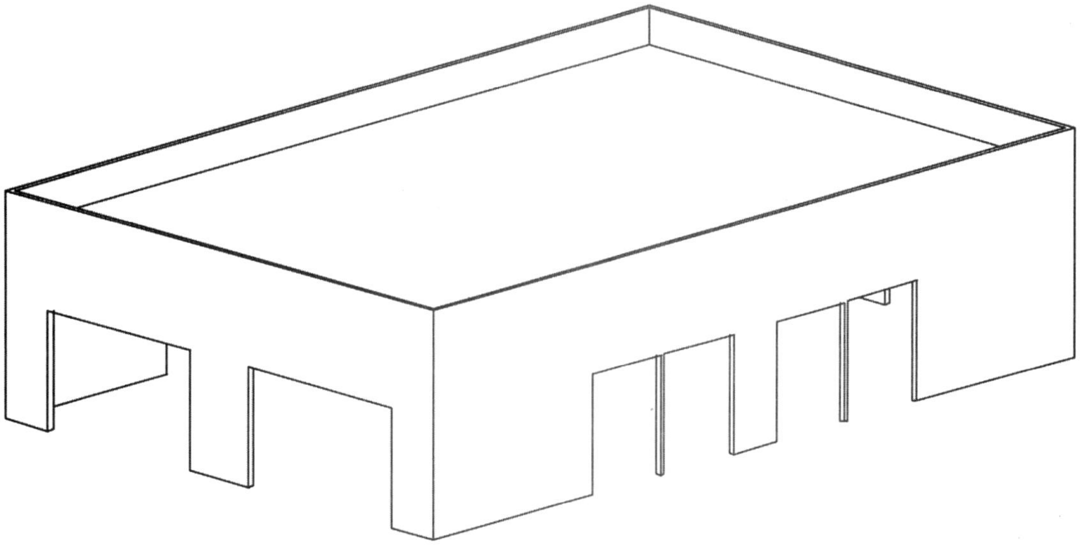

Figure 4-1. Schematic CMU building elevation

Foreword

Reinforced concrete block masonry is frequently used in one-story and lowrise construction, particularly for residential, retail, light commercial, and institutional buildings. This type of construction has generally had a good earthquake performance record. However, during the 1994 Northridge earthquake, some one-story buildings with concrete masonry unit (CMU) walls and panelized wood roofs experienced wall-roof separations similar to those of many tilt-up buildings.

This building, typical of one-story masonry buildings with wood framed roofs, is characterized as a heavy wall and flexible roof diaphragm "box building," and is shown schematically in Figure 4-1. Floor and roof plans are in Figure 4-2 and 4-3, respectively. It is a one-story bearing wall building with CMU shear walls. Roof construction consists of a plywood diaphragm over wood framing. An elevation of the building on line A is shown in Figure 4-4. A CMU wall section is shown in Figure 4-5, and a plan view of an 8-foot CMU wall/pier is shown in Figure 4-6.

This design example illustrates the strength design approach to CMU wall design for both in-plane and out-of-plane seismic forces.

Outline

Determine the following and/or design the following structural elements.

1. **Design base shear coefficient**
2. **Base shear in the transverse direction**
3. **Shear in wall on line A**
4. **Design 8′-0″ shear wall on line A for out-of-plane seismic forces**
5. **Design 8′-0″ shear wall on line A for in-plane seismic forces**
6. **Design 8′-0″ shear wall on line A for axial and in-plane bending forces**
7. **Deflection of shear wall on line A**
8. **Requirements for shear wall boundary elements**
9. **Wall-roof out-of-plane anchorage for lines 1 and 3**
10. **Chord design**

Given Information

Roof weights:

Roofing + one re-roof	7.5 psf
½-inch plywood	1.5
Roof framing	4.5
Mech./elec.	1.5
Insulation	1.5
Total dead load	17.0 psf
Roof live load	20.0 psf

Exterior 8-inch CMU Walls

75 psf (fully grouted, light-weight masonry)

$f'_m = 2500$ psi

$f_y = 60{,}000$ psi

Seismic and site data

The building is located in Oakland, California near the Hayward Fault. The values of maximum considered earthquake ground motion are read from seismic maps for periods of 0.2 seconds and 1.0 seconds. These maps are included in Chapter 22.0 and are also available in large readable format from FEMA. The maps are formatted for a typical site class B and for 5-percent damping. Ground motion values can be obtained from the USGS web site.

$S_s = 2.10g$	§11.4.1
$S_1 = 0.93g$	§11.4.1
$I = 1.0$ (Occupancy Category II)	§11.5.1
Seismic Design Category E	§11.6
Site Class D	§11.4.2

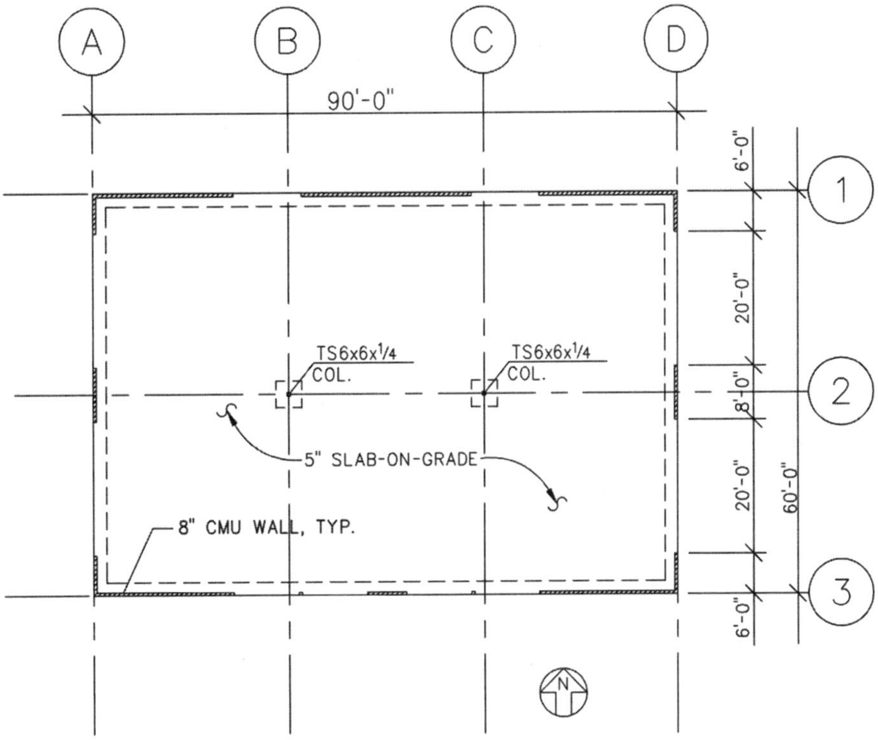

Figure 4-2. Floor plan

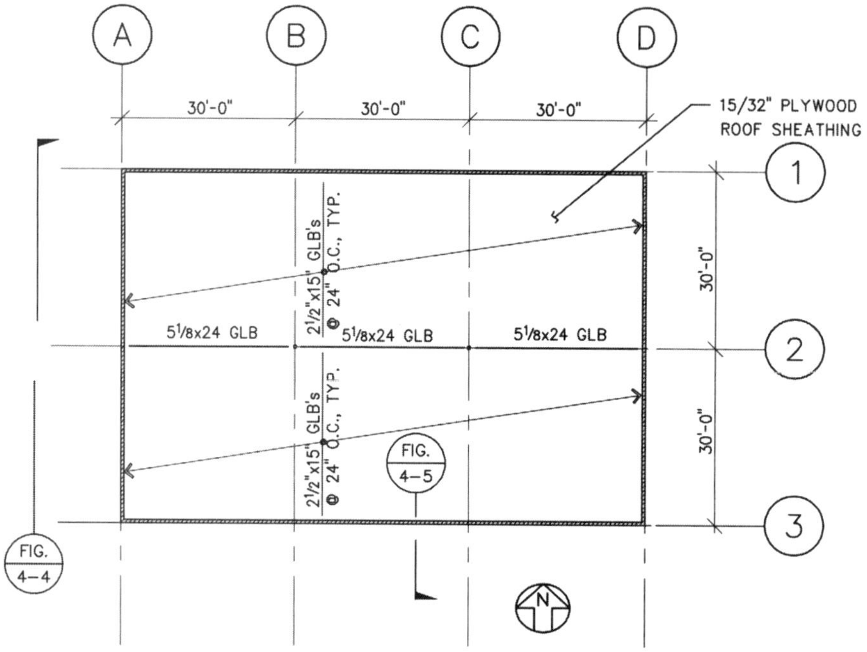

Figure 4-3. Floor plan

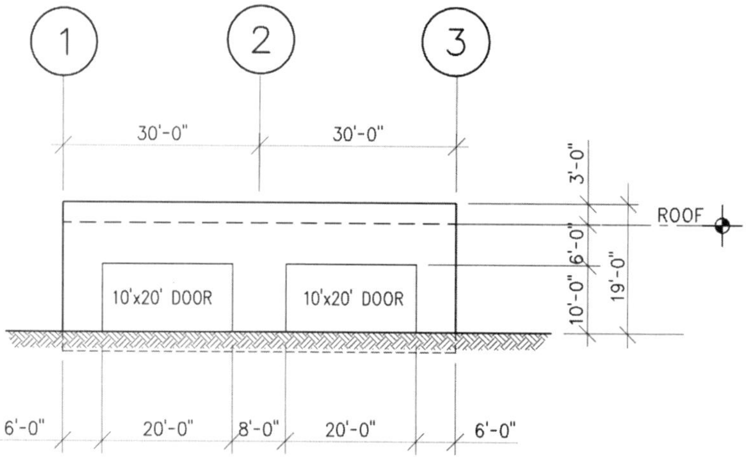

Figure 4-4. Elevation of wall on line A

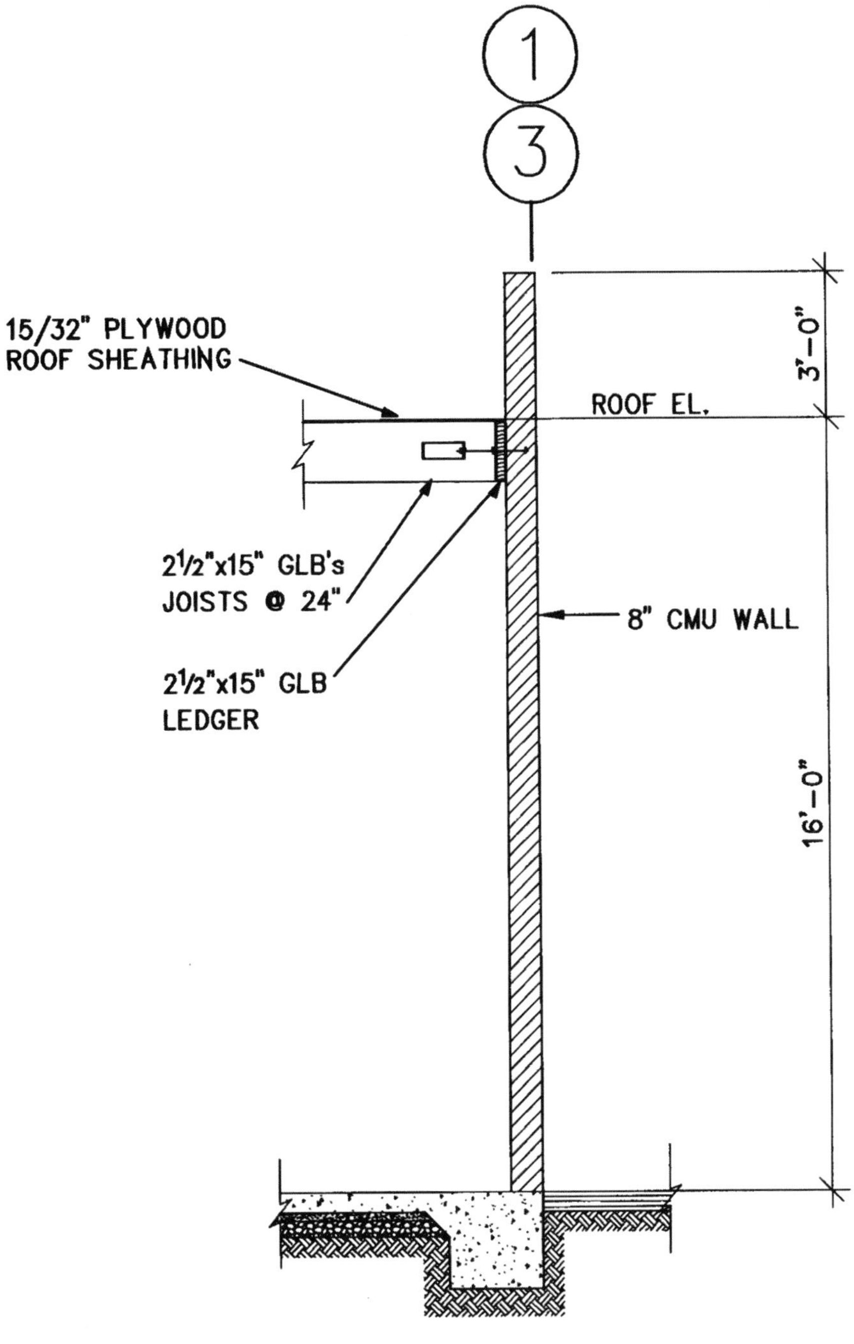

Figure 4-5. Section through CMU wall along lines 1 and 3

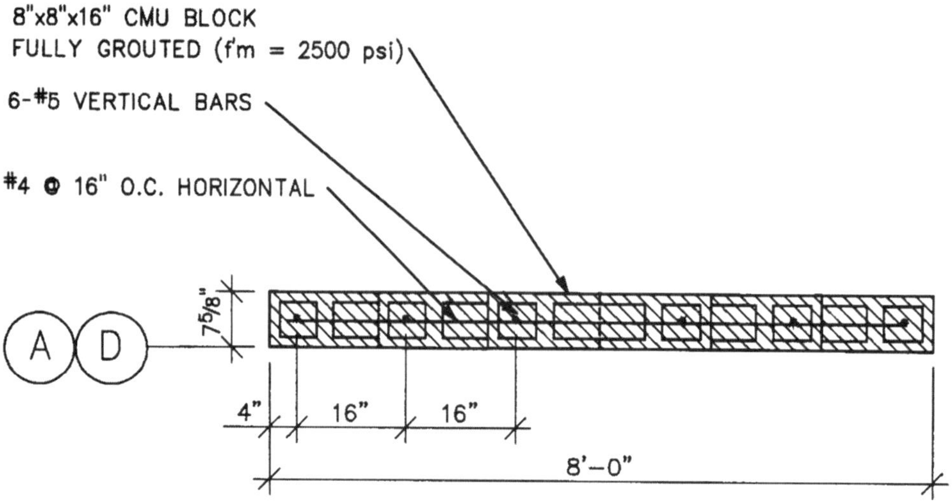

Figure 4-6. Reinforcement in 8-foot CMU shear walls on lines A and D

Calculations and Discussion	*Code Reference*

Seismic ground motion values §11.4

Period using Eq (12.8-7)

$$T = C_t(h_n)^{3/4} = 0.020(16\text{ ft})^{3/4} = 0.16\text{ sec}$$

The maximum considered earthquake spectral response acceleration, five-percent damped, for short periods S_{MS} and at one-second period S_{M1} adjusted for site class effects, shall be determined by

$F_a = 1.0$ — T 11.4-1

$F_v = 1.5$ — T 11.4-2

$S_{MS} = F_a S_S = (1.0)(2.1g) = 2.1g$ — Eq 11.4-1

$S_{M1} = F_v S_1 = (1.5)(0.93g) = 1.4g$ — Eq 11.4-2

Five-percent damped design spectral response acceleration at short periods S_{DS} and at 1-second period S_{D1} adjusted for site class effects, shall be determined by

$S_{DS} = ⅔\, S_{MS} = (⅔)(2.1g) = 1.40g$ — Eq 11.4-3

$S_{M1} = ⅔\, S_{M1} = (⅔)(1.4g) = 0.93g$ — Eq 11.4-4

1. Design base shear coefficient from the equivalent lateral force procedure. §12.8

The R coefficient for a masonry bearing wall building with special reinforced masonry shear walls is

$R = 5.0$

$\Omega_O = 2\frac{1}{2}$ — T 12.2-1

$C_d = 3\frac{1}{2}$

The seismic base shear V in a given direction shall be determined in accordance with the following

$$V = C_s W \qquad \text{Eq 12.8-1}$$

The calculation of seismic response coefficient C_S shall be

$$C_S = \frac{S_{DS}}{\frac{R}{I}} = \frac{1.40g}{\left(\frac{5.0}{1.0}\right)} = 0.280 \qquad \text{Eq 12.8-2}$$

but need not exceed

$$C_S = \frac{S_{D1}}{T\left(\frac{R}{I}\right)} = \frac{0.93g}{\left(\frac{5.0}{1.0}\right)0.16\text{ sec}} = 1.163 \text{ for } T \le T_L \qquad \text{Eq 12.8-3}$$

The value for seismic response coefficient C_s computed in Equations 12.8-2 and 12.8-3 shall not be less than the following

$$C_s = 0.01 \qquad \text{Eq 12.8-5}$$

In addition, for buildings where the 1-second spectral response S_1 is equal to or greater than 0.6g, the value of the seismic response coefficient C_s shall not be less than

$$C_s = \frac{0.5\, S_1}{\left(\frac{R}{I}\right)} = \frac{0.5(0.93g)}{\left(\frac{5.0}{1.0}\right)} = 0.093g \qquad \text{Eq 12.8-6}$$

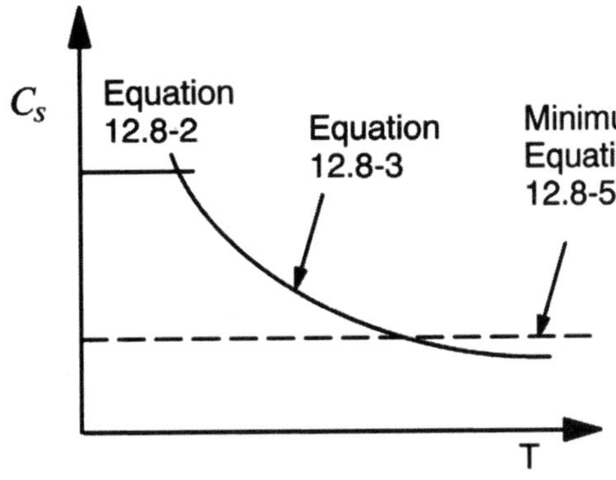

Response Spectrum

Therefore, Equation 12.8-2 controls the base shear calculation for a short period building and the seismic coefficient is

$$V = 0.280g$$

2. Base shear in transverse direction

This building has a flexible roof diaphragm and heavy CMU walls (see Figure 4-3). The diaphragm spans as a simple beam between resisting perimeter walls in both directions and will transfer 50 percent of the diaphragm shear to each resisting wall. However, in a building that is not symmetric or does not have symmetric wall layouts, the wall lines could have slightly different wall shears on opposing wall lines 1 and 3 and also on A and D.

The building weight (mass) calculation is separated into three portions: the roof, longitudinal walls, and transverse walls, for ease of application at a later stage in the calculations. The reason for separating the CMU wall masses is that masonry walls that resist ground motions parallel to their in-plane directions resist their own seismic inertia without transferring seismic forces into the roof diaphragm. This concept will be demonstrated in the example below for the transverse (N-S) direction.

For the transverse direction, the roof diaphragm resists seismic inertia forces originating from the roof diaphragm and the longitudinal masonry walls (out-of-plane walls oriented east-west) on lines 1 and 3, which are oriented perpendicular to the direction of seismic ground motion. The roof diaphragm then transfers its seismic forces to the transverse masonry walls (in-plane walls oriented north-south) located on lines A and D. The transverse walls resist seismic forces transferred from the roof diaphragm and seismic forces generated from their own weight. Thus, seismic forces are generated from three sources: the roof diaphragm, in-plane walls at lines 1 and 3, and out-of-plane walls at lines A and D.

The design in the orthogonal direction is similar and the base shear is the same; however, the proportion of diaphragm and in-plane seismic forces is different. The orthogonal analysis is similar in concept, and thus is not shown in this example.

Roof weight

$$W_{roof} = 17 \text{ psf } (5400 \text{ sf}) = 92 \text{ kips}$$

Longitudinal wall weight (out-of-plane walls)

Note that the upper half of the wall weight is tributary to the roof diaphragm. This example neglects openings in the top half of the walls.

$$W_{walls,\,long.} = 75 \text{ psf}(2 \text{ walls})(90 \text{ ft})(19 \text{ ft})\left(\frac{19 \text{ ft}}{2}\right)\left(\frac{1}{16 \text{ ft}}\right) = 75 \text{ psf } (180 \text{ ft})\frac{(19 \text{ ft})^2}{2(16 \text{ ft})} = 152 \text{ kips}$$

For forces in the transverse direction, seismic inertial forces from the transverse walls (lines A and D) do not transfer through the roof diaphragm. Therefore, the effective diaphragm weight in the N-S direction is

$$W_{trans.diaph.} = W_{roof} + W_{walls,\,long.} = 92 \text{ k} + 152 \text{ k} = 244 \text{ kips}$$

The transverse seismic inertial force (shear force), which is generated in the roof diaphragm, is calculated as follows

$$V_{trans.diaph.} = 0.280W_{trans.diaph.} = 0.280(244 \text{ kips}) = 68 \text{ kips}$$

The seismic inertial force (shear force) generated in the transverse walls (in-plane walls) is calculated using the full weight (and height) of the walls (with openings ignored for simplicity).

$$V_{trans.walls} = 0.280(75 \text{ psf})(19 \text{ ft}) (60 \text{ ft})(2 \text{ walls}) = 48 \text{ kips}$$

The design base shear in the transverse direction is the sum of the shears from the roof diaphragm shear and the masonry walls in-plane shear forces.

$$V_{trans.} = V_{trans.diaph} + V_{trans.walls} = 68 \text{ k} + 48 \text{ k} = 116 \text{ kips}$$

3. Shear wall on Line A.

The seismic shear tributary to the wall on Line A comes from the roof diaphragm (transferred at the top of the wall) and the in-plane wall inertia force

$$V_A = \frac{V_{trans.diaph.}}{2} + \frac{V_{trans.walls}}{2} = \frac{68 \text{ kips}}{2} + \frac{48 \text{ kips}}{2} = 58 \text{ kips}$$

4. Design of shear wall on Line A for out-of-plane seismic forces

In this part, the 8-foot shear wall on line A (Figure 4-4) will be designed for out-of-plane seismic forces. This wall is a bearing wall and must support gravity loads. It must be capable of supporting both gravity and out-of-plane seismic forces, and gravity plus in-plane seismic forces at different instants depending on the direction of seismic ground motion. Thus, in this part, the first of these two analyses will be performed.

The analysis will use the "slender wall" design provisions of MSJC §3.3.5. The analysis incorporates static plus $P\Delta$ deflections caused by combined gravity loads and out-of-plane seismic forces, and calculates an axial plus bending capacity for the wall under the defined load.

4a. Vertical loads.

Gravity loads from roof framing tributary to the 8-foot shear wall at line A

$$P_{DL} = (17 \text{ psf})\left(\frac{60 \text{ ft}}{2}\right)\left(\frac{30 \text{ ft}}{2}\right) = 7650 \text{ lb}$$

Live load reduction for roof loads §4.9

$$A_t = (30 \text{ ft})(15 \text{ ft}) = 450 \text{ sf}$$ Eq 4-2

$$L_r = L_O R_1 R_2 \text{ where } 12 \le L_r \le 20$$

$$R_1 = 1.2 - 0.001\, A_t \text{ for } 200 \text{ ft}^2 \le A_t \le 600\, A_t$$

$$= 1.2 - 0.001\,(450) = 0.75$$

$$R_2 = 1.0 \text{ for } F \le 4$$

$$L_r = (20 \text{ psf})\,(0.75)\,(1.0) = 15 \text{ psf}$$

The reduced live load is

$$L_r = 20 \text{ psf}\,(0.75)\,(1.0) = 15.0 \text{ psf}$$

$$P_{RLL} = (15.0 \text{ psf})\left(\frac{60 \text{ ft}}{2}\right)\left(\frac{30 \text{ ft}}{2}\right) = 6750 \text{ lb}$$

Reactions on center roof beam on CMU wall

$$P_{beam\ D+L} = 7650 \text{ lb} + 6750 \text{ lb} = 14{,}400 \text{ lb}$$

$$P_{beam\ D} = 7650 \text{ lb}$$

Wall load on 8-foot wall (at mid-height)

$$P_{wall\ DL} = (75 \text{ psf})(8 \text{ ft})\left(\frac{16 \text{ ft}}{2} + 3 \text{ ft}\right) = 6600 \text{ lb}$$

$$w_{wall\ DL} = \frac{6600 \text{ lb}}{8 \text{ ft}} = 825 \text{ plf}$$

Dead load from wall lintels

$$P_{lintel\ D} = (75 \text{ psf})(9 \text{ ft})\left(\frac{20 \text{ ft}}{2}\right) = 6750 \text{ lb}$$

The gravity loads on the 8-foot wall from the weight of the wall, the roof beam, and the two lintels are

$$\Sigma P_{DL} = (6600 \text{ lb} + 7650 \text{ lb} + 6750 \text{ lb} + 6750 \text{ lb}) = 27{,}750 \text{ lb}$$

$$\Sigma P_{RLL} = 6750 \text{ lb}$$

4b. Seismic forces

Out-of-plane seismic forces are calculated using the provisions of §12.11.1.

$$F_p = 0.40 S_{DS} I W_w$$

$$F_p = (0.40)(1.0)(1.40g) W_w$$

$$F_p = 0.56 W_w$$

$$F_p = 0.56(75 \text{ psf}) = 42 \text{ psf}$$

Calculation of wall moments due to out-of-plane forces assumes a simply-supported beam with a 3-foot cantilever.

$$W_1 = (10 + 8 + 10 \text{ ft})(42 \text{ psf}) = 1176 \text{ plf}$$

$$W_2 = 8 \text{ ft } (42 \text{ psf}) = 336 \text{ plf}$$

Using simple beam theory to calculate moment M_{oop} for out-of-plane forces, the location of maximum moment is at $h = 9.83$ feet

$$M_{oop} = 16{,}225 \text{ lb-ft} = 194{,}700 \text{ lb-in}$$

Comparison of seismic out-of-plane forces with wind (approximately 25 psf) indicate that seismic forces control the design.

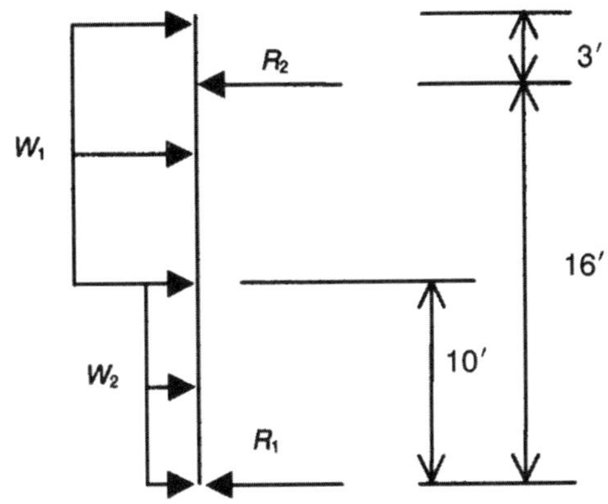

Figure 4-7. Propped cantilever loading diagram

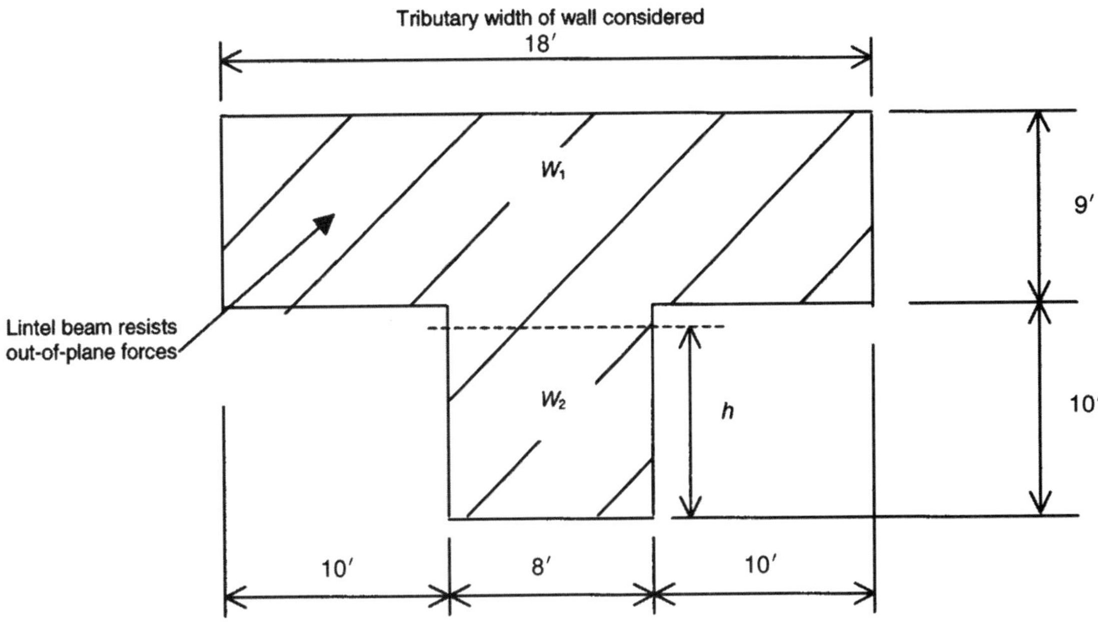

Figure 4-8. Tributary width of wall for out-of-plane seismic inertial force calculations

4c. Design for out-of-plane forces §12.11.1

The wall section shown in Figure 4-6 will be designed. The controlling load combination for dead load plus seismic load is §2.3.2

Comb. 5. $(1.2D + 1.0E) = 1.2D + (E_h + E_v)$

$E_v = 0.2S_{DS}D = 0.2(1.40g)D = 0.28D$

$P_u = 1.2\,P_D + E_v$

$P_u = 1.2(27{,}750 \text{ lb}) + (0.28)(27{,}750 \text{ lb}) = 41{,}070 \text{ lb}$

Masonry walls with an axial load of $0.2f'_m$ or less are designed under the requirements of MSJC §3.3.5.4

Check axial load vs. $0.2f'_m$ using factored loads

$$\frac{(P_{uw}+P_{uf})}{A_g} \le 0.2f'_m \qquad \text{MSJC Eq 3-23}$$

$$\therefore \frac{41{,}700 \text{ lb}}{(7.625 \text{ in})(8 \text{ ft})(12 \text{ in})} = 56 \text{ psi} \le 0.2(2500 \text{ psi}) = 500 \text{ psi} \qquad \therefore o.k.$$

Calculate equivalent area of steel area A_{se}

$$A_{se} = \frac{A_s f_y + P_u}{f_y}$$

$$A_{se} = \frac{(0.31 \text{ in}^2)(6 \text{ bars})(60{,}000 \text{ psi}) + 47{,}700 \text{ lb}}{60{,}000 \text{ psi}} = 2.55 \text{ in}^2$$

Calculate I_{cr}

$$a = \frac{P_u + A_s f_y}{0.80 f'_m b} = \frac{41{,}700 \text{ lb} + (1.86 \text{ in}^2)(60{,}000 \text{ psi})}{0.80(2500 \text{ psi})(96 \text{ in})} = 0.80 \text{ in} \qquad \text{MSJC Eq 3-28}$$

$$c = \frac{a}{0.80} = 1.0 \text{ in}$$

$$E_m = 900 f'_m = 2{,}250{,}000 \text{ psi}$$

$$n = \frac{E_s}{E_m} = \frac{29{,}000{,}000 \text{ psi}}{2{,}250{,}000 \text{ psi}} = 12.89$$

$$I_{cr} = \frac{bc^3}{3} + nA_{se}(d-c)^2 = \frac{96 \text{ in } (1.0 \text{ in})^3}{3}$$

$$+ (12.89)(2.55 \text{ in})(3.81 \text{ in} - 1.0 \text{ in})^2 = 291.5 \text{ in}^4$$

Calculate M_{cr} using the value for f_r from MSJC Table 3.1.8.2.1

$$M_{cr} = S_n f_r = \left[\frac{96 \text{ in } (7.625 \text{ in})^2}{6}\right] 145 = 134{,}886 \text{ lb-in} \qquad \text{MSJC Eq 3-32}$$

f_r = 145 psi from MSJC Table 3.1.8.2.1 for normal to bed joints in running bond with type N mortar.

Calculate I_g

$$I_g = \left[\frac{96 \text{ in } (7.625 \text{ in})^3}{12}\right] = 3546.6 \text{ in}^4$$

Calculate M_u based on MSJC Equation 3-24

First iteration for moment and deflection (note that eccentric moment at mid-height of wall is one-half of the maximum moment)

For δ_u

$$M_u = M_{oop} + M_{ecc} = E + 1.48D + 1.6(L = 0)$$

$$M_u = M_{oop} + M_{ecc} = 194{,}700 \text{ lb-in} + 1.48(7650\text{lb})(6 \text{ in})/2 \qquad \text{MSJC Eq 3-24}$$

$$M_u = 228{,}666 \text{ lb-in}$$

$$\delta_u = \frac{5M_{cr}h^2}{48E_m I_g} + \frac{5(M_u - M_{cr})h^2}{48E_m I_{cr}} \qquad \text{MSJC Eq 3-31}$$

$$\delta_u = \frac{5(134{,}886 \text{ lb-in})(192 \text{ in})^2}{48(2{,}250{,}000 \text{ psi})(3546.6 \text{ in}^4)} + \frac{5(228{,}666 \text{ lb-in} - 134{,}886 \text{ lb-in})(192 \text{ in})^2}{48(2{,}250{,}000 \text{ psi})(291.5 \text{ in}^4)}$$

$$\delta_u = 0.065 \text{ in} + 0.549 \text{ in} = 0.61 \text{ in}$$

Note: The deflection equation used is for uniform lateral loading, maximum moment at mid-height, and pinned-pinned boundary conditions. For other support and fixity conditions, moments and deflections should be calculated using established principals of mechanics. Beam deflection equations can be found in the AITC or AISC manuals, or more accurate methods can be derived.

Second iteration for moment and deflection

$$M_u = 228{,}666 \text{ lb-in} + 41{,}070 \text{ lb } (0.61 \text{ in}) = 260{,}085 \text{ lb-in}$$

$$\delta_u = 0.065 \text{ in} + \frac{5(260{,}085 \text{ lb-in} - 134{,}886 \text{ lb-in})(192 \text{ in})^2}{48(2{,}250{,}000 \text{ psi})(291.5 \text{ in}^4)}$$

$$= 0.065 \text{ in} + 0.733 \text{ in} = 0.80 \text{ in}$$

Third iteration for moment and deflection

$$M_u = 228{,}666 \text{ lb-in} + 41{,}070 \text{ lb } (0.80 \text{ in}) = 261{,}440 \text{ lb-in}$$

$$\delta_s = 0.065 \text{ in} + \frac{5(261{,}440 \text{ lb-in} - 134{,}886 \text{ lb-in})(192 \text{ in})^2}{48(2{,}250{,}000 \text{ psi})(291.5 \text{ in}^4)}$$

$$= 0.065 \text{ in} + 0.741 \text{ in} = 0.81 \text{ in}$$

Final moment (successive iterations are producing moments within 3 percent, therefore convergence can be determined)

$$M_u = 228{,}666 \text{ lb-in} + 41{,}070 \text{ lb } (0.87 \text{ in}) = 261{,}768 \text{ lb-in}$$

Calculation of wall out-of-plane strength

From MSJC Section 3.1.9

$$\phi = 0.90$$

$$\phi M_n = \phi(A_s f_y + P_u)\left(d - \frac{a}{2}\right) = \phi A_{se} f_y \left(d - \frac{a}{2}\right)$$

$$\phi M_n = 0.90(2.55 \text{ in}^2)(60{,}000 \text{ psi})\left(3.81 \text{ in} - \frac{0.81 \text{ in}}{2}\right) = 469{,}554 \text{ lb-in}$$

$$\phi M_n \geq M_u$$

$$469{,}554 \text{ lb-in} \geq 261{,}768 \text{ lb-in} \qquad \therefore \textit{o.k.}$$

Because the wall strength is greater than the demand, the wall section shown in Figure 4-4 is okay.

Note that out-of-plane deflections need to be checked using the same iteration process, but with service loads per MSJC §3.3.5.5, (i.e., $P_D = 27{,}750$ lb). Because ultimate deflections are within allowable limits, there is no need to check service deflections in this example. The limiting deflection of $0.007h$ per MSJC §3.3.5.5 is $0.007(16 \text{ ft} \times 12 \text{ in}) = 1.34$ in. The deflection from strength design is 0.81 inch. Thus the deflection is within allowable limits.

Check the maximum wall reinforcement percentage.

Check the unbraced parapet moment

$$M_u = (42 \text{ psf})(3 \text{ ft})^2/2 = 189 \text{ lb-ft} = 2268 \text{ lb-in} \leq 469{,}554 \text{ lb-in}$$

$\therefore$ Wall section is *o.k.* at parapet.

5. Design of shear wall on line A for in-plane seismic forces.

5a. Shear force distribution

The shear force on line A must be distributed to three shear wall piers (6, 8, and 6 feet in width, respectively) in proportion to their relative rigidities. This can be accomplished by assuming that the walls are fixed at the tops by the 9-foot-deep lintel. Reference deflection equations are given below for CMU or concrete walls with boundary conditions fixed top or pinned top. For this example, the fixed/fixed equations are used because the deep lintel at the wall/pier tops will act to fix the tops of wall piers.

$$\Delta_i = \frac{V_i h^3}{12 E_m I} + \frac{1.2 V_i h}{AG} \quad \text{for walls/piers fixed top and bottom}$$

$$\Delta_i = \frac{V_i h^3}{3 E_m I} + \frac{1.2 V_i h}{AG} \quad \text{for walls/piers pinned top and fixed at bottom}$$

Create modulus of rigidity

$$G = E_v = 0.4 E_m \text{ for concrete masonry per MSJC §1.8.2.2.2}$$

Thus, relative rigidity is $\frac{1}{\Delta}$ where Δ is the deflection under load V_i. Using the fixed/fixed equation, the percentage shears to each wall are as follows

Table 4-1. Distribution of line A shear to three shear walls.

Wall Length (ft)	Moment Deflection (in)	Shear Deflection (in)	Total Deflection (in)	Rigidity (1/in)	Distribution to Piers (%)	Wall Shear (k)
6	1.17E-05	3.50E-07	1.20E-05	83,283	26.6	15.4
8	6.56E-06	2.62E-07	6.82E-06	146,635	46.8	27.2
6	1.17E-05	3.50E-07	1.20E-05	83,283	26.6	15.4
Totals				313,200	100	58.0

The seismic shear force E_h to the 8-foot pier is (0.468) 58 k = 27.2 k

Calculation of reliability/redundancy factor ρ is shown below. Per Table 12.3-3, if the removal of a shear wall or wall pier with a height-to-length ratio greater than 1.0 within any story, or collector connections thereto, would not result in more than a 33% reduction in story strength, nor does the resulting system have an extreme torsional irregularity, then ρ = 1.0. §12.3.4

Each wall line resists 50% of the seismic base shear. There is no torsional response because the diaphragm is flexible. The heaviest loaded shear wall resists 47% × 50% = 24%, which is less than 35%. Therefore δ = 1.0.

The strength design shear for the 8-foot wall is

$$\therefore\ V_{8'wall} = E_h = \rho Q_E = 1.0(27.2\text{ k}) = 27.2\text{ k}$$ Eq 12.4-3

5b. Determination of shear strength

The in-plane shear strength of the wall must be determined and compared to demand. The strength of the wall is determined as follows: Vertical reinforcement is #5 @ 16 inches o/c. Try #4 @ 16 inches o/c horizontally. Note that concrete masonry cells are spaced at 8-inch centers, thus reinforcement arrangements must have spacings in increments of 8 inches (such as 8, 16, 24, 32, 40, and 48 inches). Typical reinforcement spacings are 16 and 24 inches for horizontal and vertical reinforcement.

$$V_n = V_m + V_s$$ MSJC Eq 3-18

$$V_m = \left[4.0 - 1.75\left(\frac{M_u}{V_u d_v}\right)\right] A_n \sqrt{f'_m} + 0.25 P_u$$ MSJC Eq 3-21

$\frac{M_u}{V_u d_v}$ need not be taken as greater than 1.0

$$M_u = 27.2\text{ k} \times 10\text{ ft} = 272\text{ kft}$$

$$V_u = 27.2\text{ k}$$

$$d_v = 8\text{ ft} = 96\text{ in}$$

$$\frac{M_u}{V_u d_v} = \frac{272 \text{ kft } (12 \text{ in})}{(27.2 \text{ k})(96 \text{ in})} = 1.25 \geq 1.0 \qquad \therefore = 1.0$$

$$V_m = [4.0 - 1.75(1.0)]\,(7\tfrac{5}{8} \text{ in})\,(96 \text{ in})\,\sqrt{2500 \text{ psi}} + 0.25\,(41{,}070 \text{ lb})$$

$$= 92.6 \text{ k}$$

$$V_s = 0.5\left(\frac{A_v}{s}\right) f_y d_v$$

$$= 0.5\left(\frac{0.20 \text{ in}^2}{16 \text{ in}}\right)(60{,}000 \text{ psi})(96 \text{ in})$$

$$= 36.0 \text{ k}$$

$$V_n = V_m + V_s$$

$$= 92.6 \text{ k} + 36.0\text{k} = 128.6 \text{ k}$$

Thus, using $\phi = 0.80$

$$\phi V_n = 0.80\,(128.6 \text{ k}) = 100.9 \text{ k}$$

$$\therefore\ 100.9 \text{ k} \geq 27.2 \text{ k} \qquad \therefore\ o.k.$$

$$\text{DCR} = \frac{27.2 \text{ k}}{100.9 \text{ k}} = 0.27$$

The designer should check the failure mode and, ideally, should design the shear wall to yield in bending prior to yield in shear. The method of checking the failure mode is to determine how much moment M_u is generated when the shear force is equal to shear strength V_n with $\phi = 1.0$. Then, that moment is compared with the wall P_n and M_n with a $\phi = 1.0$. If there is reserve moment capacity, there will be a shear failure. If not, there will be ductile bending yielding. Later in the example, this will be checked.

The failure mode should be checked to understand whether a brittle shear failure will occur or ductile bending yielding. Since the bending yielding is more desirable.

∴ Use #4 @ 16-inch horizontal reinforcement in the wall/pier.

6. Design shear wall on Line A for combined axial and in-plane bending actions

Part 5 illustrated the design of the wall for shear strength. This part illustrates design for wall overturning moments combined with gravity loads. A free body diagram of the wall/pier is needed to understand the imposed forces on the wall.

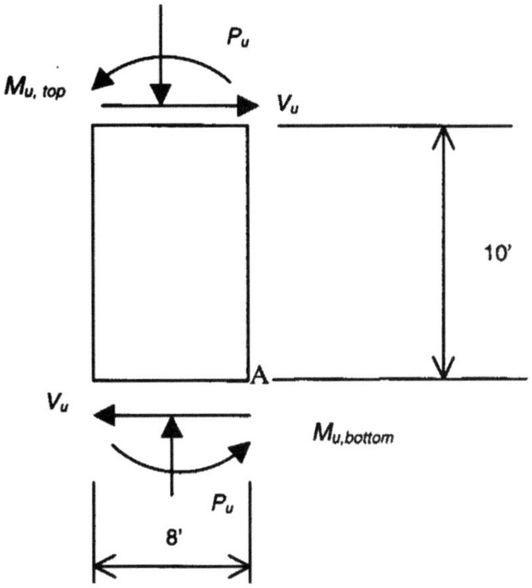

Figure 4-9. Free body diagram of 8-foot shear wall

The load combinations to be considered are specified in §12.4.2.3. These are

(Comb. 5) $1.2D + L + 1.0E = (1.2 + 0.2\,S_{DS})\,D + L + \delta Q_E$

(Comb. 7) $0.9D + 1.0E = (0.9 - 0.2\,S_{DS})\,D + \rho Q_E$

The resulting equation (5) is

$$1.2D + 0.2(1.4\text{ g})D + 1.0\,\rho Q_E = 1.48D + \rho Q_E$$

The resulting equation (7) is

$$0.9D - 0.28D - 1.0\,\rho Q_E = 0.62D - \rho Q_E$$

$$\rho Q_E = 1.0(27.2\text{ kips}) = 27.2\text{ kips}$$

Axial loads P_u are calculated as P_{u1} and P_{u2} for load combinations 5 and 7

$$P_{u1} = 1.48(27{,}750\text{ lb}) = 41.1\text{ kips}$$

$$P_{u2} = 0.62(27{,}750\text{ lb}) = 17.2\text{ kips}$$

By performing a sum of moments about the bottom corner at point A (Figure 4-9)

$$\Sigma M_A = 0 = 2M_u - V_u\,(10\text{ ft})$$

$$M_{u,top} \approx M_{u,bottom} = \frac{(27.2\text{ k }(10\text{ ft}))}{2} = 136.0\text{ kips-ft}$$

The reader is referred to an excellent book for the strength design of masonry *Design of Reinforced Masonry Structures*, by Brandow, Hart, Verdee, published by Concrete Masonry Association of California and Nevada, Sacramento, CA, Second Edition, 1997. This book describes the calculation of masonry wall/pier strength design in detail.

The *P-M* diagram for the wall must be calculated. For this, the designer must understand the controlling strain levels that define yielding and ultimate strength. At yield moment, the steel strain is the yielding strain (0.00207 inch/inch strain) and the masonry strain must be below 0.002 inch/inch (for under-reinforced sections). At ultimate strength, the masonry has reached maximum permissible strain (0.0025 inch/inch) and the steel strain is considered beyond yield strain level (see MSJC §3.3.2 for a list of design assumptions).

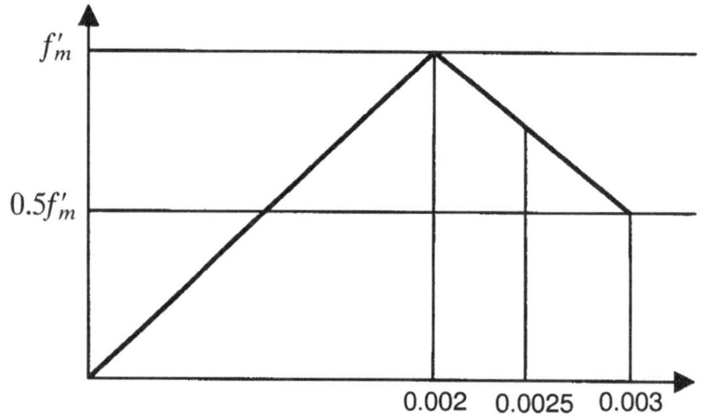

Figure 4-10. Assumed masonry compressive stress vs. strain curve

A representation of these strain states is shown in Figures 4-11 and 4-12 (the pier width is defined as *h*).

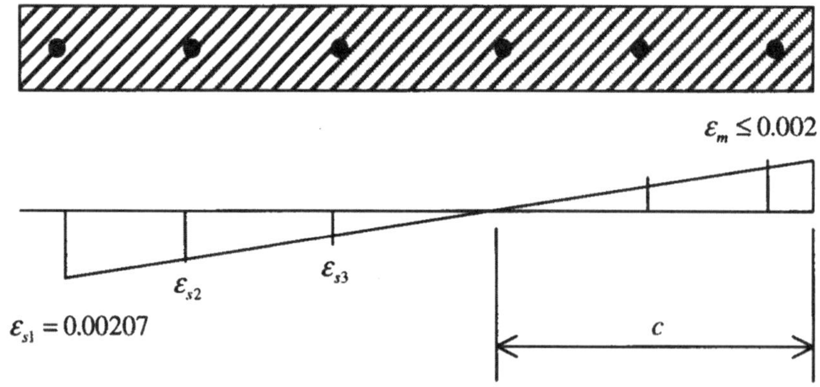

Figure 4-11. Strain diagram at yield moment. Steel strain = 0.00207 inch/inch. Masonry strain is less than yield for under-reinforced sections

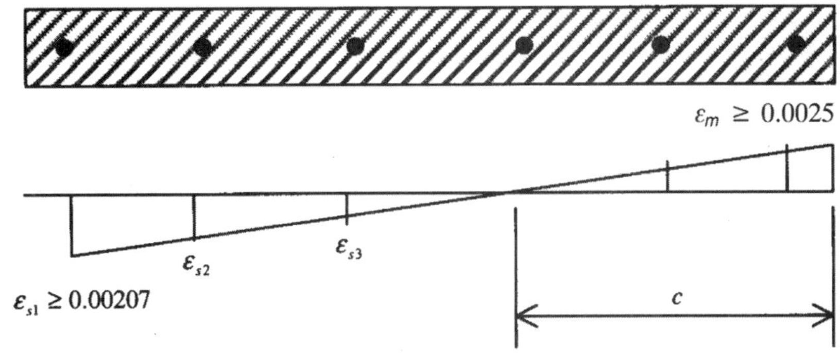

Figure 4-12. Strain diagram at ultimate moment. Masonry strain = 0.0025 inch/inch. Steel strain has exceeded 0.00207 inch/inch. The Whitney stress block analysis procedure can be used to simplify calculations

Note: Masonry strain may continue to increase with a decrease in stress beyond strains of 0.002 inch/inch at which time stresses are at f'_m. At strains of 0.0025, masonry stresses are $0.5 f'_m$. With boundary element confinement, masonry strains can be as large as 0.006 inch/inch.

By performing a summation of axial forces F, the axial load in the pier is calculated as

$$\Sigma F = P - C_1 + T_1 + T_2 + T_3$$

The corresponding yield moment is calculated as follows

$$M_y = T_1\left(d_1 - \frac{h}{2}\right) + T_2\left(d_2 - \frac{h}{2}\right) + T_3\left(d_3 - \frac{h}{2}\right) + C\left(\frac{h}{2} - \frac{c}{3}\right)$$

The ultimate moment is calculated as

$$M_u = T_1\left(d_1 - \frac{h}{2}\right) + T_2\left(d_2 - \frac{h}{2}\right) + T_3\left(d_3 - \frac{h}{2}\right) + C\left(\frac{h}{2} - \frac{a}{2}\right)$$

Strength-reduction factor for in-plane flexure $\phi = 0.90$

The balanced axial load P_b is determined as

$$P_b = f'_m b a_b$$

$$a_b = 0.80\, d\left(\frac{e_m}{e_m + \dfrac{f_y}{E_s}}\right)$$

$$P_b = 0.80(2500)(7.625 \text{ in})(0.80)(92 \text{ in})(0.0025/0.00457) = 767 \text{ kips}$$

$$\phi P_b = 0.90(767 \text{ kips}) = 690 \text{ kips}$$

A *P-M* diagram can thus be developed. The *P-M* diagrams were calculated and plotted using a spreadsheet program. By observation, the design values P_u and M_u ($P_u = 41$ k, $M_u = 136$ k-ft) are within the nominal strength limits of ϕP_n, ϕM_n values shown in Figure 4-13.

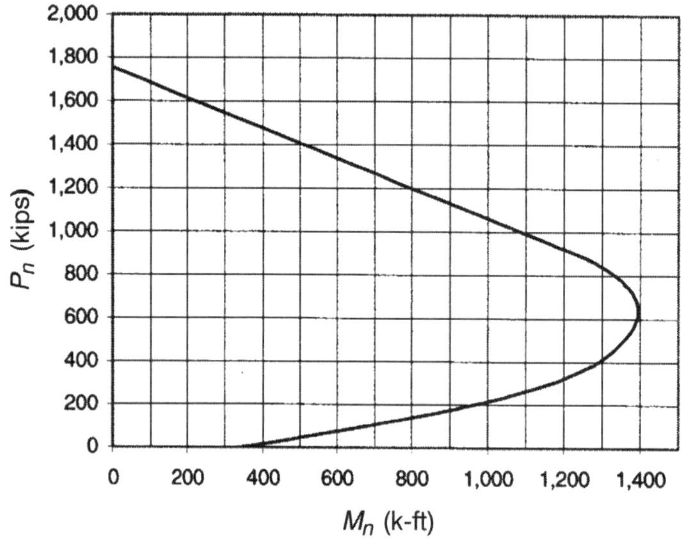

Figure 4-13. The P_n-M_n non-reduced strength curve with masonry strain at 0.0025 inch/inch

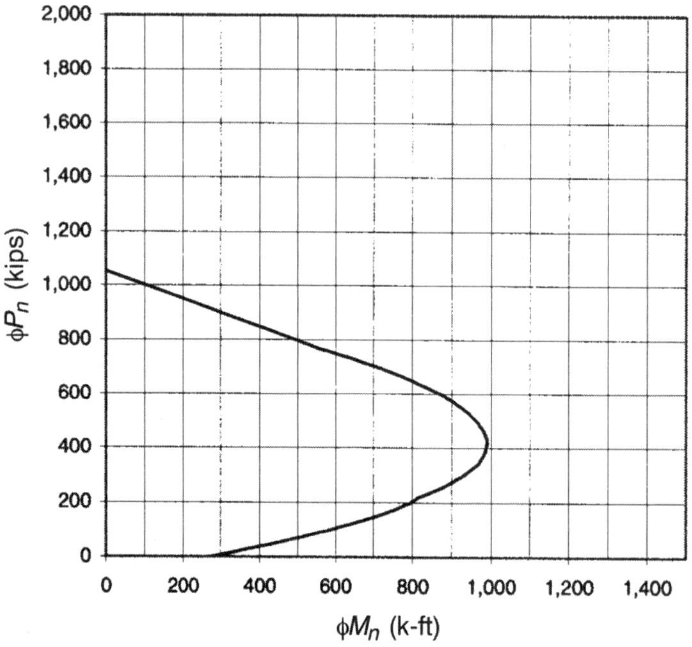

Figure 4-14. The ϕP_n-ϕM_n design strength curve with masonry strain at 0.0025 inch/inch

Check for type of wall failure by calculating wall moment at shear V_n

$$M_u = \frac{V_n(10\text{ ft})}{2} = \frac{(27.2\text{ k})(10\text{ ft})}{2} = 136\text{ k-ft}$$

$$P_u = 41.1\text{ k}$$

Looking at the P_n-M_n curve; this P_u-M_u load is inside the P_n-M_n curve. The DCR on $P + M$ is approximately 3.5. The DCR on shear is approximately 4.0. Therefore a ductile bending failure is possible.

7. Deflection of shear wall on Line A. §12.8.6

In this part, the deflection of the shear wall on line A will be determined. This is done to check actual deflections against the drift limits of §12.12.1.

Deflections based on gross properties are computed as

$$\delta_x = \frac{C_d \delta_{xe}}{I} \qquad \text{(Eq 12.8-15)}$$

$$\delta_{xe} = \frac{V_i h^3}{12E_m I} + \frac{1.2V_i h}{AG} \quad \text{for wall/piers fixed top and bottom}$$

$$\delta_{xe} = \frac{(27.2\text{ k})(120\text{ in})^3}{12(2250\text{ ksi})(562{,}176\text{ in}^4)} + \frac{1.2(27.2\text{ k})(120\text{ in})}{(732\text{ in}^2)(900\text{ ksi})} = 0.08\text{ in}$$

Assume cracked section properties and $I_{cr} = 0.3I_g$ (approximately):

$$\delta_{xe} = \frac{(27.2\text{ k})(120\text{ in})^3}{12(2250\text{ ksi})(168{,}652\text{ in}^4)} + \frac{1.2(27.2\text{ k})(120\text{ in})}{(732\text{ in}^2)(900\text{ ksi})} = 0.015\text{ in}$$

$$\delta_x = (3\tfrac{1}{2})(0.015\text{ in})/1.0 = 0.05\text{ in}$$

Thus, deflections are less than $\Delta_a = 0.010h_{sx} = 0.010(120\text{ in}) = 1.2\text{ in}$, $\therefore$ *o.k.*

8. Recommendations for shear wall boundary elements

The 1997 *Uniform Building Code* requires boundary elements for CMU shear walls with strains exceeding 0.002 inch/inch from a wall analysis with $R = 1.5$. The intent of masonry boundary elements is to help the masonry achieve greater compressive strains (up to 0.006 inch/inch) without experiencing a crushing failure.

The axial load and moment associated with this case is $P_u = 41.1$ kips

$$P_u = 41.1\text{ kips}$$

$$M_u = \frac{5.0}{1.5}(136.0\text{ kip-ft}) = 453\text{ k-ft}$$

This P, M point is not within the P-M curve using a limiting masonry strain of 0.002 inch/inch (see Figure 4-15). From an analysis it can be determined that the maximum c distance to the neutral axis is approximately 22 inches. For this example, boundary ties are required. Note that narrow shear wall performance is greatly increased by the use of boundary ties.

It is recommended that boundary elements have a minimum dimension of 3x wall thickness, which is 24 inches because of yield moments. After yield moment capacity is exceeded, the c distance is reduced. Thus, if boundary element ties are provided at each end of the wall/pier extending 24 inches inward, the regions experiencing strain greater than 0.002 inch/inch are confined. Space boundary ties at 8-inch centers. The purpose of masonry boundary ties is not to confine the masonry for compression, but to support the reinforcement during compression to prevent buckling. Tests have been performed to show that masonry walls can achieve 0.006 inch/inch compressive strains when boundary ties are present.

The *P-M* curve shown in Figure 4-15 is derived by setting masonry strain at the compression edge at 0.002 inch/inch and by increasing the steel tension strain at the opposite wall reinforcement bars. Moments are calculated about the center of the wall pier and axial forces are calculated about the cross section. *P-M* points located at the outside of the denoted *P-M* boundary element curve will have masonry strains exceeding the allowable, and thus will require boundary element reinforcement or devices.

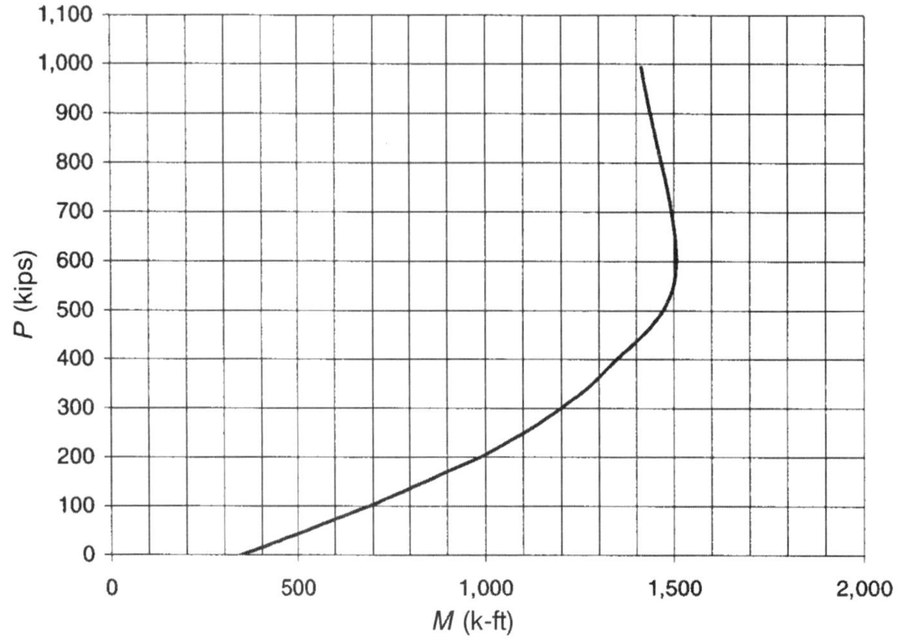

Figure 4-15. *P-M* curve for boundary element requirements with masonry strain at 0.002 inch/inch

It can be seen that boundary reinforcement is very close to being required for the point (P_u = 41 k, M_u = 453 k). To be conservative, the boundary element steel is recommended for this example.

Boundary element confinement ties may consist of #3 or #4 closed reinforcement in 10- and 12-inch CMU walls. At 8-inch CMU walls, prefabricated products such as the "masonry comb" are the best choice for boundary reinforcement because these walls are too narrow for reinforcement ties (even #3 and #4 bars). The boundary reinforcement should extend around three vertical #4 bars at the ends of the wall.

9. Wall-roof out-of-plane anchorage for Lines 1 and 3. (§1620.2.1)

CMU walls should be adequately connected to the roof diaphragm around the perimeter of the building. In earthquakes, including the 1994 Northridge event, a common failure mode was the separation of heavy walls and roofs leading to partial collapse of roofs. A SEAOC-recommended connection spacing is 8 feet maximum, however, 6 feet or 4 feet might be more appropriate and should be considered for many buildings. This anchorage should also be provided on lines A and D, which will require similar but different details at the roof framing perpendicular to wall tie condition. Section 12.11.2.2.1 requires that diaphragm struts or ties be provided, which transfer the out-of-plane anchorage forces through the roof diaphragm. Diaphragm design is presented in Design Example 5.

Per §12.11.2.1, elements of the wall out-of-plane anchorage system, in SDCs C, D, E, or F shall be designed for the forces

$$F_p = 0.8S_{DS}IW_p \qquad \text{Eq 12.11-1}$$

$$F_p = 0.8(1.0)(1.4g)W_p = 1.12W_p$$

$$F_p = 1.12(75 \text{ psf}) = 84 \text{ psf}$$

where W_p is the wall weight of 75 psf

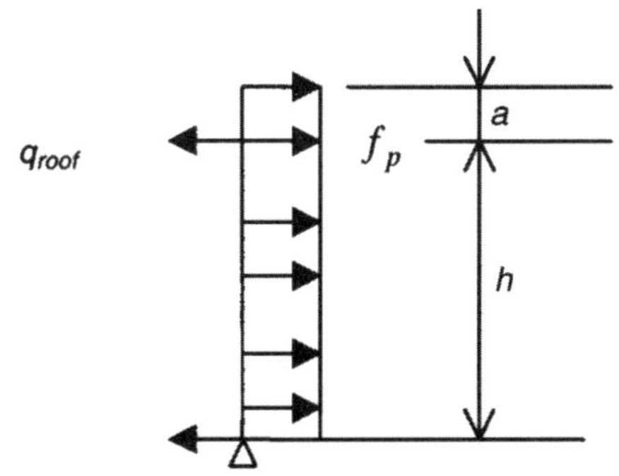

Figure 4-16. Wall-roof connection loading diagram

Calculation of the reaction at the roof level is

$$q_{roof} = \frac{w_p(h+a)^2}{2h} = \frac{(84 \text{ psf})(16 \text{ ft} + 3 \text{ ft})^2}{2(16 \text{ ft})} = 948 \text{ plf}$$

The anchorage force shall not be less than a minimum of $400S_{DS}I$ (in pounds per linear foot of wall anchorage) per §12.11.2.

$$q_{roof} \geq 400S_{DS}I = 400(1.40g)(1.0) = 560 \text{ plf}$$

$$q_{roof} = 947 \text{ plf} \geq 560 \text{ plf} \qquad \therefore o.k.$$

Additionally, the anchorage force shall not be less than 280 lb/ft.

∴ The anchorage force of 947 lb/ft is greater than 280 lb/ft.

Thus, the design anchorage reaction at different anchor spacings is

$T = C$ at 4-foot centers, $q_{roof} = 3{,}792$ lb

$T = C$ at 6-foot centers, $q_{roof} = 5{,}688$ lb

$T = C$ at 8-foot centers, $q_{roof} = 7{,}584$ lb

Therefore, choose wall-roof anchors that will develop the required force at the chosen spacing. The roof diaphragm must also be designed to resist the required force with the use of subdiaphragms (or other means). The subject of diaphragm design is discussed in Design Example 5.

For this example, a double hold-down connection spaced at 8-foot centers will be used. This type of connection must be secured into a solid roof framing member capable of developing the anchorage force.

First check anchor capacity in concrete block using the provisions of MSJC §3.1.6.2.

The required tension T for bolt embedment is $T = 7584$ lb.

The anchor bolts are spaced at 6⅝ inches center-to-center (considering purlin and hardware dimensions) and have 12-inch-diameter pull-out failure cones. Thus, the failure surfaces will overlap (Figure 4-17). In accordance with MSJC §3.1.6.2, the maximum tension of this bolt group may be determined as follows

Calculate B_{an} per bolt using the strength provisions of MSJC Equation (3-1)

$$\phi B_{an} = 4\phi A_{pt}\sqrt{f'_m} = 4(0.5)(113\text{ in}^2)(50\text{ psi}) = 11{,}300\text{ lb}$$ MSJC Eq 3-1

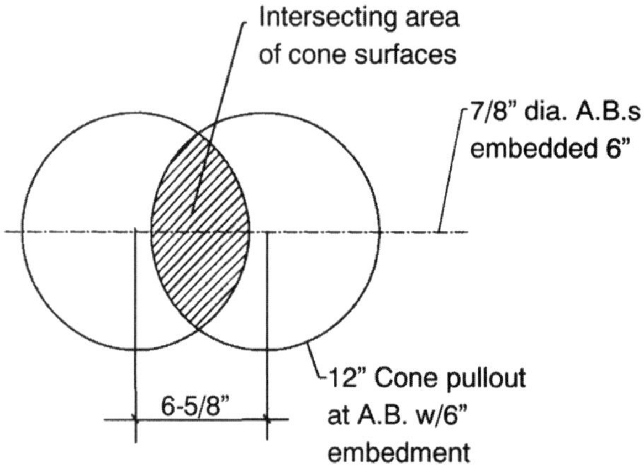

Figure 4-17. Intersection of anchor bolt tension failure cones

Calculate one-half the area of intersection of failure surfaces from two circles, each with a 6-inch radius and centers (2¹⁄₁₆ + 2½ + 2¹⁄₁₆) 6⅝ inches apart. $-A_{pt} = 37.8$ in^2. Thus the bolt group tension can be calculated as

$$2B_{an} = 4(0.5)(2 \times 113\text{ in}^2 - 2 \times 37.8\text{ in}^2/2)\text{ (50 psi)} = 18.8\text{ kips}$$

By choosing a pair of prefabricated hold-down brackets with adequate capacity for a double shear connection into a 2½-inch glued laminated framing member, the brackets are good for

$$T = 2 \times 3685 \text{ lb} = 7370 \text{ lb (ASD)} \div 0.7 = 10.5 \text{ kips (Strength)}$$

Thus, the brackets are okay.

Also check bolt adequacy in double shear hold-down connection with metal side plates (2½-inch main member, ⅞-inch bolts) per NDS Table 8.3B.

$$T = 2 \times 3060 \text{ lb} \times 1.6 = 9792 \text{ lb (ASD)} \div 0.7 = 14.0 \text{ kips (Strength)}$$

Therefore, the double shear bolts and prefabricated hold-down brackets can be used.

Thus, use two hold-down brackets on each side of a solid framing member connecting the masonry wall to the framing member with connections spaced at 8-foot centers.

Verify that the CMU wall can span laterally 8 feet between anchors. Assume a beam width of 6 feet (3-foot high parapet plus an additional 3 feet of wall below roof) spanning horizontally between wall-roof ties.

$$w = q_{roof} = 948 \text{ plf}$$

$$M_u = \frac{w\ell^2}{8} = \frac{(948 \text{ lf})(8 \text{ ft})^2}{8} = 7584 \text{ lb-ft}$$

The wall typically has #4 @ 16-inch horizontal reinforcement; therefore, a minimum four #4 bars in a 6-foot wall section.

$$a = \frac{A_s f_y}{0.80 f'_m b} = \frac{4(0.2 \text{ in}^2)(60{,}000 \text{ psi})}{0.80(2500 \text{ psi})(72 \text{ in})} = 0.333 \text{ in}$$

$$\phi M_n = \phi A_s f_y \left(d - \frac{a}{2} \right)$$

$$\phi M_n = 0.9(4)(0.20 \text{ in}^2)(60{,}000 \text{ psi}) \left(3.81 \text{ in} - \frac{0.333 \text{ in}}{2} \right) \left(\frac{1}{12 \text{ in}} \right)$$

$$= 13{,}117 \text{ lb-ft} \le 7584 \text{ lb-ft} \qquad \therefore \textit{o.k.}$$

Per §12.11.2.2.1, the wall-roof connections must be developed into the roof diaphragm with diaphragm nailing and subdiaphragm design.

The shell of the masonry unit wall next to the wood ledger should have a hole cored or drilled that allows for 1-inch grout all around the anchor bolt. Thus, for a ⅞-inch-diameter anchor bolt, the core hole is 2⅞ inches in diameter at the inside face masonry unit wall. The face shell thickness for 8-inch masonry is 1¼ inches; thus, the anchor bolt end distance to the inside face of the exterior shell is 7⅝ – 1¼ – 6 inches = ⅜ inch. It is recommended that the minimum clear dimension is ¼ inch if fine grout is used and ½ inch if coarse pea gravel grout is used (Figure 4-18).

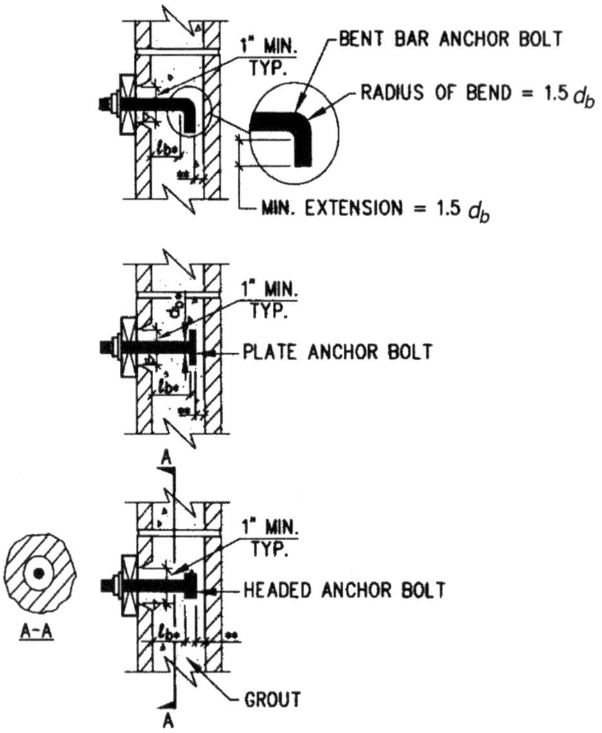

* MINIMUM EMBEDMENT LENGTH $\ell = 4d_b$ BUT ℓ_b MAY NOT BE LESS THAN 2"

** 1/4" FOR FINE GROUT
1/2" FOR COARSE (PEA GRAVEL) GROUT

Figure 4-18. Embedment of anchor bolts in CMU walls

10. Chord design

Analysis of transverse roof diaphragm chords is determined by calculation of the diaphragm simple span moment ($w\ell^2/8$) divided by the diaphragm depth based on the building base shear coefficient 0.280*W*.

$$w_{diaph.trans} = \frac{(116\text{ k})}{90\text{ ft}} = 1289\text{ plf}$$

$$M_{diaphragm} = w\ell^2/8 = 1289\text{ plf }(90\text{ ft})^2/8 = 1305\text{ k-ft}$$

$$T_u = C_u = 1305\text{ k-ft}/60\text{ ft} = 21.8\text{ kips}$$

Using reinforcement in the CMU wall for chord forces

$$A_{s,\ required} = \frac{T_u}{\phi f_y} = \frac{21.8\text{ k}}{(0.90)(60\text{ ksi})} = 0.40\text{ in}^2$$

Thus, two #5 chord bars ($A_s = 0.61\text{ in}^2$) are adequate to resist the chord forces. Place chord bars close to the roof diaphragm level. Since roof framing often is sloped to drainage, the chord placement is a matter of judgment.

The roof diaphragm must be designed for a minimum force of $0.2S_{DS}IW_{px}$ per §12.10.1.1. Also, in accordance with §12.10.1.1, the diaphragm shall be designed for a minimum force of $0.2S_{DS}IW_{px}$ or a maximum force of $0.4S_{DS}IW_{px}$. Since the diaphragm is designed for $0.28W_{px}$, the design satisfies all criteria.

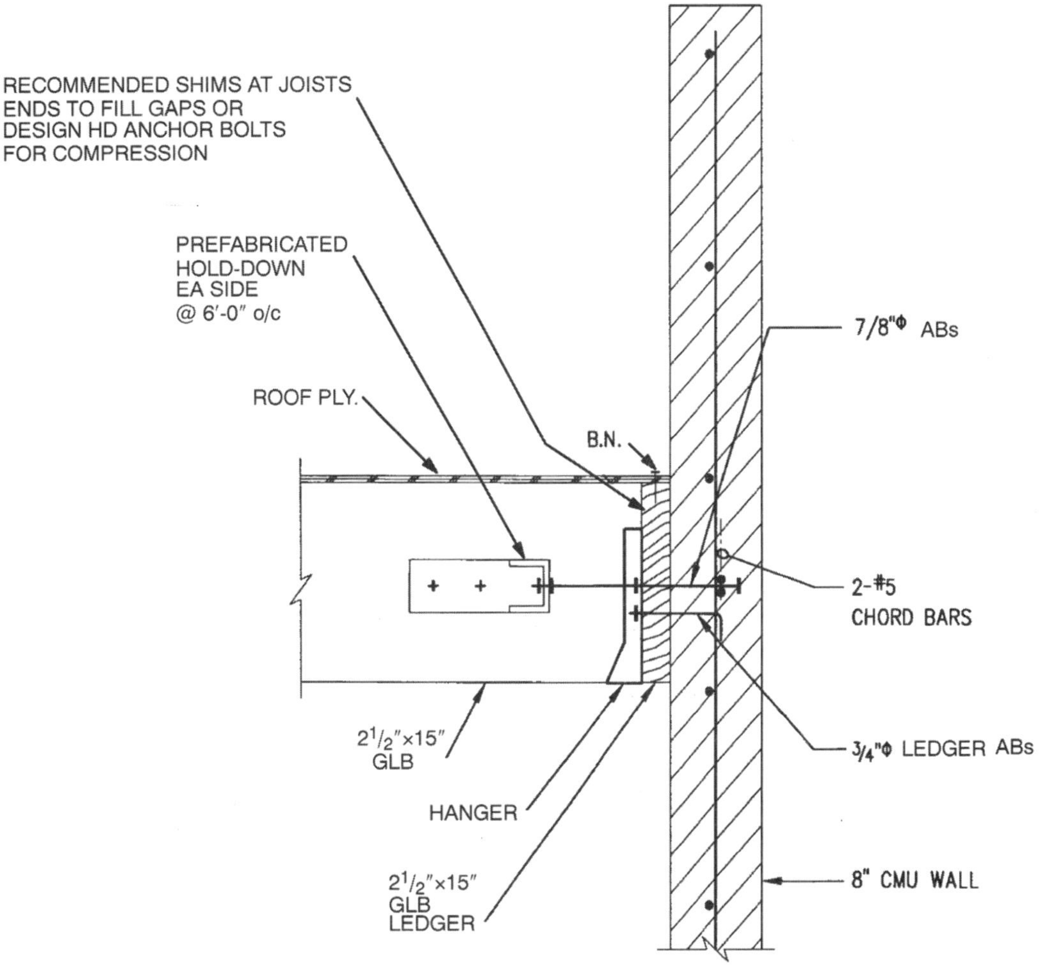

Figure 4-19. CMU wall section at wall-roof ties

References

Brandow, G. E., Hart, G., and Virdee, A., 1997, *Design of Reinforced Masonry Structures*. Concrete Masonry Association of California and Nevada, Sacramento, California.

Amrhein, J. E., 1996, *Reinforced Masonry Engineering Handbook*, 5th Edition. Masonry Institute of America, Los Angeles, California.

Paulay, T. and Priestly, M. J. N., 1992, *Seismic Design of Reinforced Concrete and Masonry Buildings*. John Wiley & Sons, Inc., New York.

Robinson, A. and Uzarski, J., 1999, CMD97, *Concrete Masonry Design to the 1997 UBC.* Computer Aided Design of Reinforced Concrete and Clay Masonry Elements in Accordance with the 1997 Uniform Building Code, Concrete Masonry Association of California and Nevada, Sacramento, California.

ACI 530-05 / ASCE 6-05 / TMS 402-99, 2005, *Building Code Requirements for Masonry Structures.* American Concrete Institute, Farmington Hills, Michigan, American Society of Civil Engineers, Reston, Virginia, The Masonry Society, Boulder, Colorado.

Design Example 5
Tilt-up Building

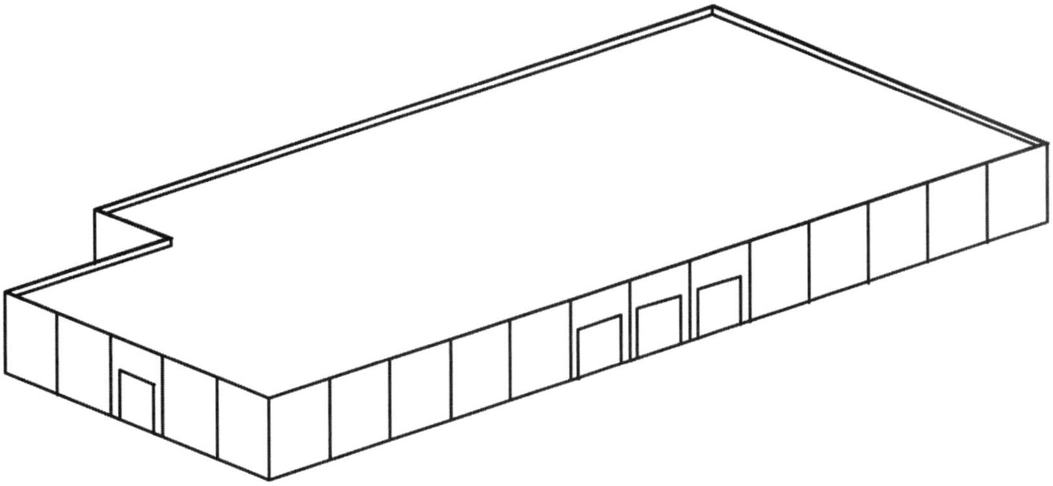

Figure 5-1. Tilt-up building

Overview

This example presents the seismic design of major components of a tilt-up building. Many tilt-up buildings have suffered severe structural damage in earthquakes, particularly during the 1971 San Fernando and 1994 Northridge events. The most common problem was wall-roof separation, with subsequent partial collapse of the roof. Since those events, the building codes have significantly improved, yet a major earthquake has yet to test the current tilt-up code provisions.

The example building is a warehouse, shown in Figure 5-1, which has tilt-up concrete walls and a panelized hybrid roof system. The hybrid roof, common in California and Nevada, consists of a panelized plywood system supported on open web steel joists. The building's roof framing plan is shown in Figure 5-2, and a typical section through the building is given in Figure 5-3. The emphasis in this design example is on the seismic design of the roof diaphragm, wall-roof anchorage, and a major collector.

Outline

This example will illustrate the following parts of the design process

1. Design base shear coefficient

2. Design the roof diaphragm

3. Required diaphragm chord for north-south seismic forces

4. Design of collector along line 3 between lines B and C

5. **Diaphragm deflection**

6. **Design shear force for north-south panel on line 1**

7. **Design wall-roof anchorage for north-south loads**

8. **Design wall-roof anchorage for east-west loads**

9. **Design typical east-west loaded subdiaphragm**

10. **Design continuity ties for east-west direction**

Given Information

Roof

dead load = 14 psf

live load (roof) = 20 psf (reducible) (T 1607.1)

Walls

thickness = 7.25 inches

height = 23 feet

normal weight concrete = 150 pcf

f'_c = 4000 psi

A615, Grade 60 rebar (F_y = 60 ksi)

Roof sheathing

Structural-I sheathing (wood structural panel)

Roof structure

Pre-engineered/pre-manufactured open-web steel joists and joist-girders with full-width nailers. All wood is Douglas-fir.

Seismic force-resisting system

Bearing wall system consisting of intermediate precast shear walls.

Seismic and site data

Mapped spectral accelerations for the site

S_s = 1.5 (Short period)

S_1 = 0.6 (1-second period)

Occupancy Category = II

Site Class = D

Wind

Assumed not to govern

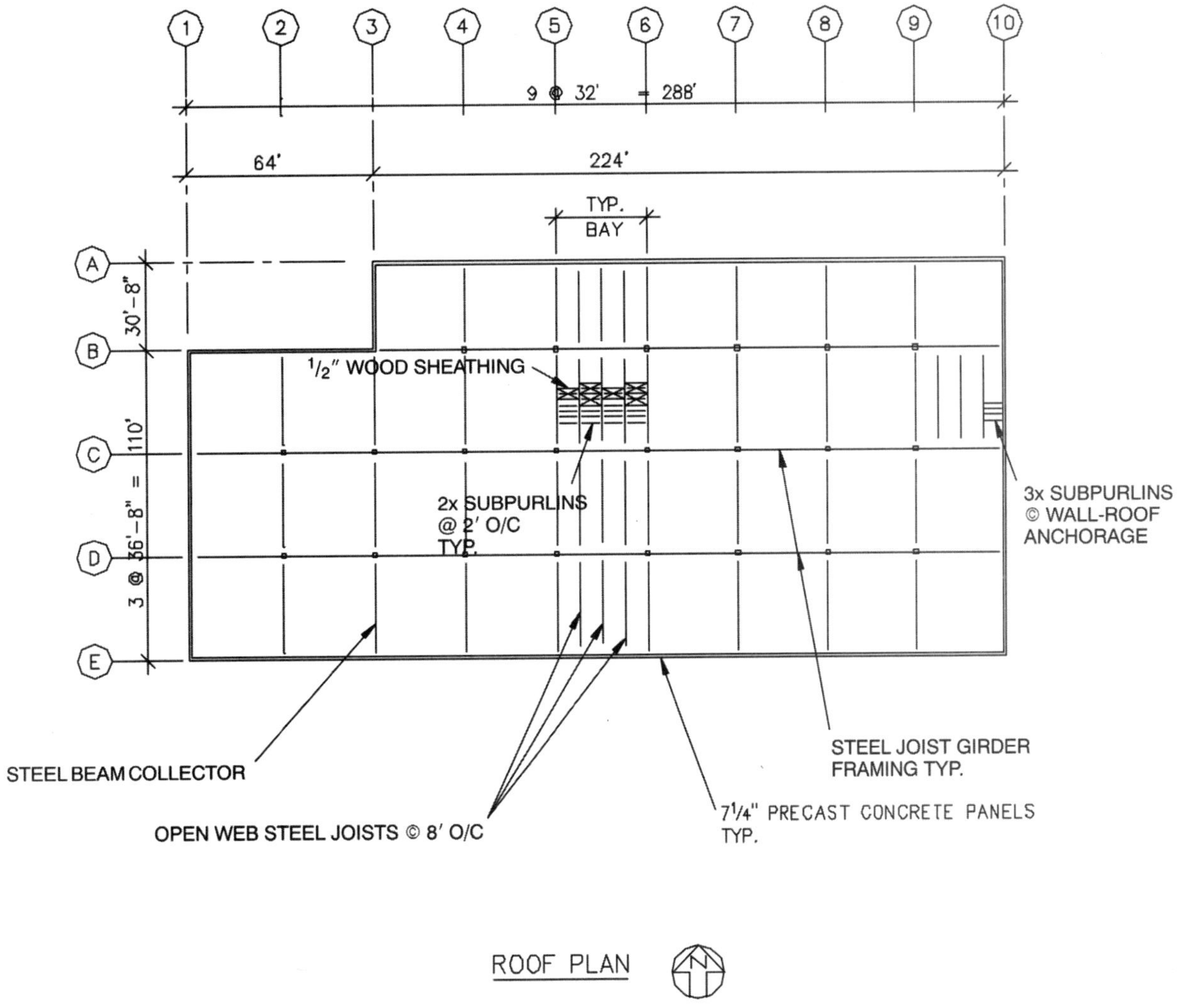

Figure 5-2. Roof framing plan of tilt-up building

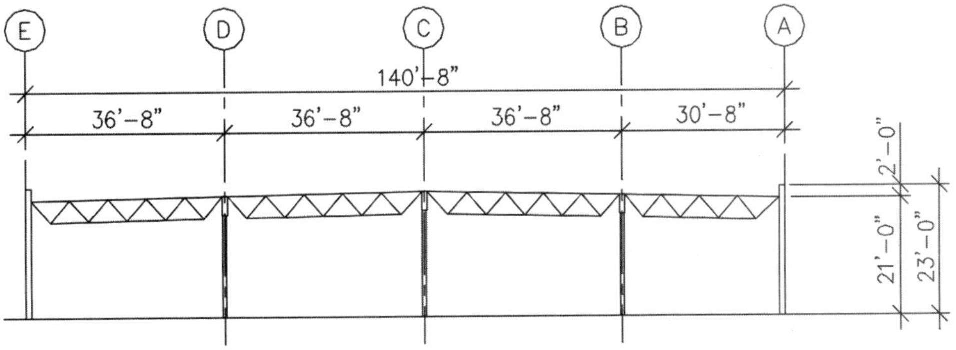

Figure 5-3. Typical cross-section

Calculations and Discussion	Code Reference

1. Design base shear coefficient

1a. Design spectral response accelerations S_{DS} and S_{D1}

The site coefficients F_a, F_v are used to modify the mapped spectral accelerations. Using the given spectral accelerations $S_s = 1.5$, $S_1 = 0.6$, and site class D, the following site coefficients are determined from IBC Tables 1613.5.3

$F_a = 1$ (short period)

$F_v = 1.5$ (1-second period)

Using these site coefficients, the site-adjusted spectral accelerations are determined

$S_{MS} = F_a S_s = 1.0(1.5) = 1.5$ (short period) (Eq 16-37)

$S_{M1} = F_v S_1 = 1.5(0.6) = 0.9$ (1-second period) (Eq 16-38)

The design spectral response accelerations are obtained as follows

$S_{DS} = ⅔ * S_{MS} = 1.0$ (short period) (Eq 16-39)

$S_{D1} = ⅔ * S_{M1} = 0.6$ (1-second period) (Eq 16-40)

Using the design spectral response accelerations and the occupancy category, the next step is to determine the appropriate seismic design category (SDC) from IBC Tables 1613.5.6. Both the short period and 1-second period design categories are level D, thus SDC D governs.

short period category = D (T 1613.5.6(1))

1-second period category = D (T 1613.5.6(2))

governing SDC = D

The appropriate analysis procedure is obtained using ASCE/SEI 7-05 §12.6 in conjunction with Table 12.6-1. Use the equivalent lateral-force procedure of §12.8 to determine the seismic base shear coefficient. For this concrete shear wall building, the approximate fundamental period T is obtained using ASCE/SEI 7-05 Equation 12.8-7 (or 12.8-9) with a $C_T = 0.020$ and an average roof height $h_n = 21$ feet.

$T_a = C_T h_n^{3/4} = 0.2$ seconds Eq 12.8-7

If this example involved a regular structure five stories or fewer in height, having a period T less than 0.5 seconds, the design spectral response acceleration, S_{DS}, need not exceed the value calculated using a value of 1.5 for S_s (§12.8.1.3). The design spectral response accelerations and SDC remain as originally calculated.

$S_{DS\ design} = 1.0$ (short period)

$S_{D1\ design} = 0.6$ (1-second period)

But this structure has a re-entrant corner irregularity per ASCE/SEI 7-05 Table 12.3-1, item 2.

1b. Base shear using the equivalent lateral-force procedure

ASCE/SEI 7-05 §12.8.1 defines the seismic base shear as

$$V = C_s W \qquad \text{where } C_s = \frac{S_{DS}}{R/I} \qquad \text{Eq 12.8-1 \& 12.8-2}$$

Because these tilt-up concrete walls will be considered load-bearing walls and intermediate precast shear walls

$R = 4$ Response modification factor — T 12.2-1

In addition, the importance factor is defined by Occupancy Category II:

$I = 1.0$ — T 11.5-1

Therefore

$$C_s = \frac{S_{DS}}{R/I} = 1.0/(4) = 0.25 \qquad \text{Eq 12.8-2}$$

Checking the maximum limit for C_s where $T \leq T_L$

$$C_{s\,max} = \frac{S_{D1}}{T\left(\frac{R}{I}\right)} = 0.75 > 0.25 \; ... \; o.k. \qquad \text{Eq 12.8-3}$$

Checking the minimum allowed value for C_s, Equations 12.8-5 and 12.8-6 are applicable. In this example, S_1 is equal to 0.6*g*, therefore Equation 12.8-6 is valid to check the minimum allowed C_s.

$$C_{s\,min} = 0.01 < 0.25 \; ... \; o.k. \qquad \text{Eq 12.8-5}$$

$$C_{s\,min} = \frac{0.5S_1}{R/I} = 0.075 < 0.25 \; ... \; o.k. \qquad \text{Eq 12.8-6}$$

The calculated value for $C_s = 0.25$ is between the maximum and minimum allowed values.

$$C_{s\,governs} = 0.25$$

Substituting into Equation 12.8-1

$$V = C_s W = 0.25W$$

1c. Base shear using the simplified alternative structural design criteria

Instead of the lengthy seismic analysis shown above, simple buildings that meet the twelve limitations of §12.14.1.1 may use the simplified analysis procedure in §12.14. Using §12.1.1, the simplified analysis procedure of §12.14 is allowed as an alternative method for designing this example's structure to resist seismic forces. This example will not follow the simplified alternative method.

2. Design the roof diaphragm

2a. Roof diaphragm shear coefficient

The roof diaphragm must be designed to resist seismic forces in each direction. The following formula is used to determine the total seismic force F_{px} on the diaphragm at a given level of a building.

$$F_{px} = \frac{\sum_{i=x}^{n} F_i}{\sum_{i=x}^{n} w_i} w_{px} \qquad \text{Eq 12.10-1}$$

Base shear for this building is $V = 0.25w$. Because it is a one-story building, Equation 12.10-1 becomes the following

$$F_{px} = 0.25w_{px}$$

F_{px} need not exceed

$$0.4S_{DS}Iw_{px} = 0.4(1.00)(1.0)w_{px} = 0.4w_{px} \qquad \text{§12.10.1.1}$$

but shall not be less than

$$0.2S_{DS}Iw_{px} = 0.2(1.00)(1.0)w_{px} = 0.2w_{px} \qquad \text{§12.10.1.1}$$

Based on the criteria given in §12.10.1.1, $F_{px} = 0.25w_{px}$

Therefore, for diaphragm design use $F_p = 0.25w_p$

2b. Roof diaphragm shears

The wood structural panel roof system is permitted to be idealized as a flexible diaphragm per §12.3.1.1 and IBC 1613.6.1. Seismic forces for the roof are computed from the tributary weight of the roof and the walls oriented perpendicular to the direction of the seismic forces. Uniform loading will be computed in each direction.

East-west direction

Because the the panelized wood roof diaphragm in this building is idealized as flexible, lines A, B, and E are considered lines of resistance for the east-west seismic forces. A collector is needed along line B to drag the tributary east-west diaphragm forces into the shear wall on line B. The loading and shear diagrams are shown below

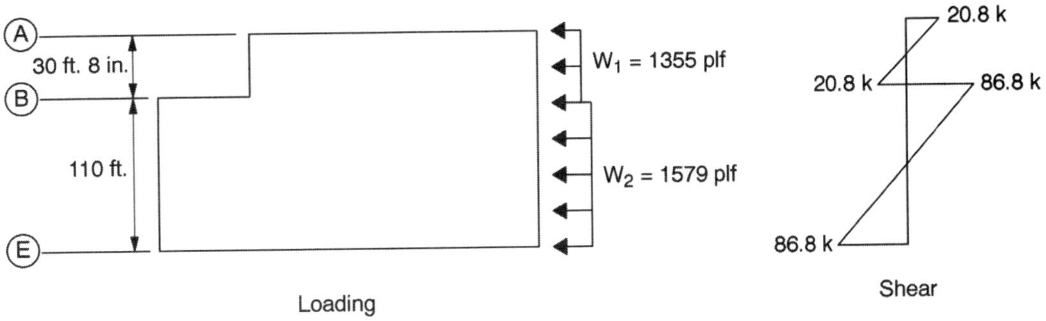

Figure 5-4. Seismic loading and shear diagrams for east-west diaphragm

The uniform loads W_1 and W_2 in the east-west direction are computed using the diaphragm lengths and wall heights.

Roof dead load = 14 psf

$$\text{Wall dead load} = \frac{7.25}{12}150 = 90.6 \text{ psf}$$

Roof height = 21 feet average

Parapet height = 2 feet average

$$W_1 = 0.25(14 \text{ psf})(224 \text{ ft}) + \left[0.25(90.6 \text{ psf})(23)\left(\frac{23}{2}\right)\left(\frac{1}{21}\right)\right]2 = 1355 \text{ plf}$$

$$W_2 = 0.25(14 \text{ psf})(288 \text{ ft}) + \left[0.25(90.6 \text{ psf})(23)\left(\frac{23}{2}\right)\left(\frac{1}{21}\right)\right]2 = 1579 \text{ plf}$$

In this example, the effect of any wall openings reducing the wall weight has been neglected. This is considered an acceptable simplification because the openings usually occur in the bottom half of the wall. In addition, significant changes in parapet height should also be considered if they occur.

Diaphragm shear at line A and on the north side of line B is

$$\frac{20{,}800 \text{ lb}}{224 \text{ ft}} = 93 \text{ plf}$$

Diaphragm shear at the south side of line B and at line E is

$$\frac{86{,}800 \text{ lb}}{288 \text{ ft}} = 301 \text{ plf}$$

North-south direction

Diaphragm forces for the north-south direction are computed using the same procedure and assumptions as the east-west direction

$$W_3 = 0.25(14)(110) + \left[0.25(90.6)(23)\left(\frac{23}{2}\right)\left(\frac{1}{21}\right)\right]2$$

$W_3 = 956$ plf

$$W_4 = 0.25(14)(140.67) + \left[0.25(90.6)(23)\left(\frac{23}{2}\right)\left(\frac{1}{21}\right)\right]2$$

$W_4 = 1{,}063$ plf

Diaphragm shear at line 1 and the west side of line 3 is

$$\frac{30{,}600 \text{ lb}}{110 \text{ ft}} = 278 \text{ plf}$$

Diaphragm shear at the east side of line 3 and at line 10

$$\frac{119{,}000 \text{ lb}}{140.67 \text{ ft}} = 846 \text{ plf}$$

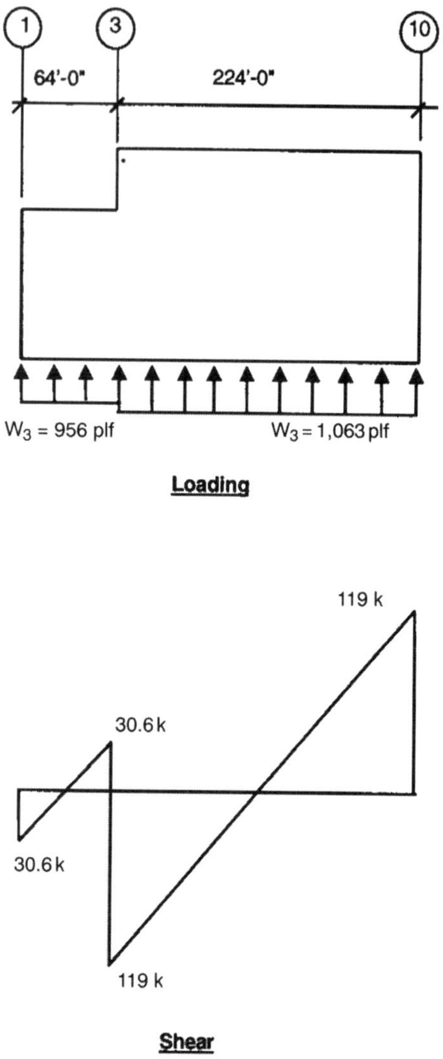

Figure 5-5. Seismic loading and shear diagram for north-south diaphragm

2c. Design of north-south diaphragm

The north-south diaphragm has been selected to illustrate the design of a wood structural panel roof diaphragm. Allowable stress design (ASD) will be used. The basic earthquake loading combinations are given in ASCE/SEI 7-05 §12.4.2.3.

The governing seismic load combination for allowable stress design is (5)

$$(1.0 + 0.14S_{DS})D + H + F + 0.7\rho Q_E \qquad \S 12.4.2.3$$

When designing the structural diaphragm, the vertical loading need not be considered in conjunction with the lateral diaphragm shear stresses. Therefore the dead load $D = 0$ in the load combinations. Additionally, $H = 0$, $F = 0$ and $L = 0$ for this example.

The redundancy factor ρ = 1.0 for typical diaphragms per §12.3.4.1. In unique situations where the diaphragm is acting to transfer forces horizontally between offsets, the redundancy factor ρ will conform to §§12.3.4 and 12.10.1.1. In this example, ρ = 1.0 for the diaphragm design. Thus, the applicable basic load combination reduces to simply $0.7Q_E$.

Assume the diaphragm is to be constructed with 15⁄32-inch Structural-I sheathing (wood structural panel) with all edges supported (blocked). Refer to IBC Table 2306.3.1 for nailing requirements. Sheathing arrangement (shown in Figure 5-2) for north-south seismic forces is Case 4. Because open web steel joist purlins with full-width wood nailers are used in this direction, the framing width in the north-south direction is greater than 3-inch nominal. However, in the east-west direction, the framing consists of 2x subpurlins, and strength is therefore limited by the 2-inch nominal width. Required nailing for panel edges for various zones of the roof (for north-south seismic only) is given in Table 5-1. Minimum intermediate (field) nailing is 10d @ 12 inches and 10d nails require 1½-inch member penetration. A similar calculation (not shown) must be done for east-west seismic forces.

Table 5-1. Diaphragm nailing capacities

Zone	Boundary and North-South Edge Nailing[1] (in)	East-West Edge Nailing[2] (in)	ASD Allowable Shear (plf)
A	10d @ 2½	4	640
B	10d @ 4	6	425
C	10d @ 6	6	320

Notes:

1. The north-south running sheet edges are the "continuous panel edges parallel to load" mentioned in IBC 2306.3.1.
2. The east-west sheet edges are the "other panel edges" in IBC 2306.3.1. Note that the nailing for east-west running diaphragm boundaries is per the tighter boundary spacing.

The diaphragm boundaries at lines 3 and 10 have a shear demand of $v = 846$ plf (see Part 2a). Converting to allowable stress design, $v_{ASD} = 0.7(846) = 592$ plf, which is less than nailing zone A's allowable stress of 640 plf.

At some location, nailing zone B (425 plf) will become acceptable as the diaphragm shears reduce farther from the diaphragm boundary. The demarcation between nailing zones A and B may be located as follows using allowable stress design:

Shear demand (ASD) = shear capacity (ASD)

$$0.7[119{,}000 \text{ lb} - (1063 \text{ plf})x] = 425 \text{ plf}(140.67 \text{ ft})$$

where

x = the demarcation distance from the diaphragm boundary.

Solving for x obtains

$$x = 31.6 \text{ ft}$$

Because a panelized wood roof system typically consists of 8-foot-wide panel modules, the demarcation is increased to the next 8-foot increment or to $x = 32$ feet.

A similar process is undertaken to determine the demarcation between zones B and C. In this situation, $x = 51.5$ ft and the demarcation is increased to 56 feet from the diaphragm boundary. The resulting diaphragm shears at these demarcation boundaries are as follows:

Table 5-2. Diaphragm nailing zone shear checks between lines 3 and 10

Nailing Zone	Distance from boundary	Maximum Shear	ASD Shear	Allowable Shear Capacity
A	0 feet	$v_{max} = 846$ plf	$v_{ASD} = 592$ plf	640 plf
B	32 feet	$v_{max} = 604$ plf	$v_{ASD} = 423$ plf	425 plf
C	56 feet	$v_{max} = 423$ plf	$v_{ASD} = 296$ plf	320 plf

The resulting nailing zones for the north-south loading are shown in Figure 5-6.

These demarcation calculations assume the full depth of the diaphragm is available for shear capacity. However, typical warehouse construction contains skylights and smoke vents that can substantially perforate the structural diaphragm. In these situations, the designer must account for these diaphragm interruptions resulting in larger shear stresses.

Comment: Plywood and other structural wood panels are common diaphragm materials in the west and parts of the south. Other parts of the nation commonly use metal deck for diaphragms in conjunction with steel roof framing. Metal deck diaphragms are approached in the same manner with a similar diaphragm table assigning various deck gauges and attachments to specific diaphragms zones depending on the shear demands.

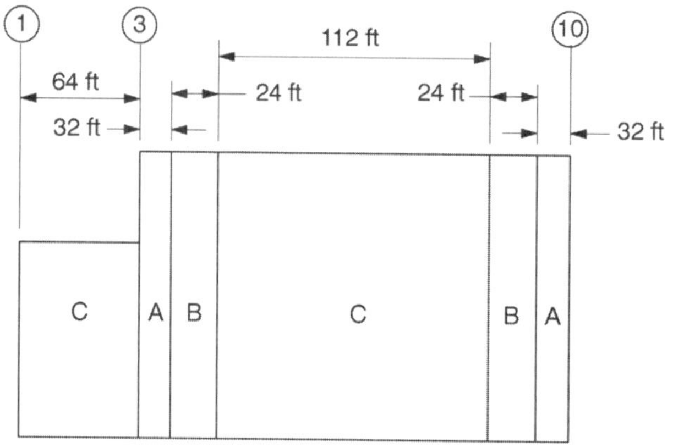

Figure 5-6. Nailing zones for north-south diaphragm

This wood diaphragm resisting seismic forces must have its aspect ratio checked against the limitations in IBC Table 2305.2.3. For blocked diaphragms of wood structural panels the maximum aspect ratio is $L/W = 4{:}1$.

For this example, $L/W = 224/140.67 = 1.6 < 4$ . . . *o.k.*

Comment: Aspect ratio limitations for metal deck diaphragms are found under the specific deck manufacturer's ICC-ES Evaluation Report. Within these reports, a table titled "Diaphragm Flexibility Limitation" provides guidance on limiting diaphragm flexibility in conjunction with diaphragm aspect ratios.

Because there is a re-entrant corner at the intersection of lines B and 3, a check for Type 2 horizontal structural irregularity must be made. Requirements for horizontal structural irregularities are given in ASCE/SEI 7-05 Table 12.3-1.

East-west direction check

$$0.15 \times 288 \text{ ft} = 43.2 \text{ ft} < 64 \text{ ft}$$

North-south direction check

$$0.15 \times (110.0 + 30.67) = 21.1 \text{ ft} < 30.67 \text{ ft}$$

Because both projections are greater than 15 percent of the plan dimension in the direction considered and the structure is SDC D or higher, a Type 2 horizontal structural irregularity exists. The requirements of ASCE/SEI 7-05 §12.3.3.4 apply, resulting in a 25-percent increase in seismic forces for connections of diaphragms to the vertical elements, and connections of diaphragms to collectors.

This 25-percent force increase is on ASCE/SEI 7-05 Equation 12.8-1, which results in diaphragm forces via Equation 12.10-1. Using the information obtained from Part 2a, the diaphragm connection forces are increased to $F_{px} = 1.25\ (0.25w_{px}) = 0.313w_{px}$. This still falls between the upper bound $0.4w_{px}$ and lower bound $0.2w_{px}$ found in Part 2a, thus $F_{px} = 0.313w_{px}$, which is a direct 25-percent increase to diaphragm connection forces.

This force increase applies to situations involving ledger and/or wood nailer bolting to shear walls, wood nailer bolting to collectors, and the row of diaphragm nailing that transfers the diaphragm shears directly to walls and collectors. The design of these elements is not a part of this example. This irregularity also affects the collector design, as will be shown in Part 4. The 25-percent force increase is not applied to out-of-plane wall anchorage forces connected to the diaphragms.

3. Required diaphragm chord for north-south seismic forces

Chords are required to carry the tension forces developed by the moments in the diaphragm. In this building, the chords are continuous reinforcement located in the wall panels at or near the roof level as shown in Figure 5-7. In this example, the chord reinforcement is below the roof ledger to facilitate the chord splice connection at the panel joint.

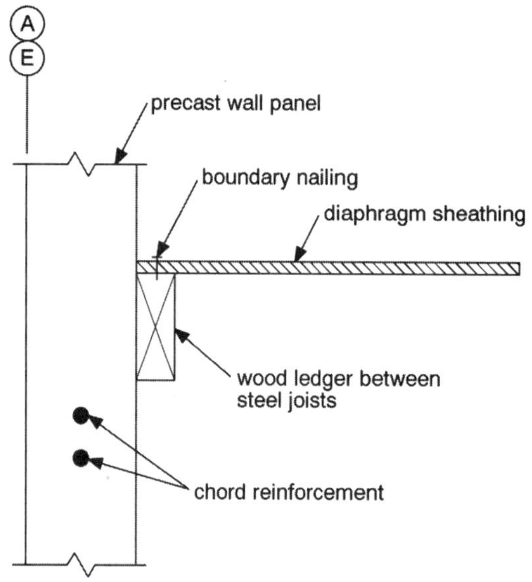

Figure 5-7. Diaphragm chord

The north-south diaphragm spans between lines 1 and 3 and lines 3 and 10. The diaphragm is idealized as flexible, and the moments in segments 1-3 and 3-10 can be computed independently assuming a simple span for each segment. In this example, the chord reinforcement between lines 3 and 10 will be determined. This reinforcement is for the panels on lines A and E.

$w = 1063$ plf from Part 2

$$M = \frac{w\ell^2}{8} = \frac{1.063\text{ klf}(224)^2}{8} = 6667\text{ kip-ft}$$

The chord forces are computed from

$$T_u = \frac{6667\text{ k-ft}}{140.67\text{ ft}} = 47.4\text{ kips}$$

The chord will be designed using strength design with ASTM A706 Grade 60 reinforcement. A706 reinforcing is used in anticipation that the steel will be welded at the panel joint splice. (See ACI §3.5.2.) The load factor is 1.0 for seismic forces. (ASCE/SEI 7-05 §2.3.2.)

$$A_s = \frac{T_u}{\phi f_y} = \frac{47.4\text{ k}}{0.9(60\text{ ksi})} = 0.877\text{ in}^2$$

$\therefore$ Use minimum two #6 bars, $A_s = 0.88\text{ in}^2 > 0.877 \ldots$ o.k.

Comment: The chord shown above consists of two #6 bars. These must be spliced at the joint between adjacent panels, typically using details that are highly dependent on the accuracy in placing the bars and the quality of the field welding. The welded reinforcing splice connection must develop at least 125 percent f_y per ACI 318 §12.14.3.4. Alternately, chords can also be combined with the ledger when steel channels or angles are used, and good quality splices can be easier to make.

4. Design collector along line 3 between lines B and C

The collector and shear wall ledger along line 3 carry one-half of the north-south roof diaphragm seismic force. The force in the collector is "collected" from the tributary area between lines B and E and transmitted to the shear wall on line 3.

4a. Determine seismic forces on the collector

From the diaphragm shear diagram for north-south seismic forces (Figure 5-5), the maximum collector load along line 3 is

$$R = 30.6\text{ k} + \left(\frac{110.0\text{ ft}}{140.67\text{ ft}}\right) 119\text{ k} = 124\text{ kips tension or compression}$$

The uniform axial load that accumulates in the collector can be approximated as the total collected load on line 3 divided by the length of the collector (110 ft) in this direction.

$$q = \frac{R}{L} = \frac{124{,}000\text{ lb}}{110\text{ ft}} = 1127\text{ plf}$$

4b. Determine the collector force in steel beam between lines B and C

Assume the collector, a W18 × 50 with wood nailer, is adequate to support dead and live loads. ASTM A992, $F_y = 50$ ksi. Calculate the seismic force at mid-span. Tributary length for collecting axial forces is

$$\ell = 110.00\text{ ft} - \frac{36.67\text{ ft}}{2} = 91.67\text{ ft}$$

$$P = q\ell = 1{,}127\text{ klf }(91.67\text{ ft}) = 103\text{ kips tension or compression in beam}$$

4c. Check steel beam collector for load combinations as required by §12.4.2.3

The governing seismic load combination for LRFD under ASCE/SEI 7-05 §12.4.2.3 is

$$(5)\ \ (1.2 + 0.25\,S_{DS})D + \rho Q_E + L + 0.25$$

For this example, $L = 0$, $S = 0$, and $S_{DS} = 1.0$. Because collectors are considered a part of the diaphragm system, the redundancy factor $\rho = 1.0$ was discussed previously in Part 2c for diaphragms. Thus, the applicable basic load combination for LRFD reduces to the following:

$$(5)\ \ 1.4D + Q_E$$

The unfactored gravity loads and moments are as follows:

$$w_D = 8 \text{ ft } (14 \text{ psf}) + 50 \text{ plf} = 162 \text{ plf}$$

$$M_D = \frac{162 \text{ plf}(36.67 \text{ ft})^2}{8} = 27{,}230 \text{ lb-ft or } 27.2 \text{ kip-ft}$$

$$L_r = L_O R_1 R_2 = (20 \text{ psf})(0.91)(1.0) = 18.2 \text{ psf} \qquad \text{(Eq 16-27)}$$

$$w_{Lr} = 8 \text{ ft } (18.2 \text{ psf}) = 146 \text{ plf}$$

$$M_{Lr} = \frac{146 \text{ plf}(36.67 \text{ ft})^2}{8} = 24{,}541 \text{ lb-ft or } 24.5 \text{ kip-ft}$$

As shown in Part 2c, this building contains a Type 2 horizontal structural irregularity, and the requirements of ASCE/SEI 7-05 §12.3.3.4 apply. This results in a 25-percent increase in seismic forces for collectors and their connections except where designed for load combinations with the overstrength factor of §12.4.3.2. The collector's axial seismic force becomes $Q_E = 1.25 \times 103$ kips = 129 kips.

AISC §H1 contains the equations for combined axial compression and bending. Because the bending is not biaxial, AISC §H1.3 is advantageous to use by checking failure about each axis independently. First, compute the available strengths P_c and M_c for use in the equations. P_c is a function of the collector's unbraced length. In this example, the lateral bracing to the collector's bottom flange is provided at the member's equal third points with use of an angle brace (design not shown) for an unbraced length of $\ell_y = 36.67/3 = 12.22$ ft. The strong axis unbraced length is simply the span $\ell_x = 36.67$ ft.

$$\frac{k\ell_x}{r_x} = \frac{1.0(36.67)12}{7.38} = 60$$

$$\frac{k\ell_y}{r_y} = \frac{1.0(12.22)12}{1.65} = 89$$

Because failure will be checked separately about each axis per AISC §H1.3, P_c corresponding with each axis will be determined:

X-axis:

$$F_{ex} = \frac{\pi^2 E}{\left(\frac{k\ell_x}{r_x}\right)^2} = \frac{\pi^2(29 \times 10^3)}{(60)^2} = 79.5 \text{ ksi} \qquad \text{AISC Eq E3-4}$$

Because $F_e \geq 0.44\, F_y$, AISC Equation E3-2 is applicable.

$$F_{cr} = [0.658^{\frac{F_y}{F_e}}]F_y = [0.658^{\frac{50}{79.5}}]50 = 38.4 \text{ ksi}$$

$$P_{nx} = F_{cr} A_g = 38.4(14.7) = 564 \text{ kips} \qquad \text{AISC Eq E4-1}$$

$$P_{cx} = \phi_c P_{nx} = 0.90(564) = 508 \text{ kips} \qquad \text{AISC §E1}$$

Y-axis:

$$F_{ey} = \frac{\pi^2 E}{\left(\frac{k\ell_y}{r_y}\right)^2} = \frac{\pi^2(29 \times 10^3)}{(89)^2} = 36.1 \text{ ksi}$$

Because $F_e \geq 0.44\ F_y$, AISC Equation E3-2 is applicable.

$$F_{cr} = [0.658^{\frac{F_y}{F_e}}]F_y = [0.658^{\frac{50}{36.1}}]50 = 28.0 \text{ ksi}$$

$$P_{ny} = F_{cr} A_g = 28.0(14.7) = 412 \text{ kips}$$

$$P_{cy} = \phi_c\, P_{ny} = 0.90(412) = 371 \text{ kips}$$

With the top flange fully supported laterally:

$$M_c = \phi_b\, M_n = \phi_b F_y Z_x = 0.90(50 \text{ ksi})(101) = 4545 \text{ in-kips}$$

$$= 379 \text{ ft-kips}$$

Determine factored P_r and M_r using basic load combination (5) 1.40 + Q_E:

$$P_r = Q_E = 129 \text{ kips (includes increase for plan irregularity)}$$

$$M_r = 1.4M_D = 1.4(27.2) = 38.1 \text{ kip-ft}$$

Per AISC §H1.3(a), the in-plane stability check uses AISC Equations H1-1. P_{cx} is the appropriate in-plane bucking strength.

$$\frac{P_r}{P_{cx}} = \frac{129}{508} = 0.25 \geq 0.20.$$

Therefore, AISC Equation H1-1a is applicable for checking combined forces.

$$\frac{P_r}{P_{cx}} + \frac{8}{9}\left(\frac{M_r}{M_{cx}}\right) = 0.25 + \frac{8}{9}\left(\frac{38.1}{379}\right) = 0.34 \leq 1.0.\ .\ .\ .\ o.k.$$

Per AISC §H1.3(b), the out-of-plane buckling check uses AISC Equation H1-2. $P_{co} = P_{cy}$.

$$\frac{P_r}{P_{co}} + \left(\frac{M_r}{M_{cx}}\right)^2 = \frac{129}{371} + \left(\frac{38.1}{379}\right)^2 = 0.36 \leq 1.0.\ .\ .\ .\ o.k.$$

Evaluating the W18 × 50 collector for combined axial tension and bending per AISC §H1.2 is not necessary, because P_c will be less and more critical for compression than for tension.

4d. Check steel beam collector for load combinations with overstrength factor per §12.4.3.2

As required by ASCE/SEI 7-05 §12.10.2.1 the steel beam (W18 × 50) must also be checked for the special load combinations of §12.4.3.2. The relevant strength design equations are

$$(5)\ (1.2 + 0.2\,S_{DS})\,D + \Omega_O\,Q_E + L + 0.25$$

$$(7)\ (0.9 - 0.2\,S_{DS})\,D + \Omega_O\,Q_E + 1.6\,H$$

$\Omega_O\,Q_E$ is an estimate of the maximum force transmitted by the collector elements in the seismic event. The horizontal seismic force Q_E is scaled by the amplification factor Ω_O for estimating E_m. The amplification factor Ω_O may be reduced by subtracting 0.5 for structures with flexible diaphragms; however, Ω_O shall not be reduced below 2.0.

$$\Omega_O = 2.5 - 0.5 = 2.0$$ T 12.2-1

Because the dead load component D is detrimental to the analysis, load combination (7) will not govern. Simplifying the remaining load combination for this example we obtain:

$$(5)\ 1.14D + 1.75Q_E\ 1.4D + 2.0Q_E$$

With this special load combination, re-analyze the W18 × 50 steel beam collector for combined axial and bending loads.

As determined earlier in Part 4, $M_D = 27.2$ kip-ft and $Q_E = 103$ kips. Notice that Q_E does not include a 1.25 factor increase for irregular buildings when considering special load combinations with overstrength per §12.3.3.4.

Because collector bending is not biaxial, AISC §H1.3 is advantageous to use by checking failure about each axis independently. Recall from Part 4c:

$$P_{cx} = 508 \text{ kips}$$

$$P_{cy} = 371 \text{ kips}$$

$$M_c = 379 \text{ ft-kips}$$

Evaluating the special load combinations with overstrength:

$$P_r = \Omega_O\,Q_E = 2.0(103 \text{ kips}) = 206 \text{ kips}$$

$$M_r = 1.4\,M_D = 1.4(27.2 \text{ kips}) = 38.1 \text{ kip-ft}$$

Per AISC §H1.3(a), the in-plane stability check uses AISC Equations H1-1. P_{cx} is the appropriate in-plane bucking strength.

$$\frac{P_r}{P_{cx}} = \frac{206}{508} = 0.41 \geq 0.20.$$

Therefore, AISC Equation H1-1a is applicable for checking combined forces.

$$\frac{P_r}{P_{cx}} + \frac{8}{9}\left(\frac{M_r}{M_{cx}}\right) = 0.41 + \frac{8}{9}\left(\frac{38.1}{379}\right) = 0.50 \le 1.0 \ldots.\ o.k.$$

Per AISC §H1.3(b), the out-of-plane buckling check uses AISC Equation H1-2. $P_{co} = P_{cy}$.

$$\frac{P_r}{P_{co}} + \left(\frac{M_r}{M_{cx}}\right)^2 = \frac{206}{371} + \left(\frac{38.1}{379}\right)^2 = 0.57 \le 1.0 \ldots.\ o.k.$$

Evaluating the W18 × 50 collector for combined axial tension and bending per AISC §H1.2 is not necessary, because P_c will be less and more critical for compression than for tension.

Thus, W18 × 50 steel beam collector is acceptable.

4e. Collector connection to shear wall

The design of the connection of the steel beam to the shear wall on line 3 is not given. This is an important connection because it transfers the large "collected" seismic force into the shear wall. The connection must be designed to carry the seismic forces from the beam, including the load combinations with overstrength per §12.10.2.1. A plan irregularity can increase the connection forces for the collector and diaphragm by 25 percent when the overstrength factor is not included. As shown in Part 2b, this building has a Type 2 horizontal structural irregularity.

Because there is also a collector along line B, there is similarly an important connection of the girder between lines 3 and 4 to the shear wall on line B. Having to carry two large tension (or compression) forces through the intersection of lines B and 3 (but not simultaneously) requires careful design consideration.

5. Diaphragm deflection

Diaphragm deflections are estimated to determine the displacements imposed on attached structural and nonstructural elements, and to evaluate the significance of the *P*-delta effects. Under IBC §2305.2.2, diaphragm deflections are limited to the amount that will permit the attached elements to maintain structural integrity and to continue supporting their prescribed loads. For structural elements, the intent here is to ensure structural stability by avoiding formation of collapse mechanisms in the vertical support system and avoiding excessive *P*-delta loading effects. For nonstructural elements, the intent of this section is to prevent failure of connections or self-integrity that could result in a localized falling hazard.

5a. Deflection of north-south diaphragm

An acceptable method of determining the horizontal deflection of a blocked wood structural panel diaphragm under lateral forces is given in AF&PA SDPWS §4.2.2. The following equation is used

$$\delta_{dia} = \frac{5vL^3}{8EAW} + \frac{0.25vL}{1000\,G_a} + \frac{\Sigma(x\Delta_c)}{2W}$$

AF&PA SDPWS Eq 4.2-1

The deflection of the diaphragm spanning between lines 3 and 10 will be computed. Values for each of the parameters in the above equation are given below

v = 846 plf (see Part 2b)

L = 224 ft

E = 29×10^6 psi

A = 2 #6 bars = $2 \times 0.44 = 0.88$ in^2

W = 140.67 ft

G_a = 20.0 k/in Zone A (see part 2b for nailing zones) AF&PA SDPWS T 4.2A

15.0 k/in Zone B

24.0 k/in Zone C

Δ_c = 0 (Assume no slip in steel chord connections)

The flexural deformation portion of the equation $\frac{5vL^3}{8EAW}$ assumes a uniformly loaded diaphragm and is computed as follows:

$$\delta_{\substack{\text{diaphragm}\\ \text{flexure}}} = \frac{5vL^3}{8EAW} = \frac{5(846\text{ plf})(224\text{ ft})^3}{8(29 \times 10^6\text{ psi})0.88(140.64\text{ ft})} = 1.66\text{ in}$$

The shear deformation portion of the equation $\frac{0.25vL}{1000\,G_a}$ is derived from a uniformly loaded diaphragm with uniform shear stiffness. Because our example has various nailing zones, and the apparent shear stiffness G_a varies by nailing zone, we will have to modify this portion of the equation. Using virtual work methods, the shear deformation of a uniformly loaded diaphragm with various shear stiffness zones is

$$\delta_{\substack{\text{diaphragm}\\ \text{flexure}}} = \Sigma \frac{0.5v_{i\,\text{ave}}L_i}{1000\,G_{ai}}$$

where

$v_{i\,\text{ave}}$ = the average diaphragm shear within each shear stiffness zone.

L_i = the length of each stiffness zone measured perpendicular to loading.

G_{ai} = the apparent shear stiffness of each shear stiffness zone being considered.

Working across the diaphragm from grid 3 to 10, the following table is helpful using information from Part 2c:

Table 5-3. Shear deformation of various nailing zones

Zone	v_{left}	v_{right}	$v_{i\,ave}$	L_i	G_a	$\frac{0.5 v_{i\,ave} L_i}{1000\, G_{ai}}$
A	846	604	725	32 ft	20	0.58 in
B	604	423	514	24 ft	15	0.41 in
C	423	0	212	56 ft	24	0.25 in
C	0	423	212	56 ft	24	0.25 in
B	423	604	514	24 ft	15	0.41 in
A	604	846	725	32 ft	20	0.58 in
						$\Sigma = 2.48$ in

$$\delta_{\substack{\text{diaphragm}\\ \text{shear}}} = 2.48 \text{ in}$$

Because the chord reinforcing bars are directly welded together at their splice, no chord slip is assumed to occur.

$$\delta_{\substack{\text{chord}\\ \text{slip}}} = \frac{\Sigma(x\Delta_c)}{2W} = 0.00 \text{ in}$$

$$\delta_{\text{dia}} = \delta_{\substack{\text{diaphragm}\\ \text{flexure}}} + \delta_{\substack{\text{diaphragm}\\ \text{shear}}} + \delta_{\substack{\text{chord}\\ \text{slip}}} = 1.66 + 2.48 + 0.00 = 4.14 \text{ in}$$

To compute the maximum expected diaphragm deflection δ_x, Equation 12.8-15 is used

$$\delta_x = \frac{C_d \delta_{xe}}{I}$$

δ_{xe} = 4.14 in (using an elastic analysis under strength forces, δ_{dia})

$C_d = 4$ T 12.2-1

$$\delta_x = \frac{4(4.14)}{1.0} - 16.6 \text{ in}$$

Note: The deflection amplification factor C_d is primarily associated with reversing the effects of applied response modification coefficient R used in determining the base shear $V = 0.25W$ and diaphragm shear coefficient $F_{px} = 0.25w$ (see Parts 1b and 2a).

Instead of using the AF&PA equation, the designer could use IBC §2305.2.2. Although the IBC method is a little more complex, it has the ability to be more accurate if properly applied. Additional information is available in the AF&PA SDPWS commentary and Skaggs, 2004.

5b. Limits on diaphragm deflection

Limits are placed on diaphragm deflection primarily for two reasons. The first reason is to separate the building from adjacent structures and property lines in accordance with §12.12.3. In this situation, d_x is computed for the shear walls and diaphragm and added together to obtain the overall deflection. Because the concrete shear wall drift is insignificant compared with the diaphragm deflection, the shear wall deformation is ignored in this example. In addition, out-of-plane wall deformaton does not need to be included.

The second reason for limiting diaphragm deflection is to maintain structural integrity under design load conditions. Diaphragm deflections are limited by IBC §2305.2.2, ASCE/SEI 7-05 §12.12.2, and AF&PA SDPWS §4.2.1.

> "Permissible deflection shall be that deflection up to which the diaphragm and any attached load distributing or resisting element will maintain its structural integrity under design load conditions, such that the resisting element will continue to support design loads without danger to occupants of the structure."

The language of this section is intentionally ambiguous, with the approach left much to the engineer's own rational judgment. The 1999 SEAOC Blue Book (§C108.2.9) states, "In lowrise concrete or masonry buildings, deflections that can cause secondary failures in structural and nonstructural walls should be considered."

The diaphragm's deflection results in the columns and perpendicular walls rotating about their bases because of the diaphragm's translation at the top. Assuming the columns and walls were modeled with pinned bases during their individual design, this base rotation is permitted to occur even if some unintentional fixity exists.

Unintentional fixity may be the result of standard column base plate anchorage or wall-to-slab anchorage. The assumption of plastic hinges forming at the base is acceptable, provided that these hinges do not result in an unstable condition.

A possible source of instability is the *P*-delta effect resulting from added diaphragm loading due to a horizontal thrust component from the axially loaded gravity columns and walls.

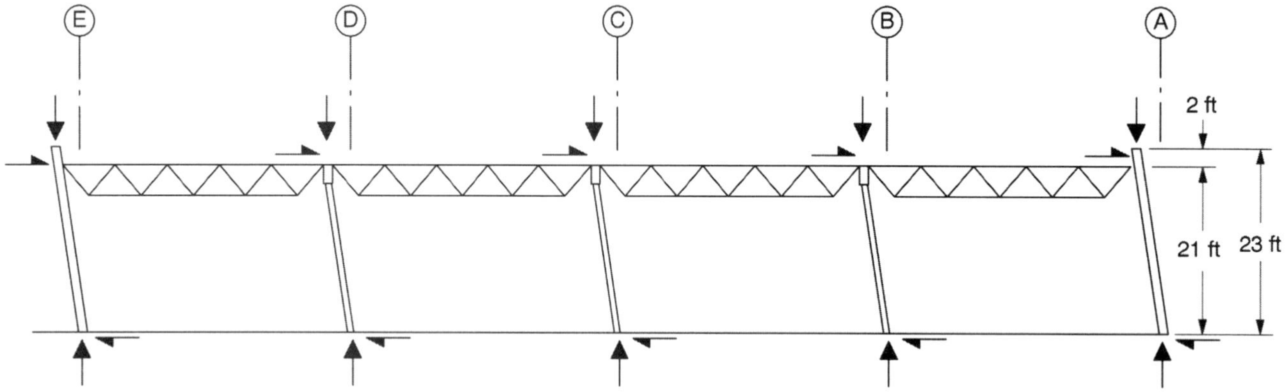

Figure 5-8. Deflected building section

Although it was not originally intended to be used to evaluate diaphragm deformations, §12.8.7 can be used as a guide to investigate stability of the roof system under diaphragm *P*-delta effects. The stability coefficient θ is defined as

$$\theta = \frac{P_x \Delta}{V_x h_{sx} C_d} \qquad \text{(Eq 12.8-16)}$$

P_x is the vertical load acting on the translating system and has two components in this example. $P_{x\,\mathrm{roof}}$ is the translating roof load, and because load combination (5) of §12.4.2.3 is applicable, no roof live load is considered. $P_{x\,wall}$ is the translating concrete wall dead load and comprises the upper half of the wall plus parapet. Load factors need not exceed 1.0.

$P_{x\,roof} = 14 \text{ psf } (224 \text{ ft})(140.67 \text{ ft}) = 441 \text{ kips}$

$P_{x\,wall} = \dfrac{7.25 \text{ in}}{12}(150 \text{ pcf})\left(\dfrac{21 \text{ ft}}{2} + 2 \text{ ft}\right)224 \text{ ft}(2 \text{ sides}) = 507 \text{ kips}$

$P_x \quad = P_{x\,roof} + P_{x\,wall} = 441 + 507 = 948 \text{ kips}$

$\Delta \quad$ = the average horizontal translation. Because this is a flexible diaphragm with an approximately parabolic deflected shape, the average translation is

$$\frac{2}{3}\delta_x = \frac{2}{3}(16.6) = 11.1 \text{ in}$$

$V_x \quad$ = the seismic shear force acting on the translating system under consideration

$V_x \quad = 1063 \text{ plf } (224 \text{ feet}) = 238 \text{ kips}$

$h_{sx} \quad = 21 \text{ ft} \times 12 = 252 \text{ in}$

$C_d \quad = 4$ T 12.2-1

Therefore:

$$\theta = \frac{948(11.1)}{238(252)4} = 0.04 < 0.10$$

Thus: *P*-delta effects on story shears, moments, and story drifts are not required to be considered.

Note: The story drift limitations of §12.12.1 are not intended to apply to flexible diaphragm deflections, but instead are intended to apply to the acting lateral-resisting wall or frame systems. These limitations on building drift were primarily developed for the classic flexible frame system with rigid diaphragm. Story drift limits are designed to ensure that the frames and walls do not excessively distort in plane. Similarly, the *P*-delta limitations of §12.8.7 are also intended to restrict in-plane movements of the vertical seismic resisting system, especially in flexible frames resisting vertical and lateral forces together while subjected to potentially large secondary moments (Tilt-up buildings generally have stiff concrete shear walls that are not impacted by secondary moments from in-plane *P*-delta effects).

6. Design shear force for north-south panel on line 1

In this part, determination of the in-plane shear force on a typical wall panel on line 1 is shown. There are five panels on line 1 (Figure 5-1). The panel with the large opening is assumed to be not effective in resisting in-plane forces, and the four panels remaining are assumed to carry the total shear.

From Part 2, the total diaphragm shear on line 1 is 30.6 kips. This force is on a strength basis and was determined by using $F_p = 0.25w_p$ for the diaphragm. The building's main lateral-force-resisting system (shear walls) is designed for a base shear of $V = 0.25W$ also (see Part 1b), thus an adjustment is not necessary to determine in-plane wall forces.

Earthquake loads on the shear walls must be modified by the redundancy factor p. For buildings of Seismic Design Category D, E, or F, this factor is either 1.0 or 1.3 depending on how much redundancy exists within the vertical lateral-force-resisting system as evaluated by §12.3.4.2. Because this building contains a horizontal structural irregularity as described in Part 2c, Table 12.3-3 must be satisfied in order to use $\rho = 1.0$. An example illustrating the computation of the redundancy factor can be found in Volume 1 of this publication's series. For the purposes of this example, it is assumed the redundancy factor $\rho = 1.0$.

Finally, seismic forces caused by panel self-weight must also be included. These are determined using the base shear coefficient 0.25 from Part 1. The panel seismic force is determined as follows:

Panel self-weight

length = 110 ft

$$W_p = 0.15\left(\frac{7.25}{12}\right)(23\text{ ft})(110\text{ ft}) = 229\text{ kips}$$

Seismic force due to panel self-weight

$$V_{\text{panel}} = 0.25W_p = 0.25(229\text{ k}) = 57.3\text{ kips}$$

The total horizontal seismic shear force on line 1 shear wall is the horizontal shear force transferred from the diaphragm and the horizontal seismic force due to the panel self-weight, both adjusted for the redundancy factor.

The wall line's horizontal shear force $V = \rho Q_E$ may be computed as

$$V_{\text{line 1}} = \rho Q_E = 1.0[30.6 + 57.3] = 87.9\text{ kips}$$

Assuming the four solid panels on line 1 have equal relative stiffnesses and the panel with the large opening is not effective, the shear force per panel is

$$\therefore\ V_{\text{panel}} = 87.9/4 = 22.0\text{ kips per panel}$$

Comment: Distribution of lateral forces along a line of resistance must consider the relative stiffnesses of the individual wall and pier elements. Unlike a masonry building or a cast-in-place concrete building, a tilt-up building has numerous panel joints that can significantly affect the force distribution within a particular wall line. The stiffnesses are affected by both flexural rigidity and shear rigidity. Flexural rigidity considers the pier element's fixity top and bottom. The shear rigidity is proportional to the wall's length and is proportionally more significant on longer solid walls.

In situations where significantly different stiffnesses occur along a wall line, the chord steel may also be required to act as a strut for distribution of forces. It is important to determine whether chord steel is governed by diaphragm chord forces or by the distribution forces.

7. Design wall-roof anchorage for north-south loads

From a historical perspective, the most critical element in tilt-up engineered buildings is the wall anchorage. Prior to the 1971 San Fernando, California earthquake, engineers in the west typically provided no positive direct tie anchoring the perimeter concrete wall panels to the supporting wood roof structure. Instead, the roof plywood sheathing was simply nailed to a wood ledger that was bolted to the inside face of the wall panels. The roof's glue-laminated beams (glulams) were supported on top of concrete pilasters and had tie connections with minimal capacity. This indirect tie arrangement relied on the wood ledger in cross-grain bending, a very weak material property of wood.

In the 1971 San Fernando earthquake, tilt-up buildings performed poorly. Many wood ledgers split in half due to cross-grain bending loads, and plywood edge nailing pulled through plywood panel edges as the result of tension loads. Partial roof collapses and wall collapses were common in the areas of strong ground motion.

Beginning with the 1973 UBC, cross-grain bending in wood was expressly prohibited and specific wall anchorage requirements were established. Over the years since then, the wall anchorage design forces have increased in response to continuing poor performance of wall anchorage during earthquakes and additional information learned from instrumented tilt-up buildings.

The current wall anchorage code requirements are a result of the 1994 Northridge earthquake. The unexpected wall anchorage damage to newer buildings was primarily attributed to inadequate connection overstrength for the roof accelerations. Research has shown that roof top accelerations may be three to four times the ground acceleration. ASCE/SEI 7-05 §§12.11.2.1 and 12.11.2.2 govern wall anchorage design for most of the tilt-up buildings in seismically active areas (Seismic Design Category C and higher for structural walls). The wall tie force of $F_p = 0.8S_{DS}IW_p$ for flexible diaphragms is double the normal wall design force in §12.11.2 and three to four times the typical tilt-up building base shear to account for the expected roof top amplification associated with flexible diaphragms.

The requirements of §13.4.2 associated with anchorage of nonstructural concrete components do not apply because all bearing walls and shear walls are classified as structural walls under §11.2. In addition, all non-structural walls supported by flexible diaphragms are also anchored per §12.11.2 per Table 13.5-1 footnote b. The design forces associated with the concrete and masonry wall anchorage at structural walls have already been factored up to maximum expected levels in comparison with material overstrengths.

7a. Forces on wall anchorage ties

In this example, the structural concrete wall anchorage forces to the flexible diaphragm are governed by Equation 12.11-1 with $S_{DS} = 1.0$ and $I = 1.0$

$$W_p = 90.6 \text{ psf}$$

$$F_p = 0.8 S_{DS} I W_p = 0.80 W_p \qquad \text{Eq 12.11-1}$$

Using statics to sum moments about the wall's base, the following calculation includes the cantilever effects of the parapet in determining the wall anchorage force.

$$W_p = 90.6 \text{ psf}(23 \text{ ft})\left(\frac{23 \text{ ft}}{2}\right)\frac{1}{21 \text{ ft}} = 1141 \text{ plf}$$

Solving for the uniform force per foot (q) at the roof level

$$F_p = 0.8W_p = 0.8(1141) = 913 \text{ plf}$$

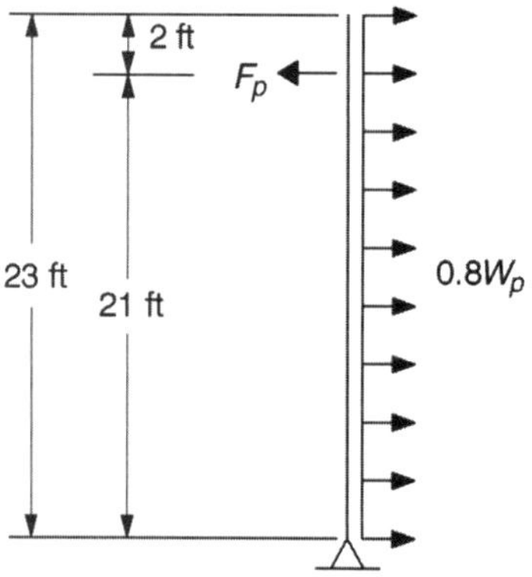

Figure 5-9. Loading diagram for wall-roof anchorage design

Check minimum wall-roof anchorage force per §12.11.2 and IBC §1604.8.2

$$913 \text{ plf} > 280 \text{ plf} \ldots \textit{o.k.} \qquad \text{§1620.1.7}$$

$$913 \text{ plf} > 400 S_{DS} I \ldots \textit{o.k.}$$

$$F_p = 913 \text{ plf} \times 8 \text{ ft} = 7304 \text{ lb}$$

Comment: When tie spacing exceeds 4 feet, §12.11.2 and IBC §1604.8.2 require that structural walls be designed to resist bending between anchors.

7b. Check concrete anchorage of typical wall-roof tie

Concrete anchorage design is in accordance with Appendix D of ACI 318, as referenced by IBC §1912.1 and modified by IBC §1908.1.16. The allowable service loads on embedded bolts in IBC Table 1911.2 are not allowed for seismic design as stated under IBC §1911.1.

The wall-roof anchorage along the north and south walls consists of a steel joist seat welded to an embedded plate with headed weld studs. (See Figure 5-10). Because the embed resists both the wall tie force and the vertical gravity reaction of the steel joist, several loads must be combined.

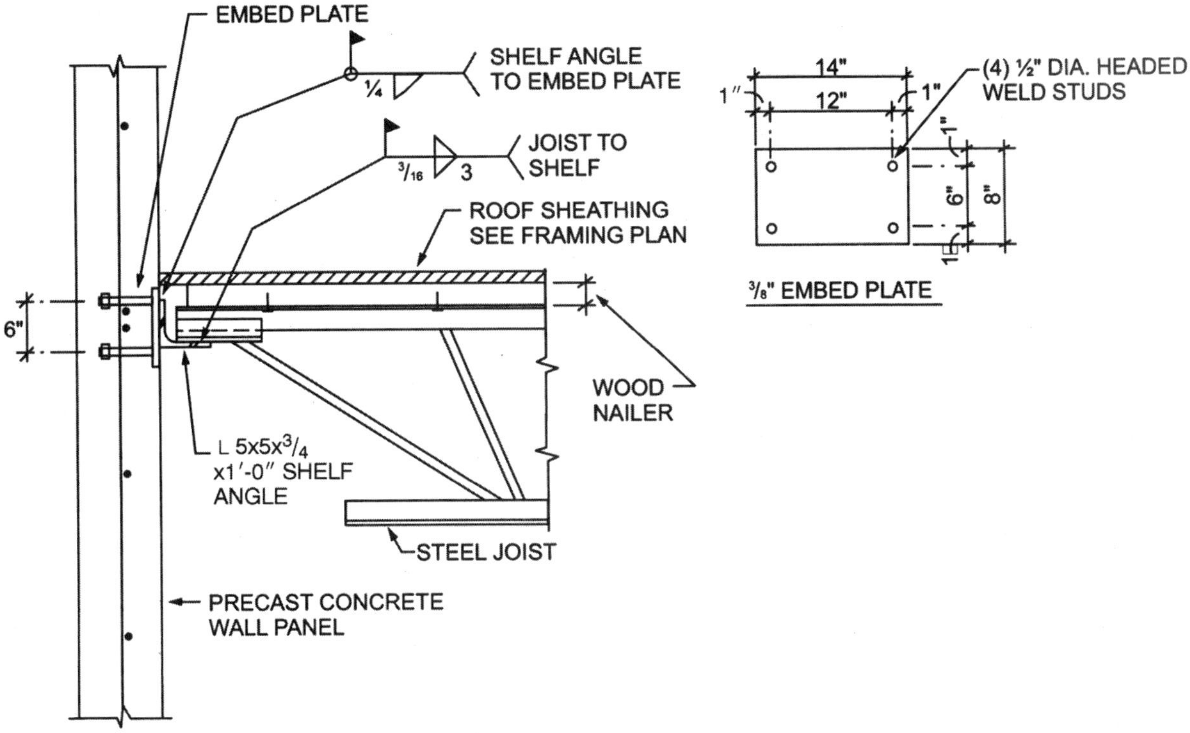

Figure 5-10. Typical steel joist wall-roof tie

The vertical gravity end reaction from the steel joist creates a prying force on the embedded plate's anchors. It will be assumed a force couple at the headed weld studs will resist the eccentric gravity load.

Calculate the joist end reaction R

$$R = (14 \text{ psf} + 20 \text{ psf})(8 \text{ ft})\left(\frac{36.67 \text{ ft}}{2}\right) = 2054 \text{ lb(dead)} + 2934 \text{ lb(live)}$$

Assuming the vertical joist reaction is acting at the edge of the shelf angle, the reaction eccentricity is 5 inches. With the 6-inch vertical spacing between the two pairs of headed weld studs, the following stud forces are determined using the load combinations of IBC §1605.2.1 and ASCE/SEI 7-05 §12.4.2.3:

Load Combination (3) (Eq 16-3)

$$1.2D + 1.6(L_r \text{ or } S \text{ or } R) + (L \text{ or } 0.8W)$$

Given $S = 0$, $R = 0$, $L = 0$, and wind is not being considered, load combination (3) reduces to

$$1.2D + 1.6L_r$$

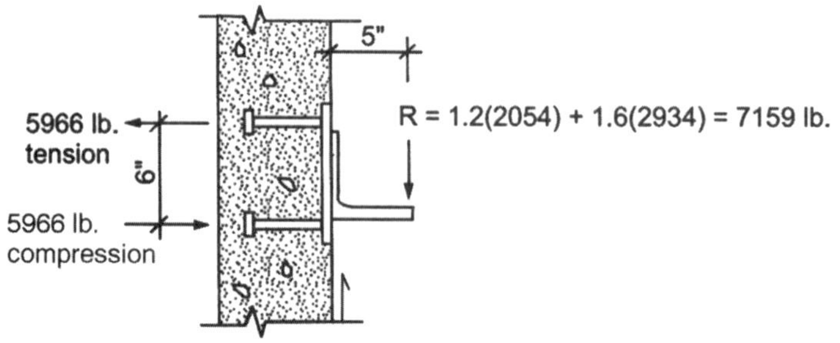

Figure 5-11. Load combination (3) force distribution

Load Combination (5) §12.4.2.3

$$(1.2D + 0.2S_{DS})D + \rho Q_E + L + 0.2S$$

Given $S_{DS} = 1.0$, $L = 0$, $S = 0$, and $\rho = 1.0$, load combination (5) reduces to

$$1.4D + Q_E$$

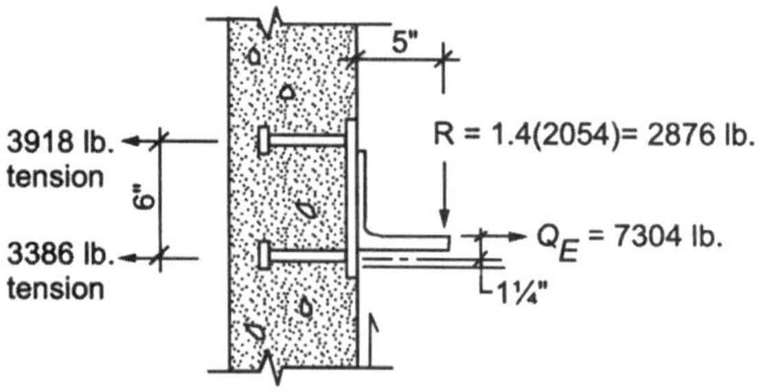

Figure 5-12. Load combination (5) force distribution

Load Combination (7) §12.4.2.3

$$(0.9\text{-}0.2S_{DS})D + pQ_E + L + 1.6H$$

Given $S_{DS} = 1.0$, $H = 0$ and $\rho = 1.0$, load combination (7) reduces to

$$0.7D + Q_E$$

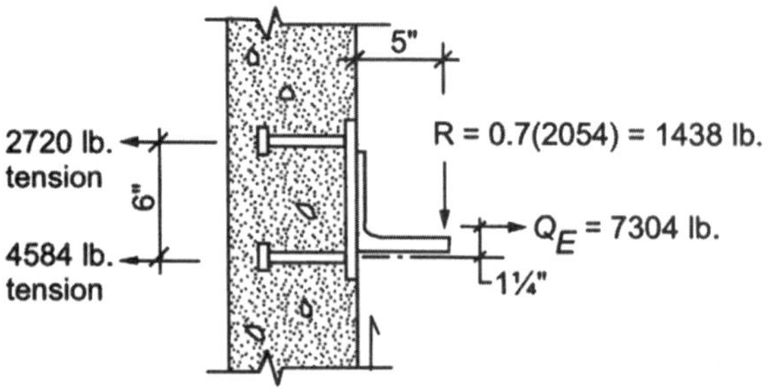

Figure 5-13. Load combination (7) force distribution

Load combination (3) results in only two weld studs loaded in tension, while load combinations (5) and (7) result in all four weld studs tension loaded. Load combination (3) is considered first

Load Combination (3) Analysis

Steel strength in tension N_{sa} — ACI §D.5.1

The nominal steel strength for two ½-inch-diameter ASTM A108 headed weld studs is computed using ACI Equation D-3.

$$N_{sa} = nA_{se}f_{uta}$$

n = 2 bolts in tension

$A_{se} = 0.196 \text{ in}^2$ (½-in-diameter shaft)

f_{uta} = 65,000 psi (AWSD1.1, Type B)

Thus, N_{sa} = 25.5 kips

Concrete breakout strength in tension N_{cbg} — ACI §D.5.2

The two top embedded weld stud anchors are spaced close enough to be considered group action. The ½-inch-diameter studs have an after-weld length of 5 inches, and with their 5⁄16-inch-thick head have an effective embedment of h_{ef} = 4.688 inches. The plate's thickness may be added to $h_{ef,}$ resulting in h_{ef} = 4.688 + 0.375 = 5.06 in. Say, h_{ef} = 5 inches.

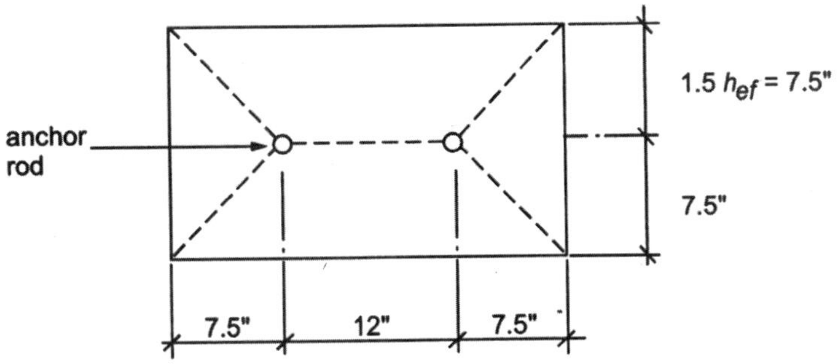

Figure 5-14. Projected failure area A_{Nc} for Load Combination (3)

$$N_{cbg} = \frac{A_{Nc}}{A_{Nco}}(\psi_{ec,N})(\psi_{ed,N})(\psi_{c,N})(\psi_{cp,N})N_b \qquad \text{ACI Eq D-5}$$

$$A_{Nc} = 2(7.5 \text{ in})\,[2(7.5 \text{ in}) + 12 \text{ in}] = 405 \text{ in}^2 \qquad \text{ACI Eq D-6}$$

Per ACI Section D.5.2.1, A_{Nc} shall not exceed nA_{Nco}

$$nA_{Nco} = 2(3\,h_{ef})^2 = 450 \text{ in}^2 > A_{Nc} \ldots o.k.$$

$$\psi_{ec,N} = 2(225)\frac{1}{\left(1+\frac{2e'_N}{3h_{ef}}\right)} \qquad \text{ACI Eq D-9}$$

where e'_N is the eccentricity of the resultant tensile force from the centroid of the bolt group acting in tension. Because there is only one row of bolts acting in tension in this load combination, the bolt group's resultant tension force aligns with the row and thus e'_N is zero.

$$e'_N = 0 \text{ in}$$

$$\psi_{ec,N} = \frac{1}{\left(1+\frac{2(0.0)}{3(5.0)}\right)} = 1.0$$

$\psi_{ed,N} = 1.0$ (no adjacent edge effects)

$\psi_{c,N} = 1.25$ (uncracked section due to short parapet)

$\psi_{cp,N} = 1.0$ (cast-in-place anchor)

$N_b = k_c\sqrt{f'_c}h_{ef}^{1.5} = 24\sqrt{4000}(5)^{1.5} = 17.0$ kips — ACI Eq D-7

$$N_{cbg} = \frac{405}{225}(1.0)(1.0)(1.25)(1.0)(17.0) = 38.3 \text{ kips}$$

Pullout strength in tension — ACI §D.5.3

$N_{pn} = \psi_{c,p}N_p$ — ACI Eq. D-14

$\psi_{c,P} = 1.4$ (assume uncracked section due to short parapet height)

$N_p = 8A_{brg}f'_c$ (where headed studs or bolts are used) — ACI Eq D-15

A_{brg} = (head area) – (shank area) = 0.785 – 0.196 = 0.589 in^2

$N_{pn} = 1.4[8(0.589)(4000 \text{ psi})] = 26.4$ kips

$nN_{pn} = 2(26.4) = 52.8$ kips

Concrete side-face blowout strength in tension — ACI §D.5.4

Because it is assumed that this concrete anchorage is not located near an edge, N_{sb} will not govern the design.

Governing tensile strength

Comparing N_{sa}, N_{cbg}, N_{pn}, and N_{sb}, the governing strength in tension is the steel strength N_{sa} = 25.5 kips. Checking ACI Equation D-1 modified by ACI §D.3.3.3

$$0.75\phi N_n = 0.75(0.75)25.5 \text{ kips} = 14.3 \text{ kips} \geq 5.97 \text{ kips} \ldots o.k.$$

where $\phi = 0.75$ for anchorage governed by ductile steel element strength per ACI §D.4.4 (Weld studs conforming to ASTM A108 Type B qualify as a ductile steel element).

Steel strength in shear V_{sa} — ACI §D.6.1

The nominal steel strength for four ½-inch-diameter ASTM A108 Type B headed weld studs is computed using ACI Equation D-19.

$$V_{sa} = nA_{se}f_{uta}$$

n = 4 bolts

$A_{se} = 0.196\text{ in}^2$ (½-in-diameter shaft)

$f_{uta} = 65{,}000$ psi

Thus, $V_{sa} = 51.0$ kips

Concrete breakout strength in shear V_{cb} — ACI §D.6.2

As previously mentioned, it is assumed in this example that the embed plate is not located near an edge of the panel. In this situation, V_{cb} will not govern (ACI §RD.6.2.1). Often, the purlin layout is not well coordinated with the concrete panel joint layout and thus conflicts are likely to occur. Where purlin embeds are located in close proximity to panel joints, V_{cb} must be evaluated. This is also true for wall panels with no parapet.

Concrete pryout strength in shear V_{cpg} — ACI §D.6.3

The nominal pryout strength for anchors in shear V_{cpg} is a function of the concrete breakout strength N_{cbg} determined earlier.

$$V_{cpg} = k_{cp}N_{cbg} \qquad \text{ACI Eq D-30}$$

$k_{cp} = 2.0$ for anchor embedments $h_{ef} \geq 2.5$ in

$N_{cbg} = 38.3$ kips

$V_{cpg} = 2(38.3) = 76.6$ kips

Governing shear strength

Comparing V_{sa}, V_{cb}, and V_{cpg} the governing strength in shear is the steel strength $V_{sa} = 51.0$ kips. Checking ACI Equation D-2 modified by ACI §D.3.3.3

$$0.75\phi V_n = 0.75(0.65)5\ 1.0\text{ kips} = 24.9\text{ kips} \geq 7.16\text{ kips} \ldots \textit{o.k.}$$

where $\phi = 0.65$ for shear anchorage governed by ductile steel strength per ACI §D.4.4.

Interaction of tensile and shear forces — ACI §D.7

Interaction equation check required if $V_{ua} < 0.2\phi V_n$. However, in Seismic Design Categories C and higher the design strength is multiplied by 0.75 per ACI §D.3.3.3. Thus in this seismic example, an interaction equation check is required if $V\upsilon\alpha < 0.2(0.75)\phi V_n$.

$$7.16\text{ kips} > 0.2(0.75)(0.65)\ 51.0 = 4.97\text{ kips}$$

Thus, interaction equation (D-31) is required to be checked. As stated in ACI §D.3.3.3, the design strength is multiplied by 0.75 in Seismic Design Categories C and higher.

$$\frac{N_{ua}}{0.75\,\phi N_n}+\frac{V_{ua}}{0.75\,\phi V_n}\le 1.2$$ ACI Eq D-31 and ACI §D.3.3.3

For the four weld stud anchorage configuration

$$\frac{5.97}{075(0.75)(25.5)}+\frac{7.16}{0.65(0.75)(51.0)}=0.42+0.29=0.71<1.2\ldots o.k.$$

In summary, the weld studs under the gravity load combination (3) are acceptable.

Load Combinations (5) and (7) Analysis

Steel strength in tension N_{sa} — ACI §D.5.1

The nominal steel strength for four ½-inch-diameter ASTM A108 headed weld studs is computed using ACI Equation D-3.

$N_{sa} = nA_{se}f_{uta}$

n = 4 bolts in tension

$A_{se} = 0.196$ in^2 (½-in-diameter shaft)

f_{uta} = 65,000 psi (AWS D1.1, Type B)

Thus, N_{sa} = 51.0 kips

Concrete breakout strength in tension N_{cbg} — ACI §D.5.2

The four embedded weld stud anchors are spaced close enough to be considered group action. The ½-inch-diameter studs have an after-weld length of 5 inches, and with their 5⁄16-inch-thick head have an effective embedment of h_{ef} = 4.688 inches. The plate's thickness may be added to h_{ef} resulting in h_{ef} = 4.688 + 0.375 = 5.06 in. Say h_{ef} = 5 inches.

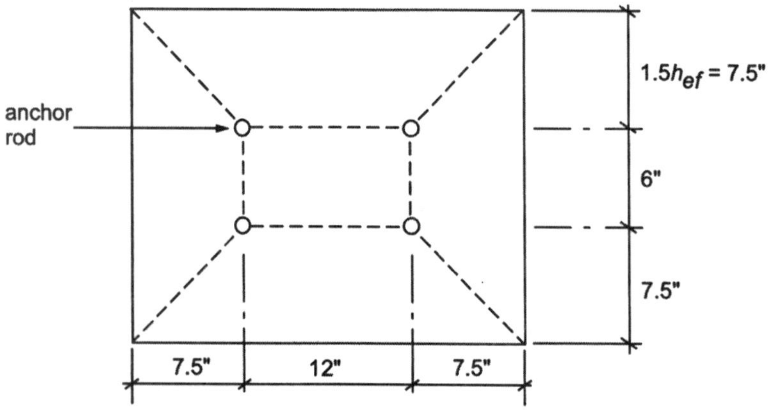

Figure 5-15. Projected failure area A_{Nc} for Load Combinations (5) and (7)

Because load combinations (5) and (7) result in all four stud anchors in tension, a larger concrete breakout projected area is used.

$$N_{cbg} = \frac{A_{Nc}}{A_{Nco}}(\psi_{ec,N})(\psi_{ed,N})(\psi_{c,N})(\psi_{cp,N})N_b \qquad \text{ACI Eq D-5}$$

$$A_{Nc} = (2(7.5\text{ in}) + 6\text{ in})[2(7.5\text{ in}) + 12\text{ in}] = 567\text{ in}^2$$

$$A_{Nco} = 9h_{ef}^2 = 9(5)^2 = 225\text{ in}^2 \qquad \text{ACI Eq D-6}$$

Per ACI §D.5.2.1, A_{Nc} shall not exceed nA_{Nco}

$$nA_{Nco} = 4(225) = 900\text{ in}^2 > A_{Nc} \ldots o.k.$$

$$\psi_{ec,N} = \frac{1}{\left(1+\frac{2e'_N}{3h_{ef}}\right)} \qquad \text{ACI Eq D-9}$$

where e'_N is the eccentricity of the resultant tensile force from the centroid of the bolt group. Using statics, e'_N is computed for both load combinations

$$e'_N = \frac{6\text{ in}}{2} - \frac{6\text{ in}(3918\text{ lb})}{7304\text{ lb}} = .022\text{ in} \qquad \text{Comb. (5)}$$

$$e'_N = \frac{6\text{ in}}{2} - \frac{6\text{ in}(2720\text{ lb})}{7304\text{ lb}} = 0.77\text{ in} \qquad \text{Comb. (7) [Governs]}$$

$$\psi_{ec,N} = \frac{1}{\left(1+\frac{2(0.77)}{3(5.0)}\right)} = 0.91$$

$$\psi_{ed,N} = 1.0 \text{ (no adjacent edge effects)}$$

$$\psi_{c,N} = 1.25 \text{ (uncracked section due to short parapet)}$$

$$\psi_{cp,N} = 1.0 \text{ (cast-in-place anchor)}$$

$$N_b = k_c\sqrt{f'_c}h_{ef}^{1.5} = 24\sqrt{4000}(5)^{1.5} = 17.0\text{ kips} \qquad \text{ACI Eq D-7}$$

$$N_{cbg} = \frac{567}{225}(0.91)(1.0)(1.25)(1.0)(17.0) = 48.7\text{ kips}$$

Pullout strength in tension — ACI §D.5.3

$$N_{pn} = \psi_{c,p}N_p \qquad \text{ACI Eq D-14}$$

$$\psi_{c,p} = 1.4 \text{ (assume uncracked section due to short parapet height)}$$

$$N_p = 8A_{brg}f'_c \text{ (where headed studs or bolts are used)} \qquad \text{ACI Eq D-15}$$

$$A_{brg} = \text{(head area)} - \text{(shank area)} = 0.785 - 0.196 = 0.589\text{ in}^2$$

$$N_{pn} = 1.4[8(0.589)(4000\text{ psi})] = 26.4\text{ kips}$$

$$nN_{pn} = 4(26.4) = 105.6\text{ kips}$$

Concrete side-face blowout strength in tension ACI §D.5.4

Because it is assumed that this concrete anchorage is not located near an edge, N_{sb} will not govern the design.

Governing tensile strength

Comparing N_{sa}, N_{cbg}, N_{pn}, and N_{sb}, the governing strength in tension is the concrete breakout $N_{cbg} = 48.7$ kips. Checking ACI Equation D-1 modified by ACI §D.3.3.3

$$0.75\phi N_n = 0.75(0.70)48.7 \text{ kips} = 25.6 \text{ kips} \geq 7.3 \text{ kips} \ldots \textit{o.k.}$$

where $\phi = 0.70$ for anchorage governed by concrete breakout per ACI §D.4.4.

Per ACI §D.3.3.4 as modified by IBC §1908.1.16, structures with SDC C or higher must show that the behavior of the anchorage or attachment is ductile or have a design strength of at least 2.5 times the connection's factored forces under seismic conditions. Because concrete breakout strength (brittle) governs over the steel strength (ductile), we must check ACI Equation D-1 modified by ACI §D.3.3.3 with the 2.5 overstrength factor

$$N_{ua} = 2.5F_p = 2.5(7.3) = 18.3 \text{ kips} \leq 25.6 \text{ kips} \ldots \textit{o.k.}$$

Because the weld stud anchorage forces are not distributed evenly among all four studs, separate checks for the steel strength N_{sa} and pullout strength N_{pn} are recommended for the heaviest loaded pair (the breakout strength equation already accounts for the uneven distribution). In load combination (7), the lower pair is the most heavily loaded.

$$N_{ua} = 2.5F_p = 2.5(4584 \text{ lb}) = 11{,}460 \text{ lb}$$

For two weld studs

$$0.75\phi N_{sa} = 0.75(0.75)51.0/2 = 14.3 \text{ kips} > 11.46 \text{ kips} \ldots \textit{o.k.}$$

$$0.75\phi N_{pn} = 0.75(0.75)2(26.4) = 29.7 \text{ kips} > 11.46 \text{ kips} \ldots \textit{o.k.}$$

Steel strength in shear V_{sa} ACI §D.6.1

The nominal steel strength for four ½-inch-diameter ASTM A108 Type B headed weld studs is computed using ACI Equation D-19.

$$V_{sa} = nA_{se}f_{uta}$$

n = 4 bolts

$A_{se} = 0.196 \text{ in}^2$ (½-in-diameter shaft)

$f_{uta} = 65{,}000$ psi

Thus, $V_{sa} = 51.0$ kips

Concrete breakout strength in shear V_{cb} — ACI §D.6.2

As previously mentioned, it is assumed in this example that the embed plate is not located near an edge of the panel. In this situation, V_{cb} will not govern (ACI §RD.6.2.1). Often, the purlin layout is not well coordinated with the concrete panel joint layout and thus conflicts are likely to occur. Where purlin embeds are located in close proximity to panel joints, V_{cb} must be evaluated.

Concrete pryout strength in shear V_{cpg} — ACI §D.6.3

The nominal pryout strength for anchors in shear V_{cpg} is a function of the concrete breakout strength N_{cbg} determined earlier.

$$V_{cpg} = k_{cp} N_{cbg} \qquad \text{ACI Eq D-30}$$

$k_{cp} = 2.0$ for anchor embedments $h_{ef} \geq 2.5$ in

$N_{cbg} = 48.7$ kips

$$V_{cpg} = 2(48.7) = 97.4\text{kips}$$

Governing shear strength

Comparing V_{sa}, V_{cb}, and V_{cpg}, the governing strength in shear is the steel strength $V_{sa} = 51.0$ kips. Checking ACI Equation (D-2) modified by ACI §D.3.3.3

$$0.75\phi V_n = 0.75(0.65)\ 51.0 \text{ kips} = 24.9 \text{ kips} \geq 2.88 \text{ kips} \ldots \textit{o.k.}$$

where $\phi = 0.65$ for shear anchorage governed by ductile steel strength per ACI §D.4.4.

Per ACI §D.3.3.4 as modified by IBC §1908.1.16, structures in SDC C or higher must show that the behavior of the anchorage or attachment is ductile or have a design strength of at least 2.5 times the connection's factored forces. Checking ACI Equation D-2 modified by ACI §D.3.3.3 with this limitation is

$$V_{ua} = 2.5(2.88 \text{ kips}) = 7.2 \text{ kips} \leq 24.9 \text{ kips} \ldots \textit{o.k.}$$

Interaction of tensile and shear forces — ACI §D.7

Interaction equation check is required if $V_{ua} < 0.2\phi V_n$. However, in Seismic Design Categories C and higher the design strength is multiplied by 0.75 per ACI §D.3.3.3. Thus in this seismic example, an interaction equation check is required if $V_{ua} < 0.2(0.75)\phi V_n$.

$$7.2 \text{ kips} > 0.2(0.75)(0.65)51.0 = 4.97 \text{ kips}$$

Thus, interaction Equation D-31 is required to be checked. As stated in ACI §D.3.3.3, the design strength is multiplied by 0.75 in Seismic Design Categories C and higher.

$$\frac{N_{ua}}{0.75\phi N_n} + \frac{V_{ua}}{0.75\phi V_n} \leq 1.2 \qquad \text{ACI Eq D-31 and ACI §D.3.3.3}$$

For the four weld stud anchorage configuration:

$$\frac{18.3}{0.75(0.70)(48.7)}+\frac{7.2}{0.75(0.65)(51.0)}=0.72+0.29=1.01<1.2\ldots o.k.$$

As discussed when checking tensile strength previously, the bottom pair of weld studs is more critically loaded under load combination (7) than the group of four weld studs under any load combination. However, $V_{ua} < 0.2(0.75)\phi V_n$ for the weld studs under load combination (7) and thus a separate interaction check is not required in this example.

Check requirements to preclude splitting failure — ACI §D.8

For the cast-in-place headed studs, the following limits are checked:

Minimum center-to-center spacing = 4 diameters = 2 inches < 6 inches
Minimum edges distance = Concrete cover per ACI Section 7.7 . . . *o.k.*

In summary, the four ½-inch-diameter × 5-inch headed weld studs are acceptable.

7c. Check shelf angle at typical wall-roof tie

In this example, the steel joist purlin sits on a steel shelf angle (L5 × 5 × ¾-in × 1 ft). Without additional information, it is assumed the load acts at the tip of the leg. The horizontal leg is subject to bending and seismic tension stresses. Evaluating the array of load combinations for strength design (ASCE §§2.3.2 and 12.4.2.3), combinations (3) and (5) potentially govern.

Simplified load combination (3) — (Eq 16-3)

$$1.2D + 1.6L_r$$

Joist reaction = 1.2(2,054 lb) + 1.6(2,934 lb) = 7159 lb

Moment arm to critical section = leg – k dimension = 5 – 1.25 = 3.75 in

$$M_r = 7{,}159 \text{ lbs } (3.75 \text{ in}) = 26{,}846 \text{ in-lb}$$

$$\text{Plastic section modulues } Z = \frac{12 \text{ in } (0.75 \text{ in})^2}{4} = 1.69 \text{ in}^3$$

Per AISC Section F11.1, the nominal flexural strength, M_n, may be checked as follows:

$$M_n = M_p = F_y Z \le 1.6M_y \qquad \text{AISC Eq F11-1}$$

$$M_p = 36{,}000 \text{ ksi } (1.69 \text{ in}^3) = 60{,}840 \text{ in-lb}$$

$$1.6M_y = 1.6F_y S = 1.6(36{,}000)\left(\frac{12 \text{ in}(0.75 \text{ in})^2}{6}\right) = 64{,}800 \text{ in-lb}$$

Thus, $M_n = 60{,}840$ in-lb

The design flexural strength is checked as follows:

$$\phi_b M_n = 0.90(60{,}840) = 54{,}756 \text{ in-lb} \geq 26{,}846 \text{ in-lb} \ldots \textit{o.k.}$$

Simplified load combination (5) — ASCE 7 §12.4.2.3

$$1.4D + Q_E$$

A combination of gravity forces with horizontal tie forces will be evaluated.

Joist gravity reaction = 1.4(2054 lb) = 2876 lb (dead load)

Moment arm to critical section = leg – k dimension = 5 – 1.25 = 3.75 in

$$M_r = 2876 \text{ lb}(3.75 \text{ in}) = 10{,}785 \text{ in-lb}$$

$$Z = \frac{12 \text{ in } (0.75 \text{ in})^2}{4} = 1.69 \text{ in}^3$$

$$M_n = M_p = F_y Z \leq 1.6M_y \qquad \text{AISC Eq F11-1}$$

$$M_p = 36{,}000 \text{ ksi}(1.69 \text{ in}^3) = 60{,}840 \text{ in-lb}$$

$$1.6M_y = 1.6F_y S = 1.6(36{,}000)\left(\frac{12 \text{ in}(0.75 \text{ in})^2}{6}\right) = 64{,}800 \text{ in-ft}$$

Thus, $M_n = 60{,}840$ in-lb

The design flexural strength is checked as follows:

$$\phi_b M_n = 0.90(60{,}840) = 54{,}756 \text{ in-lb} \geq 10{,}785 \text{ in-lb} \ldots \textit{o.k.}$$

Joist horizontal tie force = 7304 lb (from Part 7a)

Per ASCE §12.11.2.2.2, steel elements of the structural wall anchorage system (SDC C and above) are designed for strength forces with an additional 1.4 multiplier. This material-specific multiplier is based on the observed poor performance of steel straps during the Northridge earthquake. It was determined that an inadequate overstrength range existed in various steel elements to accommodate the maximum expected roof top accelerations. This 1.4 force multiplier is applied to all steel elements resisting the wall anchorage forces of §12.11 (SDC C and above) including wall connectors, subdiaphragm strapping, continuous ties and their connections. Concrete reinforcing steel, concrete anchor rods and headed weld studs, wood bolting and nailing are not subject to this force multiplier.

Required tie force $P_r = 1.4(7304 \text{ lb}) = 10{,}226$ lb — ASCE §12.11.2.2.2

Tensile area $A_g = 12 \text{ in} \times 0.75 \text{ in} = 9 \text{ in}^2$

Design tensile strength for checking combined forces per AISC §H1.2:

$$P_c = \phi_t P_n = \phi_t F_y A_g = 0.90(36{,}000)9 = 291{,}600 \text{ lb} \qquad \text{AISC Eq D2-1}$$

$\frac{P_r}{P_c} = \frac{10{,}226}{291{,}600} = 0.04 < 0.2$, therefore AISC Equation H1-1b is applicable for checking the combined forces of tension and bending flexure.

$$\frac{P_r}{2P_c} + \left(\frac{M_{rx}}{M_{cx}} + \frac{M_{ry}}{M_{cy}}\right) \le 1.0 \qquad \text{AISC Eq H1-1b}$$

$$\frac{0.04}{2} + \frac{10{,}785}{54{,}576} = 0.22 \le 1.0 \ldots o.k.$$

Therefore, the shelf angle support is adequate.

7d. Check the shelf angle weld to the embed plate

Check the use of a ¼-inch fillet weld all around the shelf angle's perimeter. Per AISC Table J2.4, the ¼-inch fillet weld meets the minimum weld size limitations for the thinner plate joined (⅜-inch embed plate), and per AISC §J2.2b the ¼-inch fillet weld meets the maximum weld size limitations for the ¾-inch edge thickness of the shelf angle.

Similar to the process in Part 7b, the force distribution to the shelf angle's upper and lower welds is shown in Figure 5-16 for the various potentially governing load combinations.

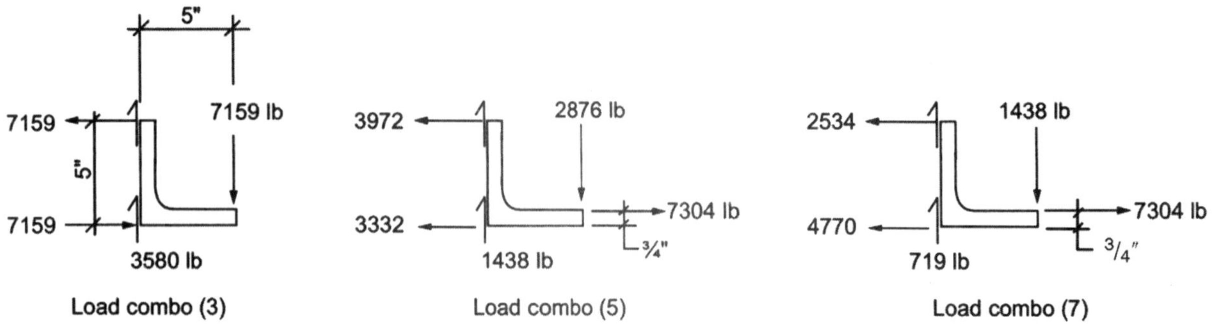

Figure 5-16. Load combination force distributions

Because load combinations (5) and (7) involve seismically induced wall anchorage force to the weld, they are subject to the 1.4 force multiplier of ASCE §12.11 (SDC C and above). Note that the dead load component of the seismic load combinations contains S_{DS} and thus both the vertical and horizontal acting forces are multiplied by 1.4. The following lists the effective results of the vertical and horizontal force vectors:

$$P_r = \sqrt{7159^2 + 3580^2} = 8004 \text{ lb} \qquad \text{Comb. (3)}$$

$$P_r = \sqrt{(1.4 \times 3972)^2 + (1.4 \times 1438)^2} = 5914 \text{ lb} \qquad \text{Comb. (5)}$$

$$P_r = \sqrt{(1.4 \times 4770)^2 + (1.4 \times 719)^2} = 6753 \text{ lb} \qquad \text{Comb. (7)}$$

In this example, the strictly gravity load combination governs at 8004 lb because the gravity load offsets a portion of the seismic anchorage force. Where larger wall anchorage loads occur, often the other load combinations govern.

Checking the strength of the ¼-inch × 12-inch-long fillet weld gives

$$\phi R_n = \phi F_w A_w = 0.75(0.6 \times 70 \text{ ksi})\left(\frac{0.25 \text{ in}}{\sqrt{2}} \times 12 \text{ in}\right) = 66.8 \text{ kips} > 8.0 \text{ kips} \ldots \textit{o.k.}$$

Therefore, the shelf angle weld to the embed plate is adequate.

7e. Check joist seat weld at typical wall-roof tie

The connection of the joist to the embed's shelf angle is through a fillet weld. Given its orientation, the steel shelf angle (L5 × 5 × ¾ in × 1 ft) has a flat run-out distance of 3¾-inches suitable for joist seat bearing.

Per the Steel Joist Institute's Standard Specification (2005), the minimum weld at the joist seat attachments is a ¼ × 2-inch-long fillet or equivalent each side of seat (LH-series joists). Because the seats in these roof systems are typically thinner than ¼ inch it is desirable to specify an equivalent 3⁄16 × 3-inch-long fillet weld. For seats of 3⁄16-inch or thicker material, this fillet weld meets maximum weld size limitations of AISC §J2.2b and the minimum weld size limitations of AISC Table J2.4.

Checking the strength of the two rows of 3⁄16 × 3-inch-long fillet weld is as follows:

$$\phi R_n = \phi F_w A_w = 0.75(0.6 \times 70 \text{ ksi})\left(\frac{0.1875 \text{ in}}{\sqrt{2}} \times 3 \text{ in}\right)2 = 25.1 \text{ kips} \qquad \text{AISC §J2.4}$$

Required tie force P_r = 1.4(7304 lb) = 10,226 lb < 25,100 lb . . . *o.k.*

Therefore, the joist seat weld to the shelf angle support is adequate.

7f. Design steel joist for typical wall-roof anchorage forces

Whether using a panelized wood sheathed roof or a metal deck roof, steel trusses or joists are the most common roof framing system now in tilt-up buildings. In the West, this trend began in the early 1990s when speculative timber prices disrupted the costs of traditional glulam wood roof systems. Specialty engineers in association with the joist manufacturer typically design the steel joist members. As required by IBC §2206.2, the building's design engineer is responsible for providing axial wall tie and continuity tie forces to the manufacturer along with information stating which load factors if any have already been applied.

In this example, it should be reported to the joist manufacture that the unfactored wall tie axial force (tension and compression) acting on the joist top chord is Fp = 7304 lb increased by the steel material overstrength factor 1.4 per §12.11.2.2.2 resulting in F_p = 7304 × 1.4 = 10,226 lb. It is necessary to indicate to the joist manufacturer that this tie force is from seismic effects so that the joist's specialty engineer is able to apply the proper load combinations of §12.4.2.3.

Though not shown in this example, the top chord axial effects of wind W must also be considered if it could lead to a governing design of the joist. Because the load combinations of §2.3.1 (strength design) and §2.4.1 (allowable stress design) contain very different formulas when considering seismic E and wind W, the design engineer cannot simply compare E and W to determine which governs. Currently, the joist industry is largely based on allowable stress design, but it is expected to transition to strength design in the future.

In conditions where axial loads are transferred through the joist seat at either the wall tie or at interior splices, it must be made clear to the manufacturer so that the seat strength will be checked also. There are limits to the amount of load that manufacturers can transfer through these joist seats, so check with the manufacturer's specialty engineer.

In Part 4 of this example, the collector member was a steel wide-flange beam. In some situations, the steel joist can resist lighter collector loads. In these situations, the building's engineer must specify an E_m collector load as well as an E wall tie load. The joist manufacturer's specialty engineer will have to check both the basic load combinations of §12.4.2.3 for$E$ as well as the basic load combinations with overstrength factor of §12.4.3.2 for E_m.

For this example, the following is the type of information to be placed on the drawings for the steel joist manufacturer to properly design his joists for lateral loadings. Note that the wall anchorage force E shown should already include the 1.4 multiplier for steel elements.

Joist Axial Forces E = 10.2 kips (unfactored)

E_m = 0.0 kips (unfactored) *Applicable only at collectors.*

W = 5.0 kips (unfactored)

Forces shall be checked in both tension and compression. Axial force shall be transferred through the joist seats where noted.

7g. Check joist-to-joist splice at the girder lines

Interconnection of elements within the building is required per ASCE/SEI 7-05 §§12.1.3 and 12.1.4. In addition, the joist axial load from the wall anchorage must be distributed across the building's main diaphragm from chord to chord per §12.11.2.2.1 using continuous ties (SDC C and above). Seismic loading in the north-south direction utilizes the steel joists as the continuous ties, and thus the joist axial load must be spliced across the interior girder lines. In Part 7c, the wall anchorage force and thus continuous tie force for the steel joists is $P_r = 1.4(7{,}304 \text{ lb}) = 10{,}226$ lb.

Per §12.1.3, the minimum interconnection force is $0.133S_{DS}W = 0.133W$, but not less than $0.05W$, where W is the dead load of the smaller portion of the building being connected together. Unlike the wall anchorage force, Win this case includes the diaphragm weight and thus could govern at the interior of buildings. The worst-case value for W is at grid line C with the following result:

$$P_r\,(\text{min}) = 0.133(14 \text{ psf})(8 \text{ ft})(30.67 \text{ ft} + 36.67 \text{ ft}) + 0.133(90.6 \text{ psf})(8 \text{ ft})(23)(23/2)/21$$

$$= 2217 \text{ lb}$$

Per §12.1.4, the minimum support connection force is 5 percent of the dead and live load reaction.

$$P_r\text{ (min)} = 0.05(14\text{ psf} + 20\text{ psf})(8\text{ ft})(36.67\text{ ft}/2) = 249\text{ lb}$$

Thus, the wall anchorage continuous tie force $P_r = 1.4(7304\text{ lb}) = 10{,}226\text{ lb}$ governs.

The splice can be accomplished with a welded cover plate from joist top chord to joist top chord (see Figure 5-17). Check the use of a ¼ × 3-in-wide cover plate with $^3/_{16}$-in fillet welds:

Check the design tensile strength per AISC §D2

$$\phi_t P_n = \phi_t F_y A_g = 0.90(36{,}000)(0.25)(3) = 24{,}300\text{ lbs} \qquad \text{AISC Eq D2-1}$$

Required tie force $P_r = 10{,}226$ lbs < 24,300 lbs . . . *o.k.*

Using two lines of $^3/_{16}$ × 2-inch-long fillet welds, check the design weld strength per AISC §J2.4

$$\phi R_n = \phi F_w A_w = 0.75(0.6 \times 70\text{ ksi})\left(\frac{0.1875\text{ in}}{\sqrt{2}} \times 2\text{ in}\right)2 = 16.7\text{ kips}$$

Required tie force $P_r = 10{,}226$ lb < 16,700 lb . . . *o.k.*

Therefore, the steel joist splice across the interior girders is adequate.

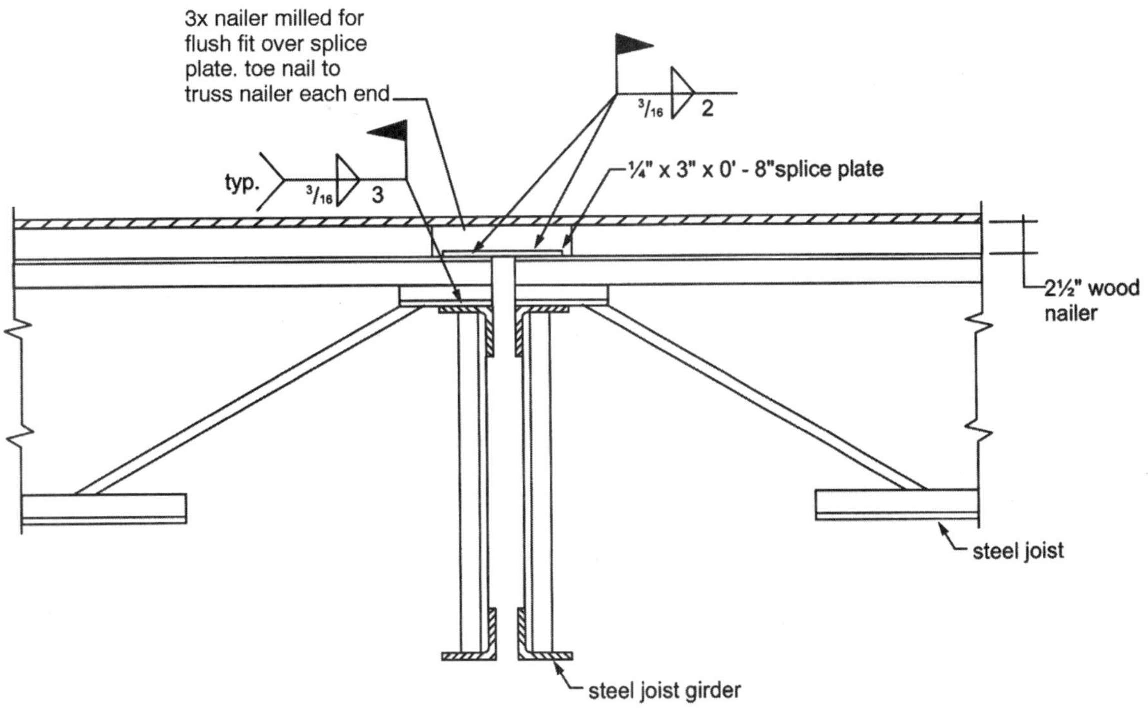

Figure 5-17. Joist-to-joist splice at joist girder

Comment: It is possible to splice the joist axial loads across the interior girders through their joist seats as is done at the wall anchorage joist end. However, this means added joist seat costs and requires the joist girder double-angle top chords to be joined together for this perpendicular force. If this is the design engineer's intent, it must be made clear to the joist manufacturer that the joist seats and joist-girders top chords are to be designed for these forces including the 1.4 overstrength factor.

8. Design wall-roof anchorage for east-west loads

On the east and west wall elevations, wall-roof ties are used to transfer out-of-plane seismic forces on the tilt-up wall panels to the subdiaphragms. Applicable requirements for connection of out-of-plane wall anchorages to flexible diaphragms are specified in §12.11.2.1.

8a. Seismic force on wall-roof tie

Seismic forces are determined using Equation 12.11-1. These are the same forces as those determined in Part 7 for the north and south walls.

$$F_p = 0.8S_{DS}IW_p = 0.8W_p = 913 \text{ plf} \qquad \text{(Eq 12.11-1)}$$

8b. Design typical wall-roof tie

Try ties at 8-foot spacing, and determine F_p

$$F_p = 8 \text{ ft} \times 913 \text{ plf} = 7304 \text{ lb}$$

Comment: When tie spacing exceeds 4 feet, §12.11.2 and IBC §1604.8.2 require that walls be designed to resist bending between anchors.

Try prefabricated metal hold-downs with two ¾-inch bolts into a 3x subpurlin and two ⅝-inch anchor rods connecting the hold-downs to the wall panel. This connection, illustrated in Figure 5-18, is designed to take both tension and compression as recommended by the SEAOS C/COLA Northridge Earthquake Tilt-up Building Task Force and the 1999 SEAOC Blue Book (§C108.2.8.1). Design of the hold-down hardware is not shown. Consult ICC-ES Evaluation Reports for the allowable load capacity of pre-manufactured hold-downs. Note that if a one-sided hold-down is used, eccentricities in the subpurlin should be considered per §12.11.2.2.6. Generally, one-sided wall-roof anchorage is not recommended in SDC C and above.

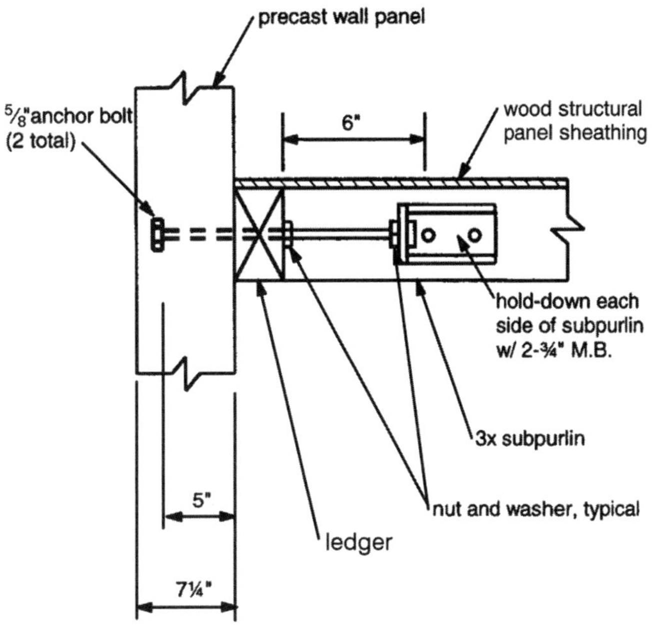

Figure 5-18. Typical subpurlin wall-roof tie

Check capacity of the two ¾-inch bolts in the Douglas Fir-Larch 3x subpurlin using 2005 AF&PA NDS Table 11G, where $C_d = 1.6$ and $C_g = 0.97$

$$(2630)(2 \text{ bolts})(1.6)(0.97) = 8164 \text{ lb} > 7304(0.7) = 5113 \text{ lb} \ldots \textit{o.k.}$$

Minimum required end distance = $7D = 7(0.75) = 5.25$ in 2005 NDS T 11.5.1B

A distance of 6 inches from the through-bolt in the hold-down to the ledger will be used. Often, there is a gap of ⅛ inch or more between the end of the subpurlin and the side of the ledger caused by panelized roof erection methods, and the use of a 6-inch edge distance will ensure compliance with the $7D$ requirement. A larger distance can be used to ensure that through-bolt tear-out does not occur in the 3x subpurlin.

Check tension capacity of two ⅝-inch ASTM F1554 (grade 36) anchor rods using LRFD

$F_t = 0.75\, F_u = 0.75(58) = 43.5$ ksi AISC T J3.2

$\phi R_n = \phi F_t A_b = 0.75(43.5)(2)(0.307) = 20.0$ kips . . . *o.k.* AISC Eq J3-1

$R_u = F_p = 7304 \text{ lb} < 20.0$ kips . . . *o.k.*

Note: The 1.4 factor normally applied to steel elements of the wall anchorage system is not applied to anchor rods per §12.11.2.2.2.

Check compression capacity of two ⅝-inch ASTM F1554 Grade 36 anchor rods using LRFD

$$P_n = F_{cr}A_g \qquad \text{AISC Eq E3-1}$$

$$A_g = A_b = 0.307 \text{ in}^2$$

$$\text{Radius of gyration of ⅝-in rod} = \frac{0.625 \text{ in}}{4} = 0.1563$$

Assume $L = 4\frac{1}{2}$ inches and $K = 1.0$

$$\frac{KL}{r} = \frac{1.0(4.5)}{0.1563} = 28.8$$

$$F_e = \frac{\pi^2 E}{\left(\frac{KL}{r}\right)^2} = 345{,}074 \text{ psi} \geq 0.44\ F_y \text{ thus AISC Equation E3-2 is applicable}$$

$$F_{cr} = \left[0.658^{\frac{F_y}{F_e}}\right]F_y = \left[0.658^{\frac{36}{345}}\right]36{,}000 = 34{,}462 \text{ psi} \qquad \text{AISC Eq E3-2}$$

$$P_n = 34{,}462(0.307)(2 \text{ rods}) = 21{,}160 \text{ lb}$$

$$\phi_c P_n = 0.90(21{,}160) = 19{,}044 \text{ lb} \geq 7304 \text{ lb} \ldots \textit{o.k.}$$

Check tension capacity of anchor rods in wall panel for concrete strength.

The tilt-up panels are exterior wall elements, but the requirements of §§13.4.2 and 13.5.3 do not apply. This is because the tilt-up panels are structural walls instead of nonstructural architectural cladding. The requirements of §12.11 are the appropriate design rules in this situation. Section 12.11.2.2.5 requires that wall anchorage using straps be attached or hooked so as to transfer the forces to the reinforcing steel. In this case, we are using cast-in-place bolts instead of straps, and the bolts are not required to be "hooked" around the wall reinforcement.

Recall that for wall anchorage, $F_p = 7304$ lb. Try a ⅝-inch-diameter ASTM F1554 Grade 36 hex headed bolt embedded in the concrete panel with 5 inches of embedment ($h_{ef} = 5$ inches). Assume that the bolt embedment is not near an edge and that the vertical shear load is negligible.

The wall's concrete anchorage needs to be checked using strength design under ACI 3 18-05 Appendix D. The vertical shear load on the anchor is very low because of the small subpurlin tributary roof load. ACI §D.7.1 allows the full tension strength to be used without reduction when the factored shear load is less than 20 percent of the nominal shear capacity of the anchorage as in this case.

ACI Equation D-1 normally requires $\phi N_n > N_{ua}$, but for structures in SDC C and above, IBC §1908.1.16 requires $0.754\ \phi N_n > N_{ua}$ when resisting seismic loads. N_n is the nominal tension strength of the anchorage. It is determined by checking the steel strength in tension N_{sa} (ACI §D.5.1), the concrete breakout strength in tension N_{cbg} (ACI §D.5.2), the pullout strength in tension N_{pn}(ACI §D.5.3), and the concrete side-face blowout strength in tension N_{sb} (ACI §D.5.4). An additional requirement for structures in SDC C and above is $N_{cbg} > N_{sa}$ (§D.3.3.4) to reduce the likelihood of brittle concrete failure. However, this may also be satisfied by providing anchors with a minimum design strength of 2.5 times the attachment's factored forces (IBC §1908.1.16).

Steel strength in tension N_{sa} — ACI §D.5.1

The nominal steel strength for ⅝-inch-diameter ASTM F1554 Grade 36 headed anchor rods is as follows. Equation D-3 is applicable

$$N_{sa} = nA_{se}f_{uta} \qquad \text{Eq D-3}$$

$n = 2$ bolts

$A_{se} = 0.226\ \text{in}^2$ (net tensile area) — AISC T 7-18

$f_{uta} = 58$ ksi — AISC T 2-5

$N_{sa} = 26.2$kips

Concrete breakout strength in tension N_{cb} — ACI §D.5.2

The two embedded anchors (one each side of subpurlin) are spaced close enough to be considered group action

$$N_{cbg} = \frac{A_{Nc}}{A_{Nco}}(\psi_{ec,N})(\psi_{ed,N})(\psi_{c,N})(\psi_{cp,N})N_b \qquad \text{Eq D-5}$$

$A_{Nc} = 2(7.5) \times (7.5 + 7.0 + 7.5)$

$= 330\ \text{in}^2 < nA_{Nco} \ldots o.k.$

$A_{Nco} = 9h_{ef}^2 = 9(5)^2 = 225\ \text{in}^2$ — ACI Eq D-6

$\psi_{ec,N} = 1.0$ (no eccentric loading)

$\psi_{ed,N} = 1.0$ (no adjacent edge effects)

$\psi_{c,N} = 1.25$ (uncracked section due to short parapet)

$\psi_{cp,N} = 1.0$

$N_b = 24\sqrt{f'_c}h_{ef}^{1.5}$ — ACI Eq D-7

$= 24\sqrt{4000}\ 5^{1.5} = 17.0$ kips

$$N_{cbg} = \frac{330}{225} \times (1.0)(1.0)(1.25)17.0 = 31.2 \text{ kips}$$

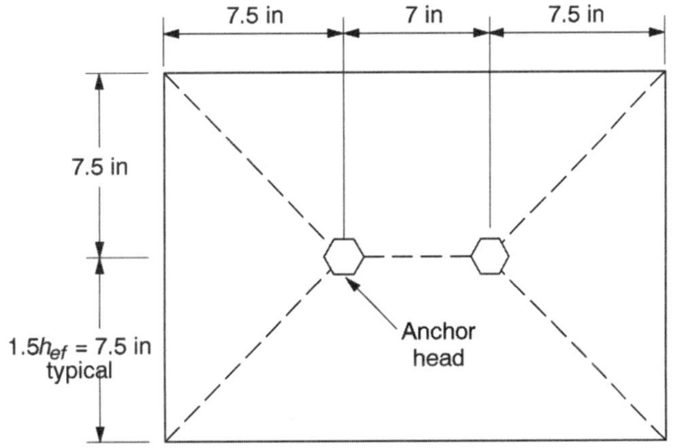

Figure 5-19. Projected failure area A_{Nc}

Pullout strength in tension §ACI D.5.3

$$N_{pn} = \psi_{c,p} N_p = \psi_{c,p}\, 8 A_{brg} f'_c$$ ACI Eq D-14, D-15

$\psi_{c,p} = 1.4$ (Assume uncracked section due to short parapet height)

A_{brg} = Bearing area of head = (Head area) – (shank area)

$$= \frac{3F^2}{2\sqrt{3}} - \text{(shank area)} = 0.761 - 0.307 = 0.454$$

$$N_{pn} = 1.4\,(0.454)8\,(4000)(2 \text{ bolts}) = 40.7 \text{ kips} > N_{sa} \ldots o.k.$$

Concrete side-face blowout strength in tension ACI §D.5.4

Since it is assumed that this concrete anchor is not located near an edge, N_{sb} will not govern the design.

Governing strength

The governing strength in tension is the steel strength N_{sa} = 26.2 kips. Checking ACI Equation D-1 modified by ACI §D.3.3.3 gives

$$0.75\phi N_n = 0.75(0.75)26.2 \text{ kips} = 14.7 \text{ kips} \geq 7.3 \text{ kips} \ldots o.k.$$

where $\phi = 0.75$ for anchorage governed by ductile steel strength per ACI §D.4.4.

Therefore, the proposed two ⅝-inch-diameter anchor rods embedded 5 inches are acceptable.

It is interesting to note that the steel rod's tensile strength obtained from the ACI procedure is lower than the tensile strength obtained earlier using the AISC-LRFD procedure. This is because ACI uses the net tensile area of the threaded fastener while AISC-LRFD uses the nominal area.

Per ACI §D.3.3.4 and IBC §1908.1.16 structures in SDC C or higher must show that the behavior of the anchorage or attachment is ductile or provide an anchorage with a minimum design strength of 2.5 times the attachment's factored forces. Because the more brittle failing N_{cbg} (31.2 kips) and N_{pn} (39.0 kips) are greater than the more ductile failing N_{sa} (26.2 kips), §D.3.3.4 is satisfied here.

<u>Compression</u>

Wall anchorage forces act in compression as well as tension. Panelized wood roof systems by their very nature are not erected tight against the perimeter wall ledger, leaving a small gap to potentially close during seismic compression forces. Strap-type wall anchors that may have yielded and stretched under tensile forces are vulnerable to buckling and low-cycle fatigue as the gaps close. Cast-in-place anchor rods used in connectors can be checked for compression, but it is important to provide an additional nut against the interior wall surface to prevent the anchor punching through the wall. A common wall-roof tie connection shown in Figure 5-20 does not offer the same compression resistance as the anchor rod scheme presented in this example. Although there have been no failures of wall panels collapsing into the building, consideration of compressive forces will maintain the integrity of the wall anchorage tie and protect the diaphragm edge nailing under the reversible seismic forces.

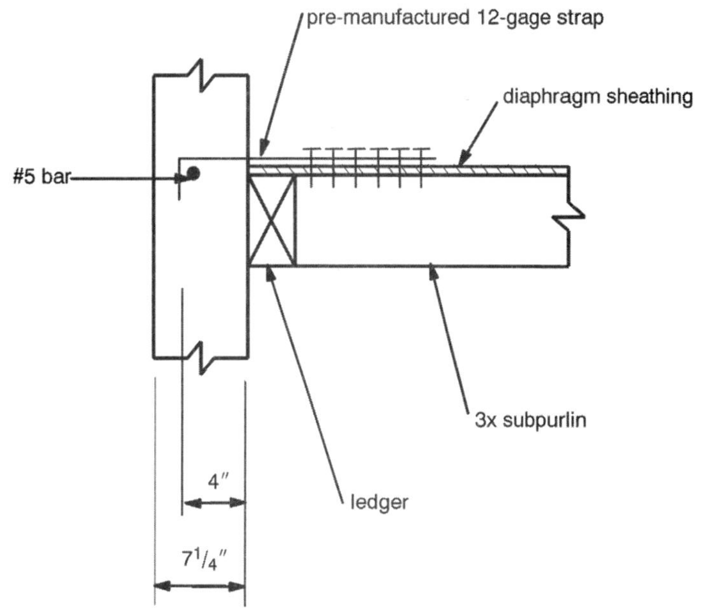

Figure 5-20. Common wall-roof strap tie

Comments about Anchorage Deformation
No prescriptive deformation limits of the wall tie system have been introduced into the IBC or ASCE/SEI 7-05, however the compatibility of the anchorage system's flexibility and the diaphragm shear nailing should be considered. Wall anchorage systems with too much flexibility will inadvertently load the wood sheathing edge nailing and either pull the nails through the sheathing edge or place the wood ledgers in cross-grain bending or tension. Pre-manufactured strap-type wall ties are designed to limit the maximum deformation to ⅛ inch at their rated allowable load, and pre-manufactured hold-down devices using anchor rods could allow even greater deformation (contact the device manufacturer for additional deformation information). This reported hold-down device flexibility is solely within the steel component itself and is additive to other sources of deformation. Additional deformation can be contributed by other anchorage components (e.g., bolts and nails) and installation practices (e.g., oversized holes).

8c. Design connection to transfer seismic force across first roof truss purlin

Under §12.11.2.2.1 for SDC C and higher, continuity ties are provided in diaphragms and subdiaphragms to distribute wall anchorage loads. Consequently, the forces used to design the wall-roof ties must also be used to design the continuity ties within the subdiaphragm. From Part 8b

$$F_p = \text{wall-roof tie load} = 7304 \text{ lb}$$

If the subdiaphragm is modeled as 32 feet deep and steel joist purlins are spaced at 8 feet, the connection at the first purlin must carry three-quarters of the wall-roof tie force.

Comment: Some engineers use the full, unreduced force, but this is not required by rational analysis.

$$\frac{(32-8)}{32} \times F_p = \frac{3}{4} \times 7304 = 5478 \text{ lb}$$

At the second and third purlins, the force to be transferred is one-half and one-fourth, respectively, of the wall-roof tie force.

$$\frac{1}{2} \times 7,304 = 3652 \text{ lb}$$

$$\frac{1}{4} \times 7,304 = 1826 \text{ lb}$$

Try 12-gage metal strap with 10d common nails. Consult ICC-ES Evaluation Reports for allowable load capacity of pre-manufactured straps and ties.

The following calculation shows determination of the number of 10d common nails into Douglas Fir-Larch required at the first connection using allowable stress design

$$\frac{(0.7)5478}{127 \text{ lb}(1.60)} = 18.9 < 19 \text{ nails}$$ 2005 NDS T 11P and T 2.3.2

∴ Use 12-gage metal strap with 19 10d nails each side

The design of the 12-gage metal strap is not presented here, but the design is based on forces increased by 1.4 times the forces otherwise required under §12.11. This requirement of §12.11.2.2.2 is a result of the early strap failures observed in the Northridge Earthquake. It was found that many steel components lacked sufficient

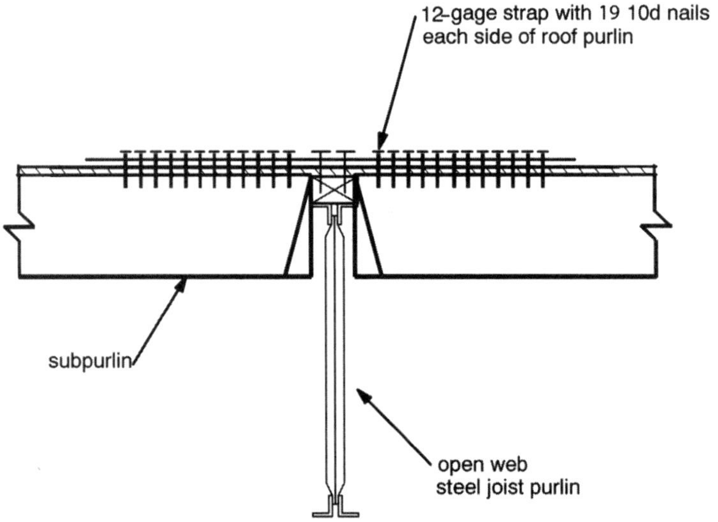

Figure 5-21. Subpurlin continuity tie at first purlin

ductility and overstrength to adequately accommodate seismic overloads. It is the intent of the 1.4 steel-material multiplier to provide sufficient overstrength to resist maximum anticipated wall anchorage forces.

Where pre-manufactured and pre-engineered straps and ties are utilized using capacity values published in ICC-ES Evaluation Reports, the engineer should compare the published capacity with the 1.4 steel increased force unless sufficient information is available to determine steel material values independently of other components.

Note that both subpurlins in Figure 5-21 likely would be 3x members because of the heavy strap nailing.

Design of the second and third connections is similar to that shown above.

Note: Additional requirements for eccentric wall anchorage and walls with pilasters are contained under §§12.11.2.2.6 and 12.11.2.2.7.

9. Design typical east-west subdiaphragm

In the 1976 UBC, the concept of subdiaphragms was introduced as an analytical device for transferring forces from the individual wall anchorage ties to the main diaphragm's continuous crossties. To transfer seismic forces from the heavy perimeter walls into the main roof diaphragm, continuous ties or crossties are necessary to drag the load uniformly across the diaphragm depth. Instead of creating a continuous tie at every wall anchorage location, continuous crossties can be placed at wider spacings using subdiaphragms. Subdiaphragms are portions of the main diaphragm that span between the continuous crossties and gather the wall anchorage loads and transfer these loads to the crossties. Once the load is collected into the continuous crossties it is distributed across the main diaphragm for further distribution to the building's shear walls and frames.

Subdiaphragms are provided for under ASCE/SEI 7-05 §12.11.2.2.1 as an analytical device to provide a rational load path for wall anchorage. Consequently, subdiaphragms are considered part of the wall anchorage system and are subject to loads per §12.11. For SDC C and above, subdiaphragm aspect ratios are limited to 2½ to 1, and this provides sufficient stiffness that the independent deflection between the subdiaphragm and the main diaphragm may be ignored.

9a. Check subdiaphragm aspect ratio

Maximum allowable subdiaphragm ratio is 2.5 to 1 §12.11.2.2.1

From Figure 5-2, the maximum north-south subdiaphragm span = $\dfrac{110\text{ ft}}{3} = 36.67\text{ ft}$

Minimum subdiaphragm depth = $\dfrac{36.67\text{ ft}}{2.5} = 14.67\text{ ft}$

Typical roof purlin spacing = 8 ft

Minimum subdiaphragm depth = 16 ft

∴ 32-foot-depth assumed . . . *o.k.*

9b. Forces on subdiaphragm

Because subdiaphragms are part of the out-of-plane wall anchorage system, they are designed under the requirements of §12.11.2.1, assuming the overall main diaphragm is flexible. Seismic forces on a typical east-west subdiaphragm are determined from Equation 12.11-1 with $S_{DS} = 1.0$ and $I = 1.0$

$$F_p = 0.8 S_{DS} I W_p = 0.80 W_p \qquad \text{Eq 12.11-1}$$

As shown in Part 7, $F_p = 913$ plf

9c. Check subdiaphragm shear

Assume a 32-foot-deep subdiaphragm as shown below. This is done for two reasons. First, the steel joist purlin along line 9 can be used as a subdiaphragm chord. Second, the deeper-than-required subdiaphragm depth (32 feet vs. 16 feet) reduces the subdiaphragm shear to manageable levels.

Shear reaction to continuity tie along lines C and D 913 plf (36.67 ft)

$$R = \frac{913 \text{ plf } (36.67 \text{ ft})}{2} = 16{,}740 \text{ lb}$$

$$\text{Maximum shear} = \frac{16{,}740 \text{ lb}}{32} = 523 \text{ plf}$$

Applying the ASD load combination

$$\text{ASD shear} = 0.7(523 \text{ plf}) = 366 \text{ plf}$$

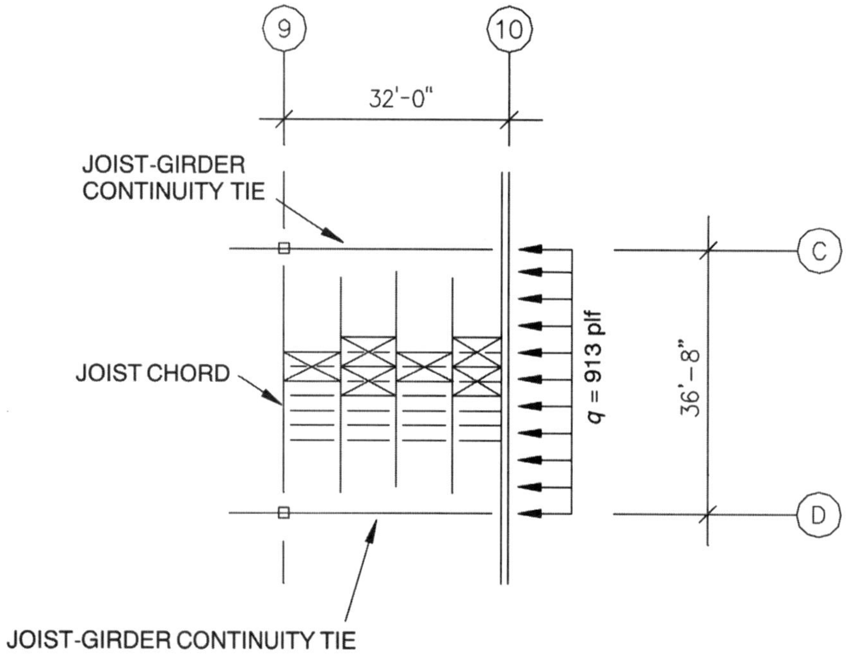

Figure 5-22. Typical subdiaphragm

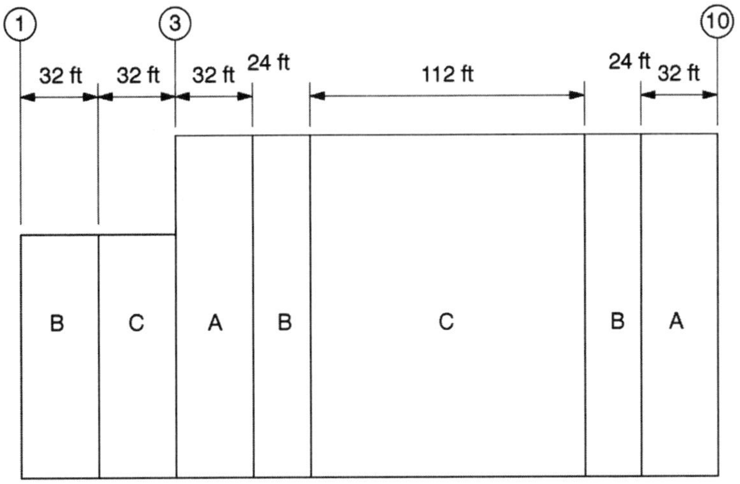

Figure 5-23. Revised nailing zones for north-south diaphragm

The 32-foot-deep subdiaphragm consists of zone A nailing (See Figure 5-6). The diaphragm's ASD shear strength 640 plf (Table 5-1) is adequate to resist the 366 plf load. On the west side of the building along line 1, a similar subdiaphragm situation exists, except the diaphragm design currently consists of the weaker zone C nailing. The first 32-feet will be revised to the stronger zone B nailing at 425 plf for purposes of the subdiaphragm.

Given the nailing of Figure 5-23, check the subdiaphragm shear.
ASD shear of the subdiaphragm = 366 plf < 425 plf. Thus, zone B nailing . . . *o.k.*

9d. Check steel joist as subdiaphragm chord

The steel joists along lines 2 and 9, and the continuous horizontal reinforcement in panels along lines 1 and 10, act as chords for the subdiaphragms. Check to see if the steel joist can carry additional seismic force.

$$\text{Chord force} \frac{913 \text{ plf} (36.67)^2}{8(32)} = 4796 \text{ lb}$$

Because the subdiaphragm chord is a steel element of the wall anchorage system, it is subject to a 1.4 force increase per §12.11.2.2.2.

$$\text{Chord Force (steel)} = 4796(1.4) = 6714 \text{ lbs}$$

This chord force is less than the wall anchorage force found in Part 7a, and thus does not govern.

Comment: In reality, the steel joist along line 9 may not act in tension as a subdiaphragm chord as shown above. It will be loaded in tension only when compressive wall anchorage forces act on the diaphragm. Under this loading, the seismic forces probably do not follow the subdiaphragm path shown above but are transmitted through the wood framing to other parts of the diaphragm. Even if subdiaphragm action does occur, the subdiaphragm may effectively be much deeper than shown. However, because it is necessary to demonstrate that there is a system to resist the out-of-plane forces on the diaphragm edge, the subdiaphragm system shown above is provided.

9e. Determine minimum chord reinforcement at exterior concrete walls

This design example assumes that there is continuous horizontal reinforcement in the walls at the roof level that acts as a chord for both the main diaphragm and the subdiaphragms.

Subdiaphragm chord force = P = 4796 lb

$$A_s = \frac{P}{\phi f_y} = \frac{4796}{0.9(60,000)} = 0.09 \text{ in}^2$$

This is a relatively small amount of reinforcement. Generally, the main diaphragm chord reinforcement exceeds this amount. In present practice, the subdiaphragm chord steel requirement is not added to the chord steel requirement for the main diaphragm. Determination of the main chord reinforcement is shown in Part 3.

10. Design continuity ties for east-west direction

In a tilt-up building, continuous ties have two functions. The first is to transmit the heavy out-of-plane wall loads into the main diaphragm. The second function is that of "tying" the interior portions of the roof together. In this example, the continuity ties on lines C and D will be designed.

10a. Seismic forces on continuity ties along lines C and D

A minimal interconnection of elements within the building is required per ASCE/SEI 7-05 §§12.1.3 and 12.1.4. Additionally, continuous ties or crossties are required per §12.11.2.2.1 (SDC C and above) to transfer seismic forces from the heavy perimeter walls into the main diaphragm. In the east/west load direction, the subdiaphragm load is collected into the continuous crossties and then distributed across the main diaphragm for further distribution to the building's shear walls and frames.

The continuous tie axial force at line 9 is the sum of both subdiaphragm reactions. Because the continuous ties are considered part of the wall anchorage system, their design force is subject to the steel material overstrength multiplier 1.4 per §12.11.2.2.2

$$P_9 = \frac{913 \text{ plf}(36.67 \text{ ft})}{2}(2)1.4 = 46,872 \text{ lb}$$

Per §12.1.3, the minimum interconnection force is $0.133S_{DS}W = 0.133W$, but not less than $0.05W$, where W is the dead load of the smaller portion of the building being connected together. Unlike the wall anchorage force, Win this case includes the diaphragm weight and thus could govern at the interior of buildings. The worst-case value for W for the continuous tie is at grid line 6 with the following result:

$$P_r\text{ (min)} = 0.133(14 \text{ psf})(36.67 \text{ ft})(4)(32 \text{ ft}) + 0.133(90.6 \text{ psf})(36.67 \text{ ft})(23)(23/2)/21$$
$$= 14,305 \text{ lb}$$

Per §12.1.4, the minimum support connection force is 5 percent of the dead and live load reaction.

$$P_r\,(\text{min}) = 0.05(14\text{ psf} + 16\text{ psf})(36.67\text{ ft})(32\text{ ft}/2) = 880\text{ lb}$$

Thus, the wall anchorage continuous tie force is governed by the subdiaphragm design

$$P_r = 46{,}872\text{ lbs}$$

Note: The continuous ties along lines C and D are not collector elements and thus are not subject to the special overstrength load combinations of §12.10.2.1. The girder line along line B functions both as a continuous tie and as a collector; therefore, both basic and overstrength load combinations must be considered.

10b. Design of joist-girders as continuity ties along lines C and D

Whether using a panelized wood sheathed roof or a metal deck roof, open web steel joist-girders are common roof girders in tilt-up buildings. Specialty engineers in association with the joist manufacturer typically design the steel joist-girder members. As required by IBC §2206.2, the building's design engineer is responsible for providing axial continuity tie forces to the manufacturer along with information stating which load factors, if any, have already been applied.

In this example, it should be reported to the joist manufacture that the unfactored wall anchorage axial force (tension and compression) acting on the joist-girder top chord is $P_r = 46{,}872$ lb. It is necessary to indicate to the joist manufacturer that this tie force is from seismic effects so that the joist-girder's specialty engineer is able to apply the proper load combinations of ASCE/SEI 7-05 §12.4.2.3.

Though not shown in this example, the top chord axial effects of wind W must also be considered if it could lead to a governing design of the joist-girder. Because the load combinations of §2.3.1 (strength design) and 2.4.1 (allowable stress design) contain very different formulas when considering seismic E and wind W, the design engineer cannot simply compare E and W to determine which governs. Currently, the joist industry is largely based on allowable stress design, but it is expected to transition to strength design in the future.

With line B acting as a collector (Figure 5-2), any joist-girders occurring there require an additional check of the overstrength load combinations of ASCE/SEI 7-05 §12.10.2.1. In this situation, the building's engineer must specify an E_m collector load as well as an E continuous tie load. The joist manufacturer's specialty engineer will have to check both the basic load combinations of §12.4.2.3 for E as well as the basic load combinations with overstrength factor of §12.4.3.2 for E_m.

The following is an example of the information to be placed on the drawings for the steel joist manufacturer to properly design his joist-girders for lateral loadings at lines C and D. Note that the wall anchorage force E shown should already include the 1.4 multiplier for steel elements.

Joist-girder
Axial Forces E = 46.9 kips (unfactored)
E_m = 0.0 kips (unfactored) *Applicable only at collectors.*
W = 13.8 kips (unfactored)
Forces shall be checked in both tension and compression.

10c. Design of joist-girders splices along lines C and D

Splicing large axial loads between joist-girder top chords is best done with a knife plate that sets down between the top chords at the joist seat (Figure 5-24). Top chords have a 1-inch gap between them, and the joist-girder manufacturer will keep this space clear if it is known in advance that a knife plate will be installed here. To facilitate installation, the knife plate should be ⅞-inch thick. The height of the knife plate is that necessary to obtain the splice welding, and often the strength of the knife plate is excessive just to accommodate installation.

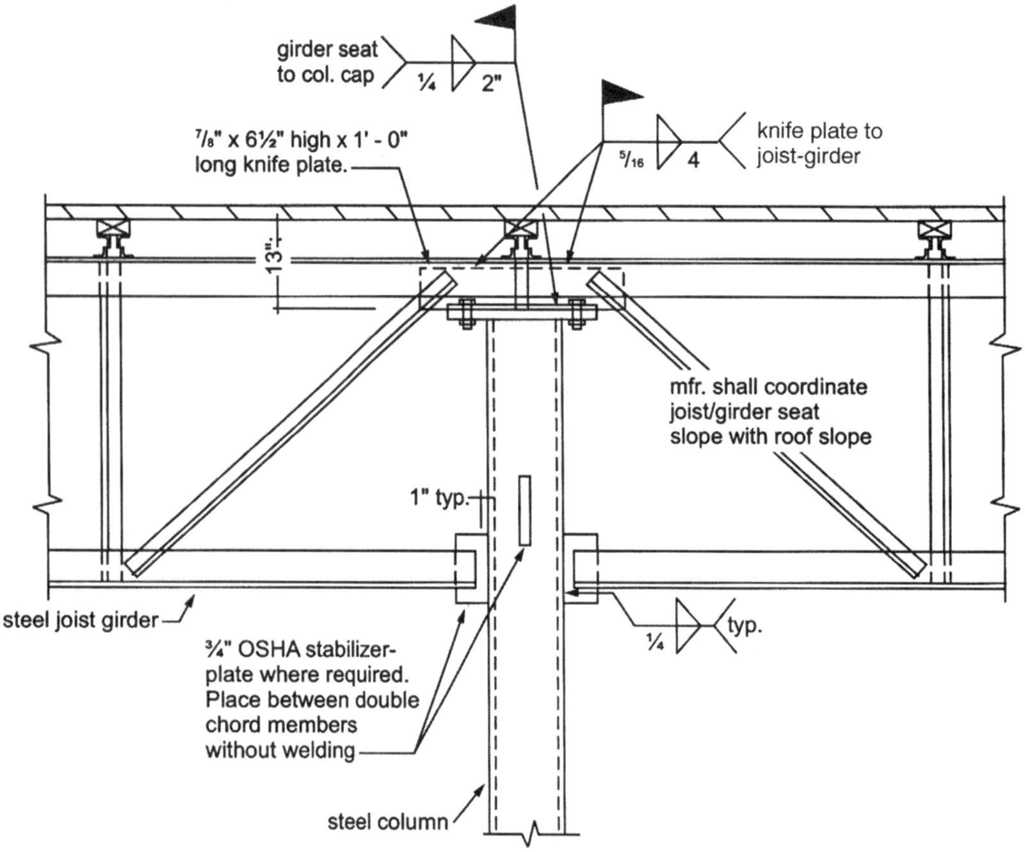

Figure 5-24. Joist-girder splice

Check the ⅞ × 6½-inch splice plate's design tensile strength per AISC §D2

$$\phi_t P_n = \phi_t F_y A_g = 0.90(36{,}000)(0.875)(6.5) = 184{,}000 \text{ lb} \qquad \text{AISC Eq D2-1}$$

Required tie force $P_r = 46{,}872$ lb < 184,000 lb . . . *o.k.*

Using two lines of 5/16 × 4-inch-long fillet welds, check the design weld strength per AISC §J2.4:

$$\phi R_n = \phi F_W A_W = 0.75(0.6 \times 70 \text{ ksi})\left(\frac{0.3125 \text{ in}}{\sqrt{2}} \times 4 \text{ in}\right)2 = 55.7 \text{ kips}$$

Required tie force $P_r = 46{,}872$ lb < 55,700 lb . . . *o.k.*

Therefore, the joist-girder splice across the columns is adequate.

10d. Comments on Metal Deck Diaphragms

Although less common in the southwest than panelized wood sheathing, flexible metal deck diaphragms (without concrete fill) are becoming more common in tilt-up construction in seismically active areas. When designed properly, metal decking can assist in providing wall anchorage and eliminate the need for subdiaphragms by acting itself as the continuous crossties. However, important detailing issues must be carefully considered.

Metal deck can only provide continuous crossties parallel to the deck span direction. ASCE/SEI 7-05 §12.11.2.2.4 specifically prohibits use of metal deck perpendicular to the direction of span for continuity, because the deck flutes will stretch out and flatten. Where the decking is spliced at the ends, a common structural member is needed to receive the attachment from both deck panels. In common steel joist systems with double top chords, it is necessary that both deck panels be attached to the same individual top chord half, otherwise crosstie loads will be inadvertently transferred through the steel joist top chord separation plate or web welding, depending on configuration. Another concern at the metal deck panel splice and direct ledger attachment is the weld tear-out through the metal deck edge. Proper deck gauge and puddle weld edge distance must be maintained for adequate wall anchorage strength.

If the metal decking is expected to carry wall anchorage forces, it must be investigated for tension *and compression* axial loads in conjunction with acting gravity loads. The axial compression loads are associated with inward wall forces and require a special axial/bending analysis of the decking. The *Standard for the Design of Cold-Formed Steel Framing* [AISI, 2004] provides design criteria for the decking, and the Structural Steel Education Council [Mayo, 2001] illustrates one approach for this wall anchorage. A more robust approach to metal deck wall anchorage is to use small steel angles or tubes that provide tension and compression wall support and distribute the load into the metal deck diaphragm.

Another challenge with metal deck diaphragms is the need for thermal expansion joints. Metal deck roof diaphragms are much more vulnerable to temperature swings than wood diaphragm systems; and with the trend toward larger roof dimensions, thermal expansion joints become very important. However, these expansion joints interrupt the continuity of the wall anchorage system and thus create several independent buildings to be analyzed separately. The wall anchorage forces must be fully developed into the main diaphragm and transferred to the applicable shear walls before reaching the expansion joint. This results in larger diaphragm shears.

10e. Design girder (continuity tie) connection to wall panel

In this example, walls are bearing walls and pilasters are not used to support the joist-girder vertically. Consequently, the kind of detail shown in Figure 5-25 must be used. This detail provides both vertical support for the girder and the necessary wall anchorage capacity. The tie force is the same as that for the wall-roof tie of Part 7a (P_{10} = 7304 lb), but not less than 5 percent of the dead plus live load reaction per ASCE/SEI 7-05 §12.1.4. The detail has the capacity to take both tension and shear forces. Details of the design are not given. The embed design is similar to that shown in Part 7.

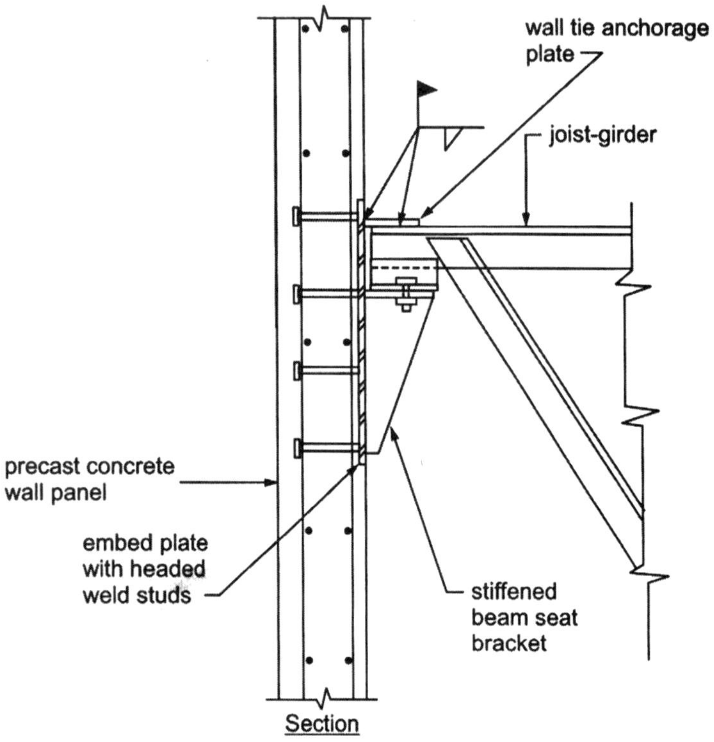

Figure 5-25. Bracket for wall-roof anchorage at joist-girder

References

ACI, 2005. 551.1R-05: *Tilt-up Concrete Construction Guide*. American Concrete Institute, P.O. Box 9094, Farmington Hills, Michigan 48333.

AISI, 2004. *Standard for Cold-Formed Steal Framing*. American Iron and Steel Institute, 1140 Connecticut Avenue, Suite 705, Washington, DC 20036.

Breyer, D. E., Fridley, K. J., and Cobeen, K. E., Pollock Jr, D. G., 2003. *Design of Wood Structures Allowable Stress Design*, Fifth edition. McGraw Hill, Inc., New York.

Brooks, Hugh, 2000. *The Tilt-up Design and Construction Manual*, Fifth edition. TCA Publications, P.O. Box 204, Mt. Vernon, IA 52314.

Fonseca, 1999. "Calculated Response of Tilt-up Structures," by F. S. Fonseca, N. M. Hawkins, and S. L. Wood. *Proceedings for 8th Canadian Conference on Earthquake Engineering*. Vancouver, British Columbia, June, 1999.

Mayo, John L., June 2001. "Metal Roof Construction on Large Warehouses or Distribution Centers," Steel Tips. Structural Education Council, P.O. Box 6190, Moraga, CA 94570.

SEAOSC/COLA, 1994. *1994 Northridge Earthquake (Structural Engineers Association of Southern California/City of Los Angeles) Special Investigation Task Force*, Tilt-up Subcommittee. Final report dated September 25, 1994.

Sheedy, 1992. "Concrete and Masonry Wall Anchorage," by P. Sheedy and C. Sheedy. *Building Standards*, July-August, 1992. International Conference of Building Officials, Whittier, California.

Skaggs, 2004. "Estimating Wood Structural Panel Diaphragm and Shear Wall Deflection," by T. D. Skaggs and Z. A. Martin. *Practice Periodical on Structural Design and Construction.* ASCE, May 2004.

Design Example 6
Tilt-up Wall Panel with Openings

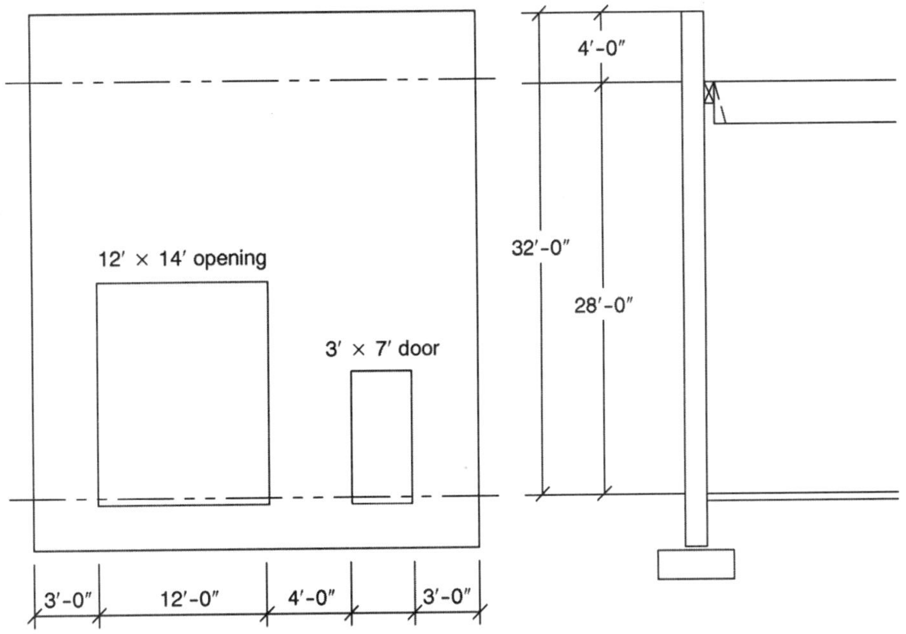

Figure 6-1. Wall elevation and section

Overview

Walls designed under the alternative slender wall method of ACI 318-05 §14.8, are typically tilt-up concrete panels that are site-cast, cured, and tilted into place. They are designed to withstand out-of-plane forces and carry vertical loads at the same time. These slender walls differ from concrete walls designed under the empirical design method (ACI 318 §14.5) and walls designed as compression members (ACI 318 §14.4) in that slender walls have greater restrictions on axial loads and must be a tension-controlled design. In addition, secondary effects of eccentricities and *P*-delta moments play an important role in analysis and design of these slender tilt-up panels.

In this example, the out-of-plane lateral design forces for a one-story tilt-up concrete slender wall panel with openings are determined, and the adequacy of a proposed reinforced concrete section is checked. The example is a single-story tilt-up concrete wall panel with two openings, site-cast, and tilted up into place. The pier between the two openings is analyzed using the slender wall design method of ACI 318 §14.8 as adopted by reference through ASCE/SEI 7-05 §14.2.1. Analysis of the wall panel for lifting stresses or other erection loads is not a part of this example.

Outline

This example will illustrate the following parts of the design process

1. **Out-of-plane lateral design forces**
2. **Primary moment from the out-of-plane forces**
3. **Primary moment from vertical load eccentricity**
4. **Total factored moment including *P*-delta effects**
5. **Nominal moment strength ϕM_n**
6. **Service load out-of-plane deflection**
7. **Special horizontal reinforcing**

Given Information

Wall material: f'_c = 3000 psi normal weight concrete

Reinforcing steel material: f_y = 60,000 psi

Wall thickness = 9¼ inches with periodic ¾-inch narrow reveals

Reinforcing steel area = seven #5 bars each face at wall section between openings

Loading data

> Roof loading to wall = uniform loading; 40-foot span of 12 psf dead load and 20 psf roof live load; no snow load
>
> Roof loading eccentricity = 4 inches from interior face of panel

Short period spectral response acceleration for design $S_{DS} = 1.0g$

Site class = *D*

Occupancy importance factor $I = 1.0$

Wind does not govern this wall panel design.

Calculations and Discussion | **Code Reference**

1. Out-of-plane lateral design forces

The wall panel is subdivided into a design strip. Typically, a solid panel is subdivided into 1-foot-wide design strips for out-of-plane design. However, for simplicity, where wall openings are involved, the entire pier width between openings is generally used as the design strip. The distributed loading accounts for the strip's self-weight, as well as the tributary loading from above each opening.

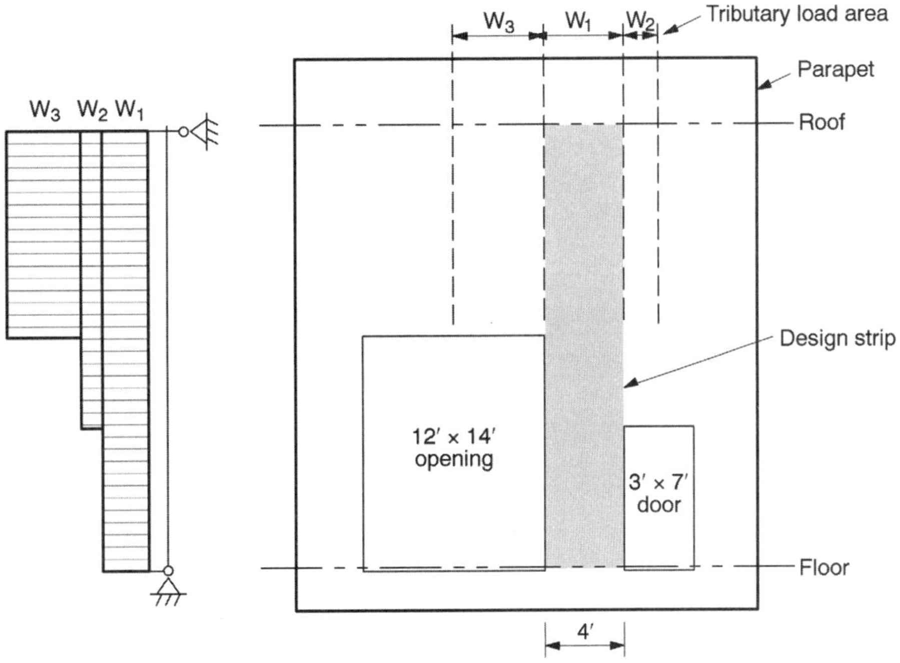

Figure 6-2. Design strip and distributed out-of-plane loading profile

1a. Seismic coefficient of wall element

The wall panel is considered a bearing wall and shear wall, thus §12.11.1 applies in determining the lateral seismic force.

$F_p = 0.40 S_{DS} I W_w$ §12.11.1

but not less than $0.10 w_w$

$I = 1.0$

$S_{DS} = 1.00g$

$F_p = 0.40(1.0)(1.00)w_w = 0.40w_w$

1b. Load combinations for strength design

For this example, the use of IBC load combination (Eq. 16-5) of §1605.2.1 is applicable, and governs for concrete strength design under seismic loading.

$$1.2D + 1.0E + L + 0.2\,S \qquad \text{(Eq 16-5)}$$

where

D = self weight of wall and dead load of roof

$L = 0$ (floor live load)

$S = 0$ (snow load)

$E = E_h + E_v = \rho Q_E + 0.20 S_{DS} D$ where $\rho = 1.0$ for wall elements §12.4.2

IBC load combination (Eq. 16-5) reduces to

$$(1.2 + 0.2S_{DS})D + 1.0Q_E \quad or \quad (1.2 + 0.20)D + 1.0Q_E \qquad \text{§12.4.2.3}$$

or $1.4D + 1.0Q_E$

1c. Lateral out-of-plane wall forces

The lateral wall forces Q_E are determined by multiplying the wall's tributary weight by the lateral force coefficient. Three different distributed loads are determined because of the presence of two door openings of differing heights. See Figure 6-2.

$$\text{Wall weight} = \frac{9.25}{12}\,150 \text{ pcf} = 116 \text{ lb/ft}^2$$

$F_{p\,wall}$ $= 0.40(116 \text{ lb/ft}^2) = 46 \text{ lb/ft}^2$

W_1 $= 46 \text{ lb/ft}^2 \times 4 \text{ ft} = 184 \text{ plf}$

W_2 $= 46 \text{ lb/ft}^2 \times 3/2 \text{ ft} = 69 \text{ plf}$

W_3 $= 46 \text{ lb/ft}^2 \times 12/2 \text{ ft} = 276 \text{ plf}$

2. Primary. moment from out-of-plane forces

Our objective is to check $\phi M_n \geq M_u$ where $M_u = M_{ua} + P_u \Delta_u$ (ACI 318 Equations 14-3 and 14-4). M_{ua} is the midheight moment due to applied factored loads and consists of two components: an out-of-plane loading moment ($M_{u\,oop}$) and a vertical eccentricity loading moment ($M_{u\,ecc}$). $P_u \Delta_u$ is a secondary moment created by P-delta effects and is investigated in Part 4.

To determine $M_{u\,oop}$, use the loading diagram in Figure 6-3.

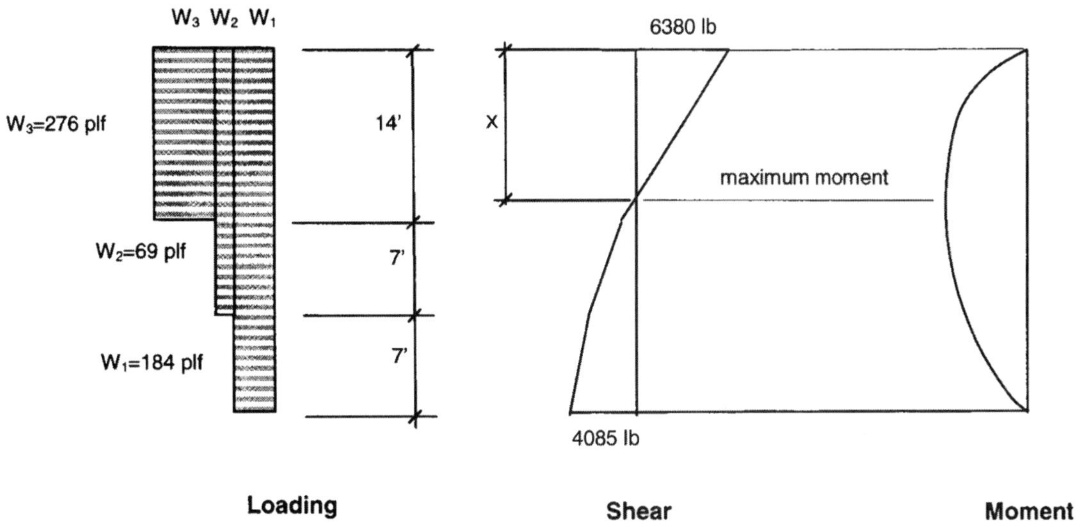

Figure 6-3. Corresponding loading, shear, and moment diagrams

ACI 318 §14.8.2.1 states, "The wall panel shall be designed as a simply supported, axially loaded member subjected to an out-of-plane uniform lateral load, with maximum moments and deflections occurring at midspan." As evident from Figure 6-3, a pier between openings has neither a uniform lateral load nor a maximum moment occurring at midspan. In this situation, it is acceptable to compute an equivalent uniform load and the more accurate maximum moment $M_{u\,oop}$ located slightly away from midspan. This is then combined with $M_{u\,ecc}$ and $P_u\Delta_u$ as computed at midspan.

Locate the point of zero shear for maximum moment $M_{u\,oop}$. Ignore the parapet's negative moment benefits in reducing the positive moment for simplicity of analysis. If the designer decides to use the parapet's negative moment to reduce the positive moment, special care should be taken to use the shortest occurring parapet height. For this analysis, the seismic coefficient for the parapet shall be the same as that for the wall below using forces based on §12.11.1. The parapet should be checked separately under §13.3.1, but is not a part of this example.

This example conservatively assumes the maximum moment occurs at a critical section width of 4 feet. In cases where the maximum moment occurs well above the doors, a more comprehensive analysis could consider several critical design sections, which would account for a wider design section at the location of maximum moment and for a narrower design section with reduced moments near the top of the doors.

2a. Determine the shear reactions at each support

R_{grade} = shear reaction at grade level for design strip

R_{roof} = shear reaction at roof level for design strip

$$R_{grade} = \left[184\frac{(28)^2}{2} + 69\frac{(21)^2}{2} + 276\frac{(14)^2}{2}\right]\frac{1}{28} = 4085 \text{ lb}$$

$$R_{roof} = [184(28) + 69(21) + 276(14)] - 4085 = 6380 \text{ lb}$$

Determine the distance of the maximum moment from the roof elevation downward (Figure 6-3)

$$X = \frac{6380}{184+69+276} = 12.1 \text{ feet to point of zero shear (maximum moment)}$$

2b. Determine $M_{u\ out\text{-}of\text{-}plane\ (oop)}$

This is the primary moment due to factored out-of-plane forces, which excludes P-delta effects and vertical load eccentricity effects

$$M_{u\ oop} = 6380(12.1) - (184+69+276)\frac{(12.1)^2}{2} = 38{,}473 \text{ lb-ft}$$

$$M_{u\ oop} = 38.5 \text{ k-ft}$$

3. Primary moment from vertical load eccentricity

Any vertical loads that act at an eccentric distance from the wall's center also apply a moment to the design wall section. In this example only the roof loads are applied to the wall with an eccentricity.

P_{roof} = gravity loads from the roof acting on the design strip

P_{roof} = (roof dead load) × (tributary width of pier) × (tributary width of roof)

$$P_{roof} = (12 \text{ psf})\left(4+\frac{3}{2}+\frac{12}{2}\right)\frac{40}{2} = 2760 \text{ lb}$$

Note: When concentrated gravity loads, such as from a girder, are applied to slender walls, the loads are assumed to be distributed over an increasing width at a slope of 2 units vertical to 1 unit horizontal down to the flexural design section height (ACI 318, §14.8.2.5).

The applicable load combination determined in Part 1 is $1.40D + 1.0Q_E$ for seismic considerations. Roof live load is not combined with seismic loads in the IBC strength design-load combinations. However, when investigating load combinations including wind design, a portion of the roof live load is included.

$$P_{u\ roof} = 1.40\ (2760) = 3864 \text{ lb}$$

The eccentric load places an applied moment at the roof level. With the base of the wall considered pinned, the resulting moment at midheight is half of the applied moment.

$$M_{u\ ecc} = P_{u\ roof}\frac{e}{2}$$

where,

$$e = 4 \text{ in} + \frac{9.25-0.75}{2} = 8.25 \text{ in}$$

$$M_{u\ ecc} = 3864\frac{8.25}{2} = 15{,}939 \text{ lb-in}$$

$$= 1.3 \text{ k-ft}$$

4. Total factored moment including *P*-delta effects

The total factored moment M_u is the applied moment M_{ua} with increase for P-delta effects. From Parts 2 and 3

$$M_{ua} = M_{u\,oop} + M_{u\,ecc}$$
$$= 38.5 + 1.3 = 39.8 \text{ k-ft}$$

M_{ua} is magnified using ACI 318 Equation 14-6

$$M_u = \frac{M_{ua}}{1 - \dfrac{5P_u \ell_c^2}{(0.75)48E_c I_{cr}}} \qquad \text{ACI Eq (14-16)}$$

This provides a direct solution without the need of an iterative calculation process. To use this equation, the wall's vertical loading and section properties must be calculated.

4a. Determine the total vertical load

$P_{total} = P_{roof} + P_{wall\,top}$

$P_{roof} = 2760$ lb (from Part 3)

$P_{wall\,top}$ = the portion of the wall's self weight above the flexural design section. It is acceptable to assume the design section is located midway between the floor and roof levels.

$$P_{wall\,top} = (116 \text{ psf})\left(4 + \frac{3}{2} + \frac{12}{2}\right)\left(\frac{28}{2} + 4\right) = 24{,}012 \text{ lb}$$

$P_{total} = P_{roof} + P_{wall\,top} = 2760 + 24{,}012 = 26{,}772$ lb

$P_u = 1.40(26{,}772) = 37{,}481$ lb

$= 37.5$ kips

4b. Determine necessary section properties

Reinforcing depth d can be based on ACI 7.7.3(a). Tilt-up panel reinforcement cover dimensions may comply with those for precast concrete, provided that the construction is similar to that normally expected under plant controlled conditions (ACI R7.7.3). With the panels normally cast on the building's concrete floor slab, reinforcement placement on chairs and form-work dimensions are able to keep to tight tolerances. For wall panels with #11 bars and smaller, the minimum cover dimension is ¾ inch.

d = thickness – reveal – cover – tie diameter – ½ bar diameter

$d = 9\frac{1}{4} - \frac{3}{4} - \frac{3}{4} - \frac{3}{8} - (\frac{1}{2})(\frac{5}{8}) = 7.06$ in

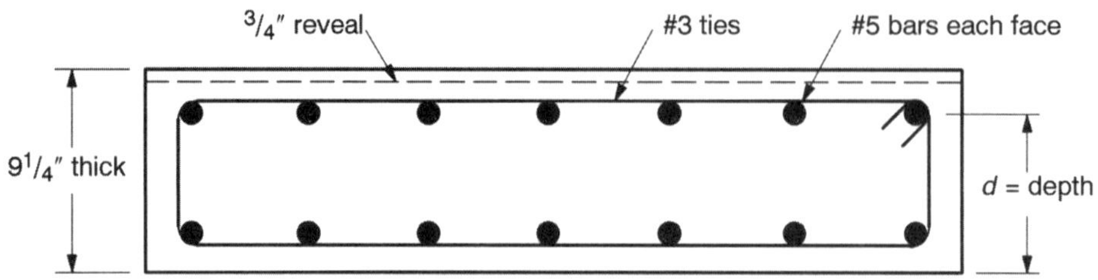

Figure 6-4. Design section

The cracked moment of inertia I_{cr} is necessary to determine the P-delta effects:

$$I_{cr} = \frac{E_s}{E_c}\left(A_s + \frac{P_u}{f_y}\right)(d-c)^2 + \frac{\ell_w c^3}{3}$$ ACI 318 (Eq 14-7)

where

$$E_s = 29{,}000 \text{ ksi}$$

$$E_c = 57\sqrt{f'_c} = 3122 \text{ ksi}$$ ACI 318 §8.5.1

$$A_{se} = \left(A_s + \frac{P_u}{f_y}\right) = 7(0.31) + \frac{37{,}500}{60{,}000} = 2.80 \text{ in}^2$$

$$a = \frac{P_u + A_s f_y}{0.85 f'_c b} = \frac{37{,}500 + 7(0.31)(60{,}000)}{0.85(3000)(48)} = 1.37 \text{ in}$$

$$c = \frac{a}{\beta_1} = \frac{1.37}{0.85} = 1.61 \text{ in}$$ ACI 318 §10.2.7

$$I_{cr} = \frac{29{,}000}{3122} 2.80\,(7.06 - 1.61)^2 + \frac{48(1.61)^3}{3} = 839 \text{ in}^4$$

4c. Determine the total factored moment magnified for *P*-delta effects

$$M_u = \frac{M_{ua}}{1 - \dfrac{5P_u\ell_c^2}{(0.75)48E_c I_{cr}}} = \frac{39.8}{1 - \dfrac{5(37.5)(28\times 12)^2}{0.75(48)3122(839)}} = 51.3 \text{ k-ft}$$ ACI 318 (Eq 14-6)

5. Check the design section's adequacy

5a. Nominal strength ϕM_n

The nominal moment strength ϕM_n is given by the following equation

$$\phi M_n = \phi A_{se} f_y\left(d - \frac{a}{2}\right);\ \text{where } \phi = 0.90 \text{ per ACI 318 §9.3.2}$$

$$= 0.90(2.80)(60{,}000)\left(7.06 - \frac{1.37}{2}\right)$$

$$= 964 \text{ k-in} = 80.3 \text{ k-ft}$$

$$M_u = 51.3 \text{ k-ft} < 80.3 \text{ k-ft}$$

$$M_u \le \phi M_n \ldots \textit{o.k.}$$

5b. Check flexural cracking moment

Verify that $M_{cr} \le \phi M_n$ to determine the acceptability of the slender wall design method (ACI 318 §14.8.2.4). M_{cr} is defined in ACI 318 §9.5.2.3.

$$M_{cr} = f_r \frac{I_g}{y_t} = \frac{7.5\sqrt{3000}(48)\dfrac{(9.25)^3}{12}}{\dfrac{9.25}{2}} = 281{,}187 \text{ lb-in} = 23.4 \text{ k-ft} \quad \text{ACI 318 (Eq 9-9)}$$

$$M_{cr} = 23.4 \text{ k-ft} \le \phi M_n = 80.3 \text{ k-ft} \ldots \textit{o.k.}$$

Reinforcing is sufficient for the use of the alternative slender wall method.

Note: For the purposes of ACI 318 §14.8.2.4, I_g and y_t are conservatively based on the gross thickness without consideration for reveal depth. This approach creates a worst-case comparison of M_{cr} to ϕM_n. In addition, the exclusion of the reveal depth in the M_{cr} calculation produces more accurate deflection values when reveals are narrow and relatively shallow.

5c. Check section is tension-controlled

ACI 318 §10.3.4 defines tension-controlled sections as those whose net tensile strain $\varepsilon_t > 0.005$ when the concrete in compression reaches its assumed strain limit of 0.003. The net tensile strain limits can also be stated in terms of the ratio c/d_t, where c is the depth of the neutral axis at nominal strength, and d_t is the distance from the extreme compression fiber to the extreme tension steel. A net tensile strain limit of $\varepsilon_t > 0.005$ is equivalent to $c/d_t < 0.375$ for grade 60 reinforcement (ACI 318 §R9.3.2.2).

$$c/d_t = 1.61/7.06 = 0.228 < 0.375 \ldots \textit{o.k.}$$

Therefore, the slender wall method is acceptable.

5d. Check the maximum vertical stress at midheight

Check the vertical stress at the midheight section to determine whether the alternative slender wall design method is acceptable (ACI 318 §14.8.2.6). ACI requires this check using strength design load levels. With only dead load D and roof live load L_r contributing to P_u, the IBC load combinations of §1605.2.1 with ASCE §12.4.2.3 reduce to the following:

IBC (Eq. 16-1) $1.4(D + F)$	$= 1.4D$
IBC (Eq. 16-2) $1.2(D + F + T) + 1.6(L + H) + 0.5(L_r \text{ or } S \text{ or } R)$	$= 1.2D + 0.5L_r$
IBC (Eq. 16-3) $1.2D + 1.6(L_r \text{ or } S \text{ or } R) + (L \text{ or } 0.8W)$	$= 1.2D + 1.6L_r + 0.8W$
IBC (Eq. 16-4) $1.2D + 1.6W + L + 0.5(L_r \text{ or } S \text{ or } R)$	$= 1.2D + 1.6W + 0.5L_r$
IBC (Eq. 16-5) $(1.2 + 0.2S_{DS})D + 1.0Q_E + L + 0.2S$	$= 1.4D + 1.0Q_E$
IBC (Eq. 16-6) $0.9D + 1.6W + 1.6H$	$= 0.9D + 1.6W$
IBC (Eq. 16-7) $(0.9\text{-}0.2S_{DS})D + 1.0Q_E + 1.6H$	$= 0.7D + 1.0Q_E$

From inspection of the load combinations above, only combinations (16-1), (16-3) and (16-5) can govern vertical load. As determined in Part 4a, the total vertical dead load D is 26,772 lbs. The roof live load L_r is determined as follows

$$L_r = 20 \text{ psf} \times 40/2 \times (4+3/2+12/2) = 4600 \text{ lbs.}$$

Load combinations (16-1), (16-3) and (16-5) result in the following P_u vertical loads

IBC (Eq. 16-1) $1.4D = 1.4(26{,}772) = 37{,}481$ lb

IBC (Eq. 16-3) $1.2D + 1.6L_r + 0.8W = 1.2(26{,}772) + 1.6(4600) = 39{,}486$ lb (governs)

IBC (Eq. 16-5) $1.4D + Q_E = 1.4(26{,}772) = 37{,}481$ lb

Vertical stress $P_u/A_g = 39{,}486/(48 \times (9.25 - 0.75)) = 96.8 \text{ psi} < 0.06(3000) = 180 \text{ psi}$ . . . *o.k.*

The compressive stress is low enough to use the alternative slender wall method; otherwise a different method, such as the empirical design method (ACI 318, §14.5) or the compression member method (ACI 318, §14.4), would be required along with their restrictions on wall slenderness.

6. Service load out-of-plane deflection

In the process of incorporating provisions for slender wall design, ACI 318 included UBC limits for service load deflection Δ_s (including P-delta effects) to a maximum of $\ell_c/150$ (ACI 318 §14.8.4).

$$\Delta_s = \frac{5M\ell_c^2}{48E_cI_e} \qquad \text{ACI 318 (Eq 14-8)}$$

where

$$M = \frac{M_{sa}}{1 - \dfrac{5P_s\ell_c^2}{48E_cI_e}} \qquad \text{(Eq 14-9)}$$

$$I_e = \left(\frac{M_{cr}}{M_a}\right)^3 I_g + \left[1 - \left(\frac{M_{cr}}{M_a}\right)^3\right] I_{cr} \le I_g;\ \text{and } M_a = M \qquad \text{ACI 318 (Eq 9-8)}$$

Unfortunately, during this incorporation no clear direction was given for the service-level load combinations expected to be used in evaluating Δ_s. ACI 318 §8.2.2 simply refers to the "general building code," but with a general transition to strength-based design and the wide variety of load combinations currently in the IBC and ASCE/SEI 7-05 codes, there is no clear direction as to the proper load combination for evaluating service-level seismic deflection Δ_s.

ASCE/SEI 7-05 Appendix C provides a brief discussion on serviceability considerations, and the Appendix C Commentary (added later as errata) provides some guidance for a service-level *wind* load combination. However, no specific discussion on service-level *seismic* load is found in ASCE/SEI 7-05.

As mentioned earlier, many of the slender wall provisions are from the Uniform Building Code. Under the UBC, service-level deflection checks were intended to be determined using the UBC allowable stress load combinations. Those original UBC load combinations are very similar to those currently found in IBC §1605.3.2 *"Alternative basic load combinations."* Until service-level seismic load combinations are more clearly defined in the IBC, ASCE/SEI 7-05, or ACI 318, it is appropriate to use the load combinations in IBC §1605.3.2. For evaluating service-level deflections, IBC Eq. 16-20 will govern.

$$D + L + S + E/1.4$$ IBC (Eq. 16-20)

where

$$E = \rho Q_E + 0.2S_{DS}D$$ §12.4.2

Thus

$$D + L + S + (\rho Q_E + 0.2S_{DS}D)/1.4$$

or

$$(1 + 0.14S_{DS})D + L + S + \rho Q_E/1.4$$

With $L = 0$, $S = 0$, $\rho = 1.0$, and $S_{DS} = 1.0$, the applicable load combination for service-level seismic loads reduces to the following:

$$1.14D + Q_E/1.4$$

6a. Determine service level moment

M_{sa} is the applied service-level moment, and comprises $M_{s\,oop}$ (out-of-plane) and $M_{s\,ecc}$

$$M_{sa} = M_{s\,oop} + M_{s\,ecc}$$

Because $M_{s\,oop}$ is solely caused by seismic loads Q_E,

$$M_{s\,oop} = \frac{M_{u\,oop}}{1.4} = \frac{(38.5)}{1.4} = 27.5 \text{ k-ft}$$

Additionally

$M_{s\,ecc} = P_{roof}(e/2) = 1.14(2760)8.25/2 = 12{,}979$ lb-in (See Part 3) = 1.1 k-ft

$M_{sa} = 27.5 + 1.1 = 28.6$ k-ft

$P_s = 1.14\,D = 1.14(26{,}772 \text{ lbs}) = 30{,}520$ lb = 30.5 k (from Part 4a)

$M_{cr} = 23.4$ k-ft (from Part 5b)

$I_{cr} = 839 \text{ in}^4$ (from Part 4b)

$$I_g = 48\frac{(9.25)^3}{12} = 3166 \text{ in}^4$$

I_g is based on gross thickness, without substracting for the architectural reveal depth, because this produces more accurate results when the reveals are narrow and relatively shallow.

First iteration

Because M and I_e are dependent on each other, some iterations between ACI 318 Equations 9-8 and 14-9 are necessary to obtain an accurate deflection Δ_s. Begin with $M = M_{sa}$

$$I_e = \left(\frac{23.4}{28.6}\right)^3 3166 + \left[1 - \left(\frac{23.4}{28.6}\right)^3\right] 839$$

$$I_e = 2114 \text{ in}^4 \leq I_g \ldots o.k.$$

$$M = \frac{28.6}{1 - \dfrac{5(30.5)(28 \times 12)^2}{48(3122)(2114)}} = 30.2 \text{ k-ft}$$

Second iteration

$$I_e = \left(\frac{23.4}{30.2}\right)^3 3166 + \left[1 - \left(\frac{23.4}{30.2}\right)^3\right] 839$$

$$I_e = 1921 \text{ in}^4 \leq I_g \ldots o.k.$$

$$M = \frac{28.6}{1 - \dfrac{5(30.5)(28 \times 12)^2}{48(3122)(1921)}} = 30.4 \text{ k-ft}$$

Third iteration

$$I_e = \left(\frac{23.4}{30.4}\right)^3 3166 + \left[1 - \left(\frac{23.4}{30.4}\right)^3\right] 839$$

$$I_e = 1900 \text{ in}^4 \leq I_g \ldots o.k.$$

$$M = \frac{28.6}{1 - \dfrac{5(30.5)(28 \times 12)^2}{48(3122)(1900)}} = 30.4 \text{ k-ft} \ldots \text{converged}$$

6b. Check service load deflection — ACI 318 §14.8.4

$$\Delta_s = \frac{5M\ell_c^2}{48E_c I_e}$$

$$= \frac{5(30.4)(28 \times 12)^2 12}{48(3122)(1900)}$$

$$= 0.72 \text{ in} < \frac{\ell_c}{150} = 2.24 \text{ in} \quad .. o.k.$$

Therefore the proposed slender wall section is acceptable using the alternative slender wall method.

7. Special horizontal reinforcing

7a. Determine the horizontal reinforcing required above the largest wall opening for out-of-plane loads

The portion of wall above the 12-foot-wide door opening spans horizontally to the vertical design strips on each side of the opening. This wall portion will be designed as a 1-foot unit horizontal design strip and subject to the out-of-plane loads computed earlier in this example.

$$F_{p\,wall} = 0.40(116 \text{ lb/ft}^2) = 46 \text{ lb/ft}^2$$

The moment is based on a simply supported horizontal beam

$$M_u = F_p \frac{(\text{opening width})^2}{8} = 46\frac{12^2}{8}$$

$$= 828 \text{ lb-ft} = 0.83 \text{ k-ft}$$

Try using #5 bars at 18-inch spacing to match the bar size being used vertically at the maximum allowed spacing for wall reinforcing.

$$\phi M_n = \phi A_s f_y \left(d - \frac{a}{2}\right)$$

where

$$A_s = 0.31\left(\frac{12}{18}\right) = 0.21 \text{ in}^2$$

$$a = \frac{A_s f_y}{0.85 f'_c b} = \frac{(0.21)60{,}000}{0.85(3000)(12)} = 0.41 \text{ in}$$

$$c = \frac{a}{\beta_1} = \frac{0.41}{0.85} = 0.48 \text{ in}$$

Assume the reinforcing above the opening is a single curtain with the vertical steel located at the center of the wall's net section. The horizontal reinforcing in concrete tilt-up construction is typically placed over the vertical reinforcing when assembled on the ground.

$$d = \tfrac{1}{2}(\text{thickness} - \text{reveal}) - \text{bar diameter}$$

$$d = \tfrac{1}{2}(9\tfrac{1}{4} - \tfrac{3}{4}) - \tfrac{5}{8} = 3.63 \text{ in}$$

Determine ϕ per ACI 318 §R9.3.2.2.

$$\frac{c}{d_t} = \frac{0.48}{3.63} = 0.132 \le 0.375$$

Therefore, it is a tension-controlled section and $\phi = 0.9$

$$\phi M_n = 0.9(0.21)(60)\left(3.63 - \frac{0.41}{2}\right) = 38.8 \text{ k-in}$$

$$= 3.24 \text{ k-ft} \geq 0.83 \text{ k-ft} \ldots \textit{o.k.}$$

$$= \phi M_n \geq M_u \ldots \textit{o.k.}$$

Therefore, the horizontal reinforcing is acceptable.

7b. Typical reinforcing around openings

Two #5 bars are required around all window and door openings per ACI §14.3.7. The vertical reinforcing on each face between the openings provides two bars along each jamb of the openings, and thus satisfies this requirement along vertical edges. Horizontally, two bars above and below the openings are required. In addition, it is common to add diagonal bars at the opening corners to assist in limiting the cracking that often occurs because of shrinkage stresses (Figure 6-5).

7c. Horizontal (transverse) reinforcing between the wall openings

The style and quantity of horizontal (transverse) reinforcing between the wall openings is typically dependent on the in-plane shear wall design. For intermediate precast structural walls, ACI 318 §21.13.3 and ASCE/SEI 7-05 §14.2.2.14 provide special reinforcements.

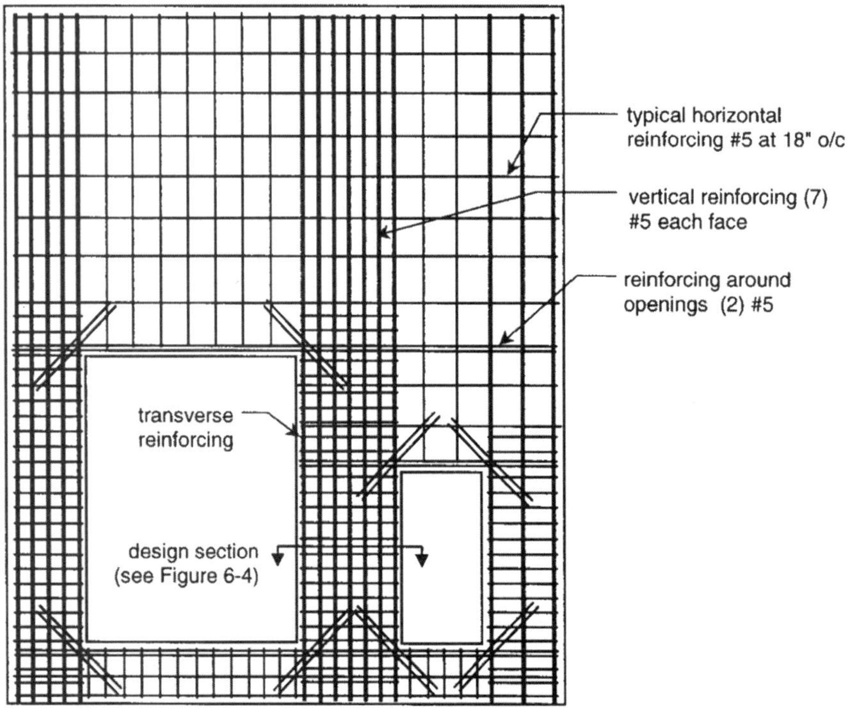

Figure 6-5. Typical wall reinforcing

In this example, two curtains of vertical reinforcing are provided for out-of-plane loads. In this situation, the horizontal reinforcing is often provided in the form of hoops or ties to assist in supporting both layers of the vertical reinforcing during construction even if two curtains of horizontal reinforcing are not required by analysis. See Figure 6-5.

Commentary

The ACI 318 section on the alternative design of slender walls made its debut in the 1999 edition. It is generally based on the 1997 Uniform Building Code, which incorporated the equations, concepts, and full-scale testing developed by the Structural Engineers Association of Southern California and published in the *Report of the Task Committee on Slender Walls* in 1982.

In the process of converting the 1997 UBC slender wall equations to ACI format, the equation for Δ_s was significantly revised. It has been reported that the current ACI procedure overestimates the panel's service-level stiffness compared with the test results of the 1980s.

The calculation for M_u using ACI Eq. 14-6 provides a direct solution for second-order effects including *P*-delta moments, instead of the interative process of ACI Eq. 14-5. Various software programs on the market today still use an iterative second-order approach or, in some cases, have no second-order analysis. Software program results can have significant errors when improper input assumptions are made. The designer is cautioned to ensure a proper second-order analysis is utlized with proper panel stiffness assumptions.

Tilt-up wall construction has become very popular because of its versatility and its erection speed. Failures of the concrete wall section out-of-plane are extremely rare; however, wall anchorage failures at the roofline have occurred during earthquakes. In response to these failures, the current anchorage design forces and detailing requirements are significantly more stringent than they were under older codes.

References

Recommended Tilt-up Wall Design, Structural Engineers Association of Southern California, 1979.5360 Workman Mill Road, Whittier, CA 90601 (562) 908–6131.

Report of the Task Committee on Slender Walls, Southern California Chapter American Concrete Institute and the Structural Engineers Association of Southern California, 1982.

Tilt-up Construction and Engineering Manual, TCA Publications, Sixth Edition, 2004. P.O. Box 204, Mount Vernon, IA 52314. (319) 895–6911

Tilt-up Concrete Construction Guide, ACI 551R-05, American Concrete Institute, 2005. P.O. Box 9094, Farmington Hills, MI 48333 (248) 848–3700.

Design Example 7
Wind Load Examples

The following are illustrative examples of the simplified wind load procedure from ASCE/SEI 7-05. They cover buildings of differing roof angles to show how the tables work with interpolation, etc.

Example 7A

Calculate the wind loads on the following building:

Dimensions: 100 ft wide by 120 ft long by 25 ft high (2 stories: 13 ft and 12 ft).

Wind Speed: Located in Minneapolis, Minnesota. From Fig. 6-1 and local building department—**90 mph zone.**

Importance: The facility is an office building with no special functions—Therefore, the Occupancy Category in Table 1-1 is **II** and, per Table 6-1, $I = 1.0$.

Exposure: Suburban office park surrounded by trees and typical suburban construction on all sides—Therefore the exposure category is **B.**

Enclosure: The building has no unusual openings in the envelope, nor is it in a hurricane region, so no concerns for wind-borne debris—Classify as **Enclosed.**

Topography: Height of adjacent hills is less than 60 ft and wind speed-up effects are not a concern $K_{zt} = 1.0$ (ASCE/SEI 7-05 §6.5.7.1.5)

Structure: The structure is an X-braced steel frame with evenly distributed braces on all four exterior walls. The second floor is concrete slabs on metal form deck on steel joists on steel floor beams. The roof is metal roof deck on steel joists on steel joist girders. The frame has no expansion joints or other structural separations. All the exterior walls are framed with studs that span vertically. Because it is also under 60 feet, is wider in both directions than it is tall (low rise) and has symmetrical cross-sections, the structure meets all the conditions to be a simple diaphragm building. For a building with well-distributed MWFRS, torsional load case in Note 5 of Figure 6-10 will not govern the design. **Therefore: design by §6.4, Method 1.**

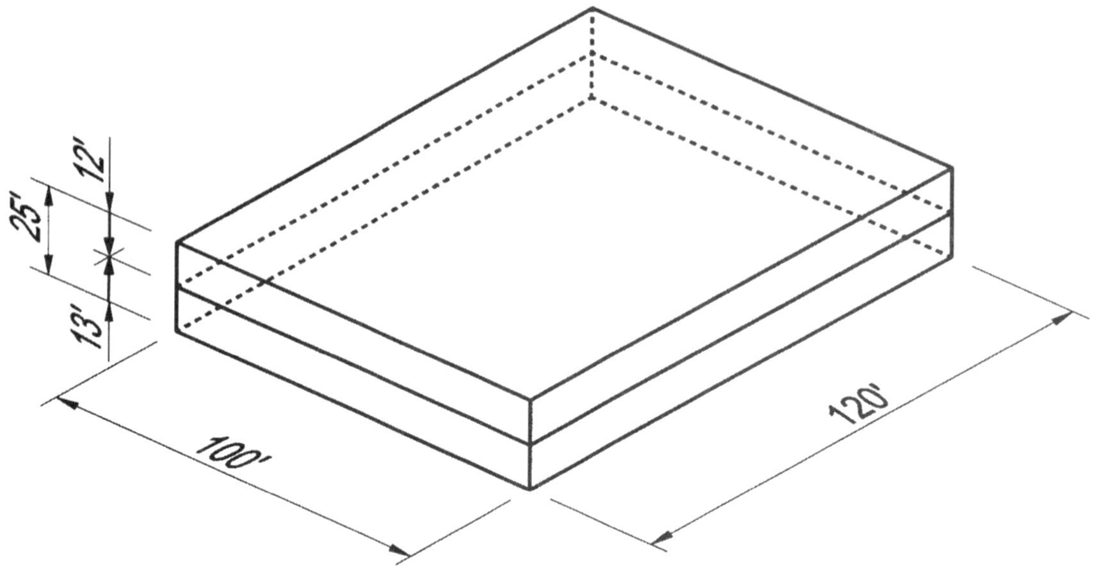

Main Windforce-resisting System-MWFRS
(Lateral Load Structural Frame)

The wind pressures on a simple diaphragm building p_s is the product of the base simplified design pressure p_{s30} taken from Figure 6-2 and multiplied by the Height and Exposure Adjustment Factor (λ) from Figure 6-2, the Topographic Factor (K_{zt}) from § 6.5.7, and by the Importance Factor (I) from Table 6-1.

Calculate the MWFRS End Zone width [See Figure 6-2]

End zone = $2a$, so first calculate a, the edge strip width as defined in Note 10, Figure 6-2.

Edge strip = a = Lesser of:
- 10% of the least horizontal dimension = 0.10 × 100 ft = 10 ft
- 40% of the eave height = 0.40 × 25 ft = 10 ft

But not less than:
- 4% of the least horizontal dimension = 0.04 × 10.0 ft = 4 ft
- 3 ft

Therefore a = 10 ft, so the end zone = $2a$ = 2 × 10 ft = 20 ft

Calculate the MWFRS design wind pressures

Using Equation 6-1: $p_s = \lambda\, K_{zt}\, I\, p_{s30}$
Look up the base pressures directly from Figure 6-2, then modify for height, exposure, topography, and importance factors. No interpolation is required since the flat roof angle falls in the row of 0° to 5°. With the mean roof height of 25 ft and exposure being B, the Height and Exposure Adjustment Factor from Figure 6-2, $\lambda = 1.0$. For a level building site, from §6.5.7, $K_{zt} = 1.0$. For a building Occupancy Category II, from Table 6-1 the Importance Factor $I = 1.0$.

Longitudinal and Transverse MWFRS – 90 mph, Exposure B, Height = 25 ft

Type	Zone	Surface	Label	p_{s30} Roof Angle 0° to 5°	λ Ht. & Exp. Factor	K_{zt} Topographic Factor	I Importance Factor	p_s Design Pressure
Horiz	End	Wall	A	12.8	× 1.00	× 1.00	× 1.00	12.8 psf
		Roof	B	No Roof Projection for Flat Roofs or Longitudinal Directory				
	Int	Wall	C	8.5	× 1.00	× 1.00	× 1.00	8.5 psf
		Roof	D	No Roof Projection for Flat Roofs or Longitudinal Directory				
Vert	End	Wind	E	−15.4	× 1.00	× 1.00	× 1.00	−15.4 psf
		Lee	F	−8.8	× 1.00	× 1.00	× 1.00	−8.8 psf
	Int	Wind	G	−10.7	× 1.00	× 1.00	× 1.00	−10.7 psf
		Lee	H	−6.8	× 1.00	× 1.00	× 1.00	−6.8 psf

Apply the pressures to the building as described in Figure 6-2. The designations of *Transverse* and *Longitudinal* are keyed to the direction of the MWFRS being evaluated. When the resisting system being designed is perpendicular to the ridge line of the gable or the hip roof, its direction is classified as *Transverse*. When it is parallel to the ridge, it is classified as *Longitudinal*. When the roof is flat (slope ≤5°), and thus has no ridge line, the loading diagrams become the same in each direction, as shown in the following diaphragm. The loading diagrams shown should be mirrored about each axis of the building until each of the four corners has been the "reference corner" as shown for each load case.

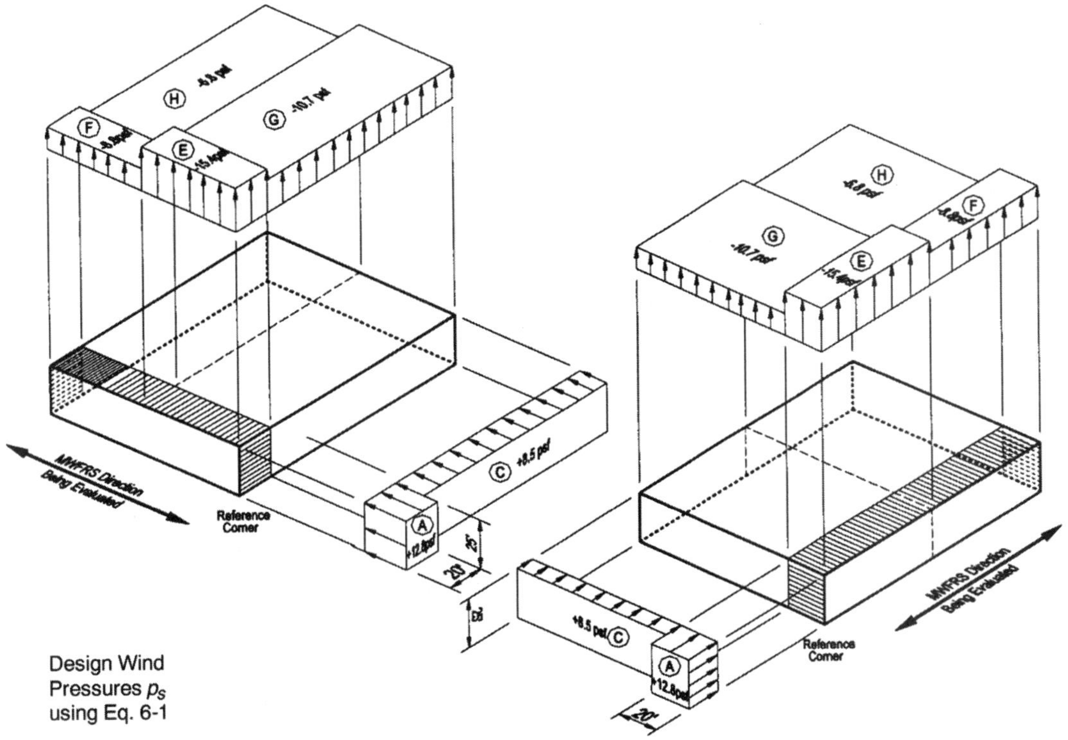

Design Wind Pressures p_S using Eq. 6-1

In addition, the minimum load case from §6.4.2.1.1 must also be checked. Apply a load of 10 psf on the building projection on a vertical plane normal to the wind. In other words, create a load case with all horizontal zones equal to 10 psf, and all vertical zones equal to 0. Check this load case as an independent case. Do not combine with the case from §6.4.2.1. It should be applied in each direction as well.

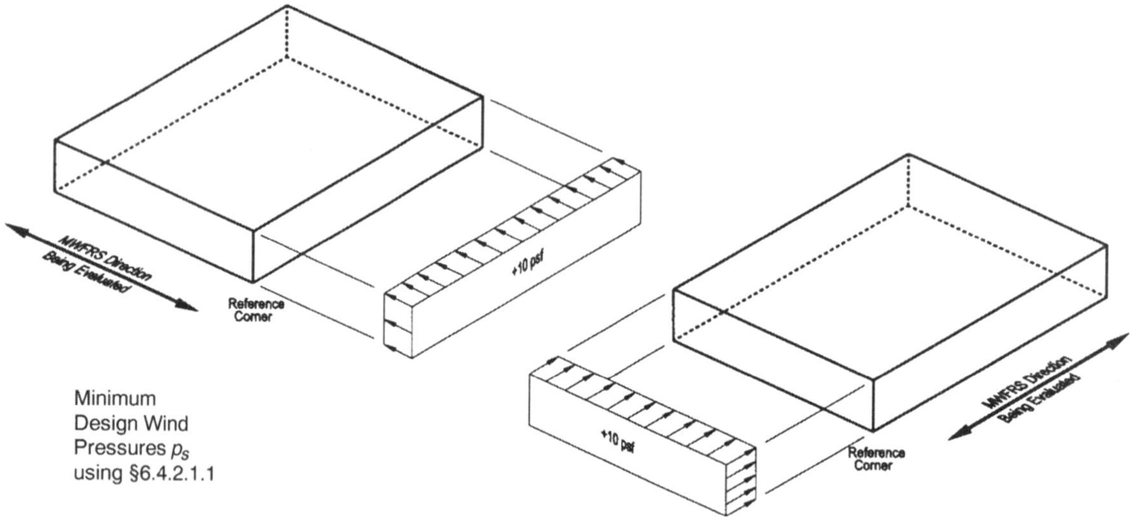

Minimum Design Wind Pressures p_S using §6.4.2.1.1

Components and Cladding (C&C) (Everything except the MWFRS)

According to §6.1.1, all "buildings, structures, and parts thereof" must be designed for wind loads. Therefore, all parts of the exterior building envelope and any load paths that are not part of the MWFRS, should be designed as C&C. For buildings such as this that qualify under §6.4.1.2, the C&C can be designed using §6.4.2.2 Equation 6-2.

Calculate the edge strip, *a*

Previously calculated in the MWFRS calculations, $a = 10$ ft

Calculate the design wind pressure on several components

For this example we will calculate the wind pressures on five example components. For this building assume the roof deck is screwed to the joists each flute, with flutes spaced 8 inches o/c. The roof joists are spaced 12 ft o/c with a span of 20 ft. Exterior wall studs are spaced at 16 inches o/c with horizontal siding.

Effective wind areas:

Deck screws: 8 in o/c × 12-ft joist span = 8 sf < 10 sf (use 10 sf)

Roof deck: 1-ft tributary width × 12-ft span = 12 sf

Roof joist = 12-ft tributary width × 20-ft span = 240 sf > 100 sf (use 100 sf)

Siding = Spanning stud to stud < 10sf

Studs = 13-ft-high studs @ 16 in o/c = 17.3 sf

Using Equation 6-2: $p_{net} = \lambda\, K_{zt}\, Ip_{net30}$
Look up the base pressures directly from Figure 6-3, then modify for height, exposure, topography, and importance factors. With the mean roof height of 25 ft and exposure being

B, the Height and Exposure Adjustment Factor from Figure 6-3, $\lambda = 1.0$. For a level building site, from §6.5.7, $K_{zt} = 1.0$. For a building Occupancy Category II, from Table 6-1, the Importance Factor $I = 1.0$.

C&C – 90 mph, Exposure B, Height = 25 ft

Type	Zone	Item	Eff Wind Area	Direction	Interpolation	p_{net30} Base Press	λ Ht & Exp. Factor	K_{zt} Topo-graphic Factor	I Impor-tance Factor	p_{net} Design Pressure
Roof 0° to 7°	Int (1)	Deck Screw	<10 sf	Positive	None Required	5.9	× 1.00	× 1.00	× 1.00	5.9*
				Negative	None Required	−14.6	× 1.00	× 1.00	× 1.00	−14.6
		Roof Deck	12 sf	Positive	10 sf / 20 sf / 12 sf: +5.9 / +5.6 / +5.8	5.8	× 1.00	× 1.00	× 1.00	5.8*
				Negative	10 sf / 20 sf / 12 sf: −14.6 / −14.2 / −14.5	−14.5	× 1.00	× 1.00	× 1.00	−14.5
		Joist	>100 sf	Positive	None Required	4.7	× 1.00	× 1.00	× 1.00	4.7*
				Negative	None Required	−13.3	× 1.00	× 1.00	× 1.00	−13.3
	Edge (2)	Deck Screw	<10 sf	Positive	None Required	5.9	× 1.00	× 1.00	× 1.00	5.9*
				Negative	None Required	−24.4	× 1.00	× 1.00	× 1.00	−24.4
		Roof Deck	12 sf	Positive	See Int (1)	5.8	× 1.00	× 1.00	× 1.00	5.8*
				Negative	10 sf / 20 sf / 12 sf: −24.4 / −21.8 / −23.9	−23.9	× 1.00	× 1.00	× 1.00	−23.9
		Joist	>100 sf	Positive	None Required	4.7	× 1.00	× 1.00	× 1.00	4.7*
				Negative	None Required	−15.8	× 1.00	× 1.00	× 1.00	−15.8
	Corner (3)	Deck Screw	<10 sf	Positive	None Required	5.9	× 1.00	× 1.00	× 1.00	5.9*
				Negative	None Required	−36.8	× 1.00	× 1.00	× 1.00	−36.8
		Roof Deck	12 sf	Positive	See Int (1)	5.8	× 1.00	× 1.00	× 1.00	5.8*
				Negative	10 sf / 20 sf / 12 sf: −36.8 / −30.5 / −35.5	−35.5	× 1.00	× 1.00	× 1.00	−35.5
		Joist	>100 sf	Positive	None Required	4.7	× 1.00	× 1.00	× 1.00	4.7*
				Negative	None Required	−15.8	× 1.00	× 1.00	× 1.00	−15.8
Wall	Int (4)	Siding	<10 sf	Positive	None Required	14.6	× 1.00	× 1.00	× 1.00	14.6
				Negative	None Required	−15.8	× 1.00	× 1.00	× 1.00	−15.8
		Stud	17.3 sf	Positive	10 sf / 20 sf / 17.3 sf: +14.6 / +13.9 / +14.1	14.1	× 1.00	× 1.00	× 1.00	14.1
				Negative	10 sf / 20 sf / 17.3 sf: −15.8 / −15.1 / −15.3	−15.3	× 1.00	× 1.00	× 1.00	−15.3
	Edge (5)	Siding	<10 sf	Positive	None Required	14.6	× 1.00	× 1.00	× 1.00	14.6
				Negative	None Required	−19.5	× 1.00	× 1.00	× 1.00	−19.5
		Stud	17.3 sf	Positive	See Int (4)	14.1	× 1.00	× 1.00	× 1.00	14.1
				Negative	10 sf / 20 sf / 17.3 sf: −19.5 / −18.2 / −18.6	−18.6	× 1.00	× 1.00	× 1.00	−18.6

*Note: A minimum pressure of 10 psf is required per §6.4.2.2.1.

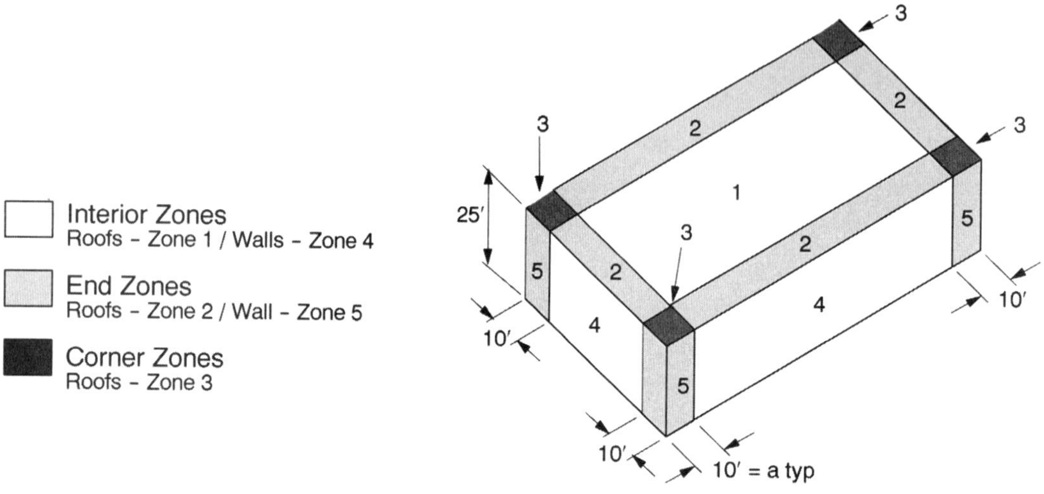

Example 7B

The building owner decides to repeat the building in Example 7A, however the site for his new, duplicate facility is located in a prairie type terrain; i.e., open grassland. Calculate the difference in loads.

MWFRS

Everything remains the same except for the Height and Exposure Adjustment Factor from Figure 6-2. With a mean roof height of 25 feet and the Exposure being C (refer to §6.5.6.2 surface roughness "C" and §6.5.6.3), the adjustment factor $\lambda = 1.35$. The Topographic Factor K_{zt} and Importance Factor I remain 1.00.

Longitudinal and Transverse MWFRS – 90 mph, Exposure C, Height = 25 ft

Type	Zone	Surface	Label	p_{s30} Roof Angle 0° to 5°	λ Ht. & Exp. Factor	K_{zt} Topographic Factor	I Importance Factor	p_s Design Pressure
Horiz	End	Wall	A	12.8	× 1.35	× 1.0	× 1.00	17.3 psf
		Roof	B	No Roof Projection for Flat Roofs				
	Int	Wall	C	8.5	× 1.35	× 1.0	× 1.00	11.5 psf
		Roof	D	No Roof Projection for Flat Roofs				
Vert	End	Wind	E	−15.4	× 1.35	× 1.0	× 1.00	−20.8 psf
		Lee	F	−8.8	× 1.35	× 1.0	× 1.00	−11.9 psf
	Int	Wind	G	−10.7	× 1.35	× 1.0	× 1.00	−14.4 psf
		Lee	H	−6.8	× 1.35	× 1.0	× 1.00	−9.2 psf

Because the adjustment factor λ is 1.35 versus the 1.00 used in Exposure B, each pressure is **35% greater** on the new site. Thus the base shear and uplift forces will each be **35% greater**.

The same logic applies to the C&C Loads. They will all be **35% greater** than the results in Example 7A.

Example 7C

Modify the building in Example 7A to have a 4-in-12 gable roof. All other parameters remain the same.

Dimensions: 100 ft wide by 120 ft long by 41.67 ft high (2 stories: 13 ft and 12 ft, plus 16.67-ft gable): Mean roof height = (25 + 41.67)/2 = 33.33 ft

Wind Speed: **90 mph zone**

Importance: **Occupancy Category II,** $I = 1.0$

Exposure: **Category B**

Typography: **Flat** $K_{zt} = 1.0$

Enclosure: **Enclosed**

Structure: Since mean roof height is less than 60 ft, building still qualifies for the simplified procedure: **design by §6.4, Method 1**

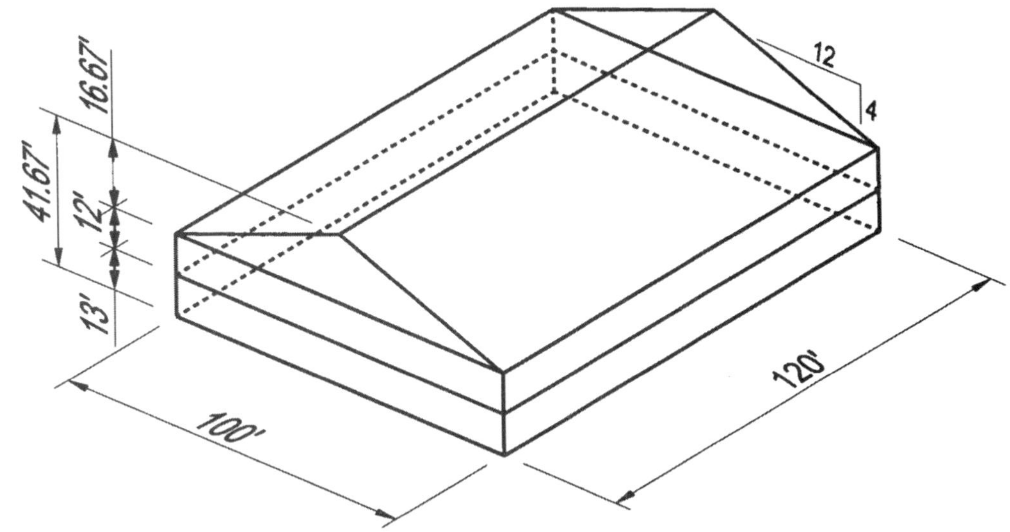

MWFRS (Lateral Load Structural Frame)

Using Equation 6-1: $p_s = \lambda\, K_{zt}\, I_{ps30}$
The wind pressures on a simple diaphragm building are taken from Figure 6-2 and multiplied by the Height and Exposure Adjustment Factor from Figure 6-2, by the Topographic Factor K_{zt} from §6.5.7, and by the Importance Factor from Table 6-1.

Calculate the MWFRS End Zone width

End zone = $2a$, so first calculate a, the edge strip width.

Edge strip = a = Lesser of:
- 10% of the least horizontal dimension = 0.10 × 100 ft = 10 ft
- 40% of the eave height = 0.40 × 25ft = 10 ft

But not less than:
- 4% of the least horizontal dimension = 0.04 × 10.0 ft = 4 ft
- 3 ft

Therefore: $a = 10$ ft, so the End Zone = $2a = 2 \times 10$ ft = 20 ft

Calculate the MWFRS design wind pressures

Using Equation 6-1: $p_s = \lambda K_{zt} I_{ps30}$
A 4-in-12 roof slope equals a roof angle of 18.4°. That falls between the 15° and 20° lines so interpolation is required per footnote 6 of Figure 6-2. Look up the base pressures from Figure 6-2 for both 15° and 20°: interpolate, then modify for height, exposure, topography, and importance factors. With the mean roof height of 33.33 ft and exposure being B, the Height and Exposure Adjustment Factor from Figure 6-2, $\lambda = 1.03$. For a level building site, from §6.5.7, $K_{zt} = 1.0$. For a building Occupancy Category II, from Table 6-1, the Importance Factor $I = 1.0$.

Transverse MWFRS – 90 mph, Exposure B Height = 33.33 ft

Type	Zone	Surface	Label	p_{s30} Roof Angle Interpolation 15°	20°	18.4°	λ Ht. & Exp. Factor	K_{zt} Topography Factor	I Import. Factor	p_s Design Pressure
Horiz	End	Wall	A	16.1	17.8	17.3	× 1.03	× 1.00	× 1.00	17.8 psf
		Roof	B	−5.4	−4.7	−4.9	× 1.03	× 1.00	× 1.00	0* psf
	Int	Wall	C	10.7	11.9	11.5	× 1.03	× 1.00	× 1.00	11.9 psf
		Roof	D	−3.0	−2.6	−2.7	× 1.03	× 1.00	× 1.00	0* psf
Vert	End	Wind	E	−15.4	−15.4	−15.4	× 1.03	× 1.00	× 1.00	−15.9 psf
		Lee	F	−10.1	−10.7	−10.5	× 1.03	× 1.00	× 1.00	−10.8 psf
	Int	Wind	G	−10.7	−10.7	−10.7	× 1.03	× 1.00	× 1.00	−11.0 psf
		Lee	H	−7.7	−8.1	−8.0	× 1.03	× 1.00	× 1.00	−8.2 psf

*Not less than zero, Note 7 Figure 6-2.

Longitudinal MWFRS – 90 mph, Exposure B, Height = 33.33 ft

Type	Zone	Surface	Label	p_{s30}* Base Press	λ Ht. & Exp. Factor	K_{zt} Topography Factor	I Importance Factor	p_s Design Pressure
Horiz	End	Wall	A	12.8	× 1.03	× 1.00	× 1.00	13.2 psf
		Roof	B	No Roof Projection in Longitudinal Direction				
	Int	Wall	C	8.5	× 1.03	× 1.00	× 1.00	8.8 psf
		Roof	D	No Roof Projection in Longitudinal Direction				
Vert	End	Wind	E	−15.4	× 1.03	× 1.00	× 1.00	−15.9 psf
		Lee	F	−8.8	× 1.03	× 1.00	× 1.00	−9.1 psf
	Int	Wind	G	−10.7	× 1.03	× 1.00	× 1.00	−11.0 psf
		Lee	H	−6.8	× 1.03	× 1.00	× 1.00	−7.0 psf

*$\theta = 0°$ For longitudinal MWFRS.

Apply the pressures to the building as described in Figure 6-2. The designations of *Transverse* and *Longitudinal* are keyed to the direction of the MWFRS being evaluated. When the resisting system being designed is perpendicular to the ridge line of the gable or hip roof, its direction is classified as *Transverse*. When it is parallel to the ridge, it is classified as *Longitudinal*. The loading diagrams shown below should be mirrored about each axis of the building until each of the four corners has been the "reference corner" as shown for each load case.

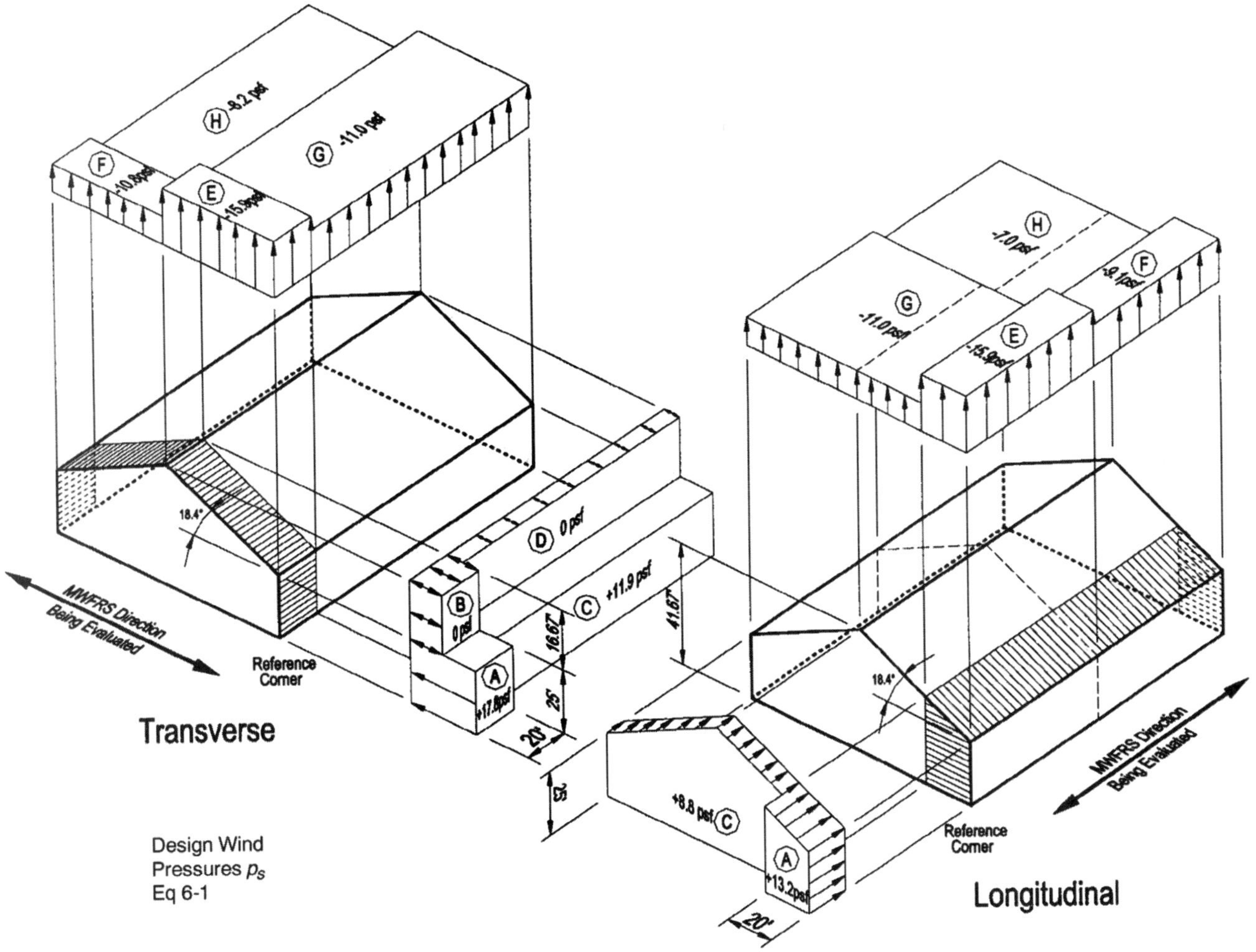

In addition, the minimum load case from §6.4.2.1.1 must be checked. Apply a load of 10 psf on the building projection on a vertical plane normal to the wind. In other words, create a load case with all horizontal zones equal to 10 psf and all vertical zones equal to 0. Check this load case as an independent case. Do not combine with the case from §6.4.2.1. It should be applied in each direction as well.

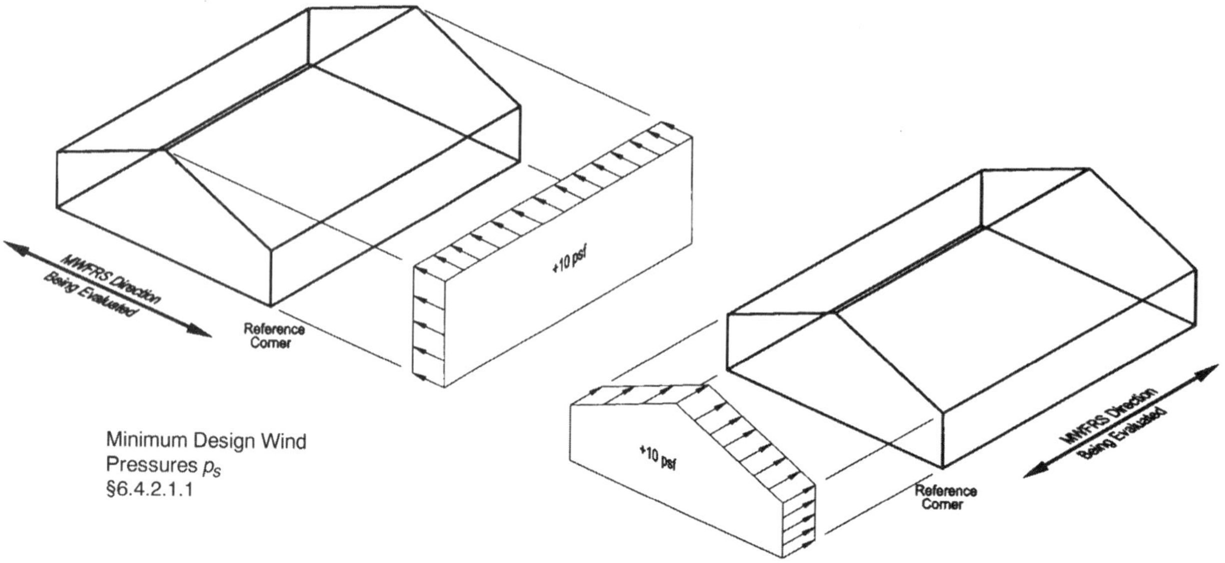

Minimum Design Wind Pressures p_S
§6.4.2.1.1

Components and Cladding (C&C) (Everything except the MWFRS)

According to §6.1.1, all "buildings, structures, and parts thereof" must be designed for wind loads. Therefore, all parts of the exterior building envelope and any load paths, that are not part of the MWFRS, should be designed as C&C. For buildings such as this that qualify under §6.4.1.2, the C&C can be designed using §6.4.2.2, Equation 6-2.

Calculate the edge strip, *a*

Previously calculated in the MWFRS calculations, $a = 10$ ft

Calculate the design wind pressure on several components:

Using Equation 6-2: $p_{net} = \lambda\, K_{zt}\, I\, p_{net30}$
Look up the base pressures directly from Figure 6-3, then modify for height, exposure, topography, and importance factors. With the mean roof height of 33.33 ft and exposure being B, the Height and Exposure Adjustment Factor from Figure 6-3 $\lambda = 1.03$. For a level building site, from §6.5.7, $k_{zt} = 1.0$. For a building Occupancy Category II, from Table 6-1, the Importance Factor $I = 1.00$.

C&C – 90 mph, Exposure B, Height = 33.33 ft

Type	Zone	Item	Eff Wind Area	Direction	Interpolation	p_{net30} Base Press	λ Ht. & Exp. Factor	K_{zt} Topo-graphic Factor	I Impor-tance Factor	p_{net} Design Pressure
Roof 7° to 27°	Int (1)	Deck Screw	<10 sf	Positive	None Required	8.4	× 1.03	× 1.00	× 1.00	8.7*
				Negative	None Required	−13.3	× 1.03	× 1.00	× 1.00	−13.7
		Roof Deck	12 sf	Positive	10 sf | 20 sf | 12 sf +8.4 | +7.7 | +8.3	8.3	× 1.03	× 1.00	× 1.00	8.5*
				Negative	10 sf | 20 sf | 12 sf −13.3 | −13.0 | −13.2	−13.2	× 1.03	× 1.00	× 1.00	−14.6
		Joist	>100 sf	Positive	None Required	5.9	× 1.03	× 1.00	× 1.00	6.1*
				Negative	None Required	−12.1	× 1.03	× 1.00	× 1.00	−12.5
	Edge (2)	Deck Screw	<10 sf	Positive	None Required	8.4	× 1.03	× 1.00	× 1.00	8.7*
				Negative	None Required	−23.2	× 1.03	× 1.00	× 1.00	−23.9
		Roof Deck	12 sf	Positive	See Int (1)	8.3	× 1.03	× 1.00	× 1.00	8.5*
				Negative	10 sf | 20 sf | 12 sf −23.2 | −21.4 | −22.8	−22.8	× 1.03	× 1.00	× 1.00	−23.5
		Joist	>100 sf	Positive	None Required	5.9	× 1.03	× 1.00	× 1.00	6.1*
				Negative	None Required	−17.0	× 1.03	× 1.00	× 1.00	−17.5
	Corner (3)	Deck Screw	<10 sf	Positive	None Required	8.4	× 1.03	× 1.00	× 1.00	8.7*
				Negative	None Required	−34.3	× 1.03	× 1.00	× 1.00	−35.3
		Roof Deck	12 sf	Positive	See Int (1)	8.3	× 1.03	× 1.00	× 1.00	8.5*
				Negative	10 sf | 20 sf | 12 sf −34.3 | −32.1 | −33.9	−33.9	× 1.03	× 1.00	× 1.00	−34.9
		Joist	>100 sf	Positive	None Required	5.9	× 1.03	× 1.00	× 1.00	6.1*
				Negative	None Required	−26.9	× 1.03	× 1.00	× 1.00	−27.7
Wall	Int (4)	Siding	<10 sf	Positive	None Required	14.6	× 1.03	× 1.00	× 1.00	15.0
				Negative	None Required	−15.8	× 1.03	× 1.00	× 1.00	−16.3
		Stud	17.3 sf	Positive	10 sf | 20 sf | 17.3 sf +14.6 | +13.9 | +14.1	14.1	× 1.03	× 1.00	× 1.00	14.5
				Negative	10 sf | 20 sf | 17.3 sf −15.8 | −15.1 | −15.3	−15.3	× 1.03	× 1.00	× 1.00	−15.8
	Edge (5)	Siding	<10 sf	Positive	None Required	14.6	× 1.03	× 1.00	× 1.00	15.0
				Negative	None Required	−19.5	× 1.03	× 1.00	× 1.00	−20.1
		Stud	17.3 sf	Positive	See Int (4)	14.1	× 1.03	× 1.00	× 1.00	14.5
				Negative	10 sf | 20 sf | 17.3 sf −19.5 | −18.2 | −18.6	−18.6	× 1.03	× 1.00	× 1.00	−19.2

*Note: A minimum pressure of 10 psf is required per §6.4.2.2.1.

The C&C pressures should be applied as described in Figure 6-3 and as shown in the following diagram.

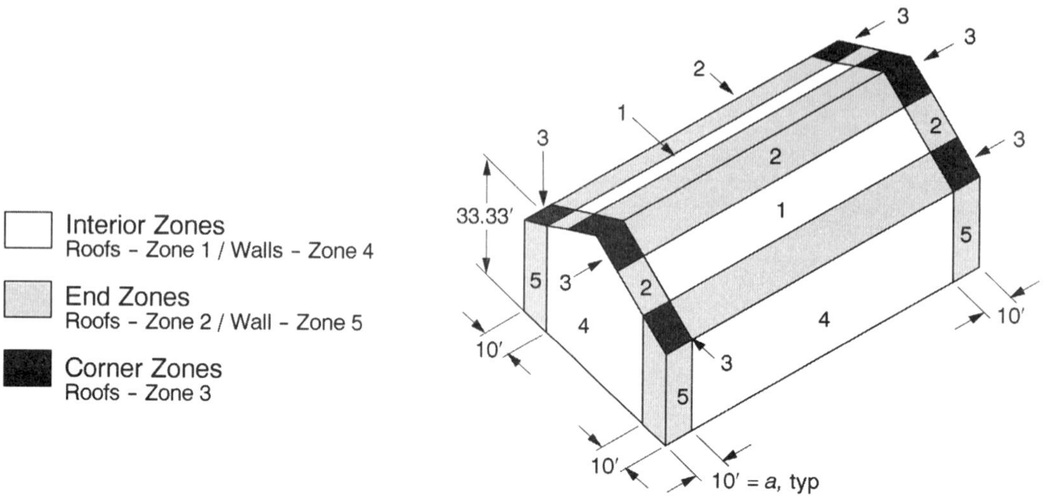

Example 7D

The building owner decides that the building in Example 7C will be designated as an emergency shelter. Calculate the difference in loads.

MWFRS

Everything remains the same except the occupancy category from Table 6-1. The building becomes Occupancy Category IV, therefore the Importance Factor *I* becomes 1.15.

Transverse MWFRS – 90 mph, Exposure B, Height = 33.33 ft

Type	Zone	Surface	Label	p_{s30} Roof Angle Interpolation 15°	20°	18.4°	λ Ht. & Exp. Factor	K_{zt} Topographic Factor	*I* Import. Factor	p_s Design Pressure
Horiz	End	Wall	A	16.1	17.8	17.3	× 1.03	× 1.00	× 1.15	20.5 psf
		Roof	B	–5.4	–4.7	–4.9	× 1.03	× 1.00	× 1.15	0* psf
	Int	Wall	C	10.7	11.9	11.5	× 1.03	× 1.00	× 1.15	13.7 psf
		Roof	D	–3.0	–2.6	–2.7	× 1.03	× 1.00	× 1.15	0* psf
Vert	End	Wind	E	–15.4	–15.4	–15.4	× 1.03	× 1.00	× 1.15	–18.2 psf
		Lee	F	–10.1	–10.7	–10.5	× 1.03	× 1.00	× 1.15	–12.4 psf
	Int	Wind	G	–10.7	–10.7	–10.7	× 1.03	× 1.00	× 1.15	–12.7 psf
		Lee	H	–7.7	–8.1	–8.0	× 1.03	× 1.00	× 1.15	–9.4 psf

*Not less than zero, Note 7 Figure 6-2.

Longitudinal MWFRS – 90 mph, Exposure B, Height = 33.33 ft

Type	Zone	Surface	Label	p_{s30}* Base Press	λ Ht. & Exp. Factor	K_{zt} Topography Factor	I Importance Factor	p_s Design Pressure
Horiz	End	Wall	A	12.8	× 1.03	× 1.00	× 1.15	15.2 psf
		Roof	B	No Roof Projection in Longitudinal Direction				
	Int	Wall	C	8.5	× 1.03	× 1.00	× 1.15	10.1 psf
		Roof	D	No Roof Projection in Longitudinal Direction				
Vert	End	Wind	E	−15.4	× 1.03	× 1.00	× 1.15	−18.2 psf
		Lee	F	−8.8	× 1.03	× 1.00	× 1.15	−10.4 psf
	Int	Wind	G	−10.7	× 1.03	× 1.00	× 1.15	−12.7 psf
		Lee	H	−6.8	× 1.03	× 1.00	× 1.15	−8.1 psf

*$\theta = 0°$ For longitudinal MWFRS.

Since the importance factor is 1.15 versus the 1.00 used in Example 7C, each pressure is **15% greater.** Thus the base shear and uplift forces will each be **15% greater.**

The same logic applies to the C&C Loads. They will all be **15% greater** than the results in Example 7C.

Example 7E

Modify the building in Example 7C to have a 6-in-12 gable roof. All other parameters remain the same.

Dimensions: 100 ft wide by 120 ft long by 50 ft high (2 stories: 13 ft and 12 ft, plus 25-ft gable). Mean roof height = (25 + 50)/2 = 37.5 ft

Wind Speed: **90 mph zone**

Importance: **Occupancy Category II,** $I = 1.0$

Exposure: **Category B**

Topography: **Flat** $K_{zt} = 1.0$

Enclosure: **Enclosed**

Structure: Since mean roof height is less than 60 ft, building still qualifies for the simplified procedure: **design by §6.4**

MWFRS (Lateral Load Structural Frame)

Using Equation 6-1: $p_s = \lambda\, K_{zt}\, I\, p_{s30}$
The wind pressures on a simple diaphragm building are taken from Figure 6-2 and multiplied by the Height and Exposure Adjustment Factor from Figure 6-2, by the Topographic Factor K_{zt} from §6.5.7, and by the Importance Factor from Table 6-1.

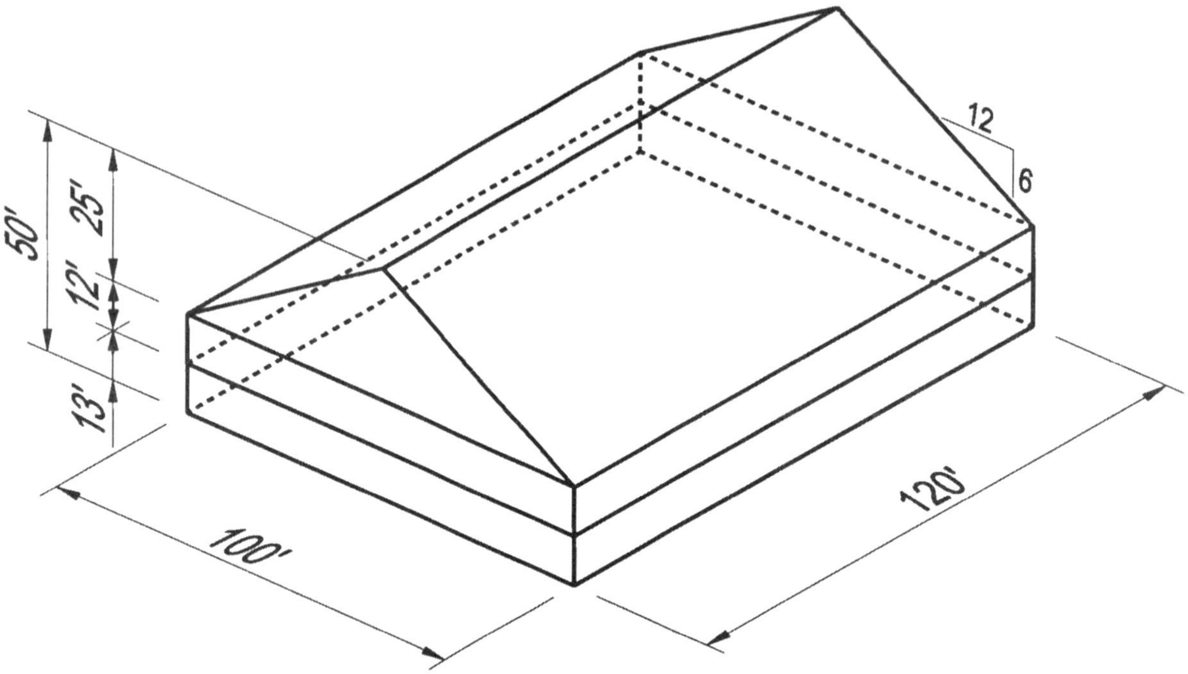

Calculate the MWFRS End Zone width

End zone $= 2a$, so first calculate a, the edge strip width

Edge strip $= a =$ Lesser of: • 10% of the least horizontal dimension $= 0.10 \times 100$ ft $= 10$ ft

• 40% of the eave height $= 0.40 \times 25$ ft $= 10$ ft

But not less than: • 4% of the least horizontal dimension

• 3ft

Therefore: $a = 10$ ft, so the End Zone $= 2a = 2 \times 10$ ft $= 20$ ft

Calculate the MWFRS design wind pressures

Using Equation 6-1: $p_s = \lambda\, K_{zt}\, I\, p_{s30}$
A 6-in-12 roof slope equals a roof angle of 26.6°. That falls between the 25°and 30°-to-45° lines so interpolation is required per footnote 6 of Figure 6-2. Look up the base pressures from Figure 6-2 for both 25° and 30°-to-45°, interpolate, then modify for height, exposure, topography, and importance factors. With the mean roof height of 37.5 ft and exposure being B, the Height and Exposure Adjustment Factor from Figure 6-2, $\lambda = 1.07$. For a level building site, from §6.5.7, $K_{zt} = 1.0$. For a building Occupancy Category II, from Table 6-1, the Importance Factor $I = 1.0$.

Transverse MWFRS – 90 mph, Exposure B, Height = 37.5 ft

Type	Zone	Surface	Label	p_{s30} Roof Angle Interpolation 2°	30 to 45°	26.6°	λ Ht. & Exp. Factor	K_{zt} Topographic Factor	I Import. Factor	p_s Design Pressure
Horiz	End	Wall	A	16.1	14.4	15.6	× 1.07	× 1.00	× 1.00	16.6 psf
		Roof	B	2.6	9.9	4.9	× 1.07	× 1.00	× 1.00	5.3 psf
	Int	Wall	C	11.7	11.5	11.6	× 1.07	× 1.00	× 1.00	12.5 psf
		Roof	D	−2.7	7.9	4.3	× 1.07	× 1.00	× 1.00	4.6 psf
Vert	End	Wind	E (1)	−7.2	1.1	−4.54	× 1.07	× 1.00	× 1.00	−4.9 psf
			E (2)	−2.7	5.6	−0.04	× 1.07	× 1.00	× 1.00	0.0 psf
		Lee	F (1)	−9.8	−8.8	−9.48	× 1.07	× 1.00	× 1.00	−10.1 psf
			F (1)	−5.3	−4.3	−4.98	× 1.07	× 1.00	× 1.00	−5.3 psf
	Int	Wind	G (1)	−5.2	0.4	−3.41	× 1.07	× 1.00	× 1.00	−3.6 psf
			G (2)	−0.7	4.8	1.06	× 1.07	× 1.00	× 1.00	1.1 psf
		Lee	H (1)	−7.8	−7.5	−7.70	× 1.07	× 1.00	× 1.00	−8.2 psf
			H (2)	−3.4	−3.1	−3.30	× 1.07	× 1.00	× 1.00	−3.5 psf

Note: (1) and (2) represent load cases 1 and 2 per note 4 of Figure 6-2.

Longitudinal MWFRS – 90 mph, Exposure B, Height = 37.5 ft

Type	Zone	Surface	Label	p_{s30} Base Press	λ Ht. & Exp. Factor	K_{zt} Topographic Factor	I Importance Factor	p_s Design Pressure
Horiz	End	Wall	A	12.8	× 1.07	× 1.00	× 1.00	13.7 psf
		Roof	B	No Roof Projection in Longitudinal Direction				
	Int	Wall	C	8.5	× 1.07	× 1.00	× 1.00	9.1 psf
		Roof	D	No Roof Projection in Longitudinal Direction				
Vert	End	Wind	E	−15.4	× 1.07	× 1.00	× 1.00	−16.5 psf
		Lee	F	−8.8	× 1.07	× 1.00	× 1.00	−9.4 psf
	Int	Wind	G	−10.7	× 1.07	× 1.00	× 1.00	−11.4 psf
		Lee	H	−6.8	× 1.07	× 1.00	× 1.00	−7.3 psf

Apply the pressures to the building as described in Figure 6-2. The designations of *Transverse* and *Longitudinal* are keyed to the direction of the MWFRS being evaluated. When the resisting system being designed is perpendicular to the ridge line of the gable or hip roof, its direction is classified as *Transverse*. When it is parallel to the ridge, it is classified as *Longitudinal*. The loading diagrams shown below should be mirrored about each axis of the building until each of the four corners has been the "reference corner" as shown for each load case.

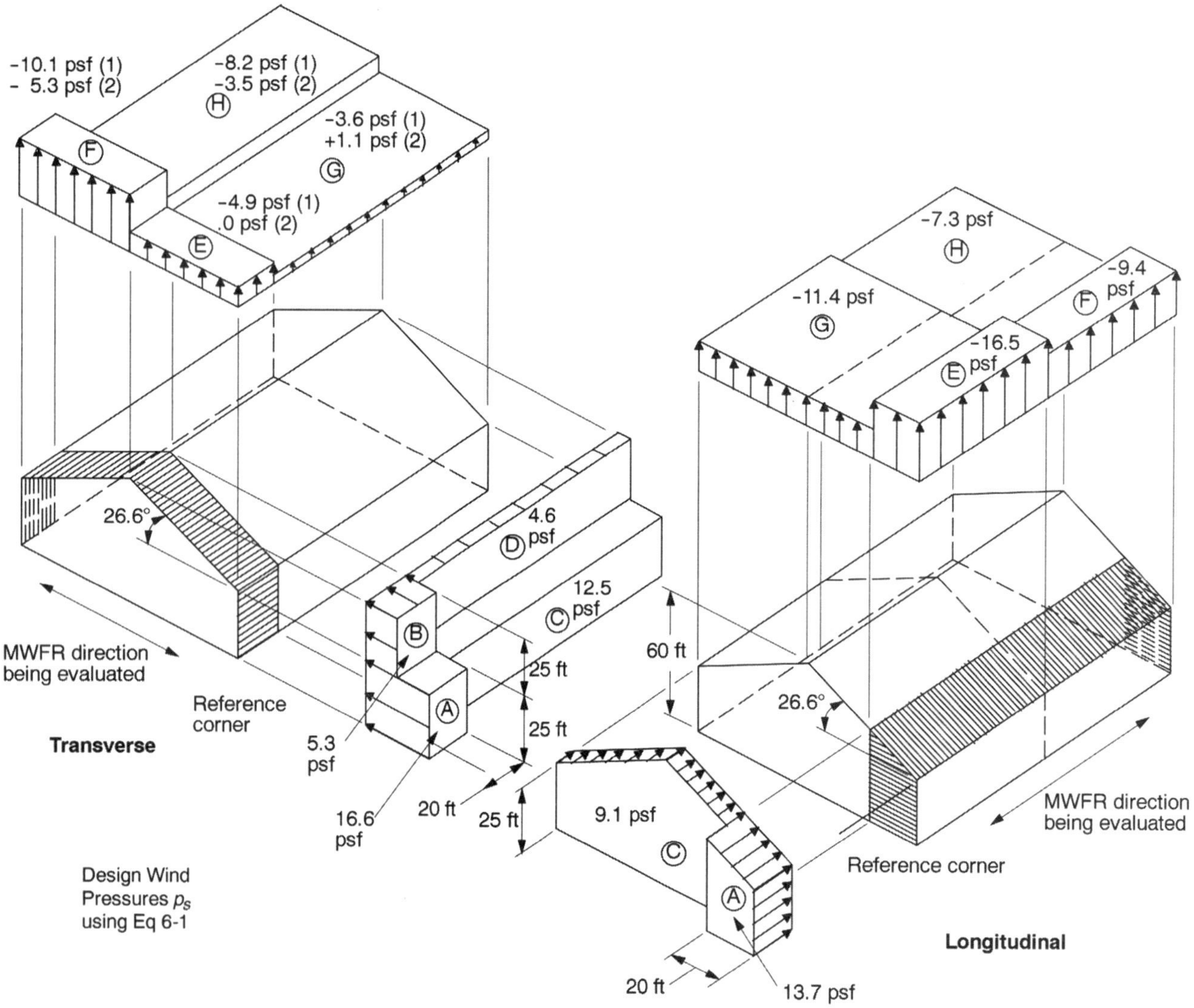

In addition, the minimum load case from §6.4.2.1.1 must be checked. Apply a load of 10 psf on the building projection on a vertical plane normal to the wind. In other words, create a load case with all horizontal zones equal to 10 psf, and all vertical zones equal to 0. Check this load case as an independent case. Do not combine with the case from §6.4.2.1. It should be applied in each direction as well.

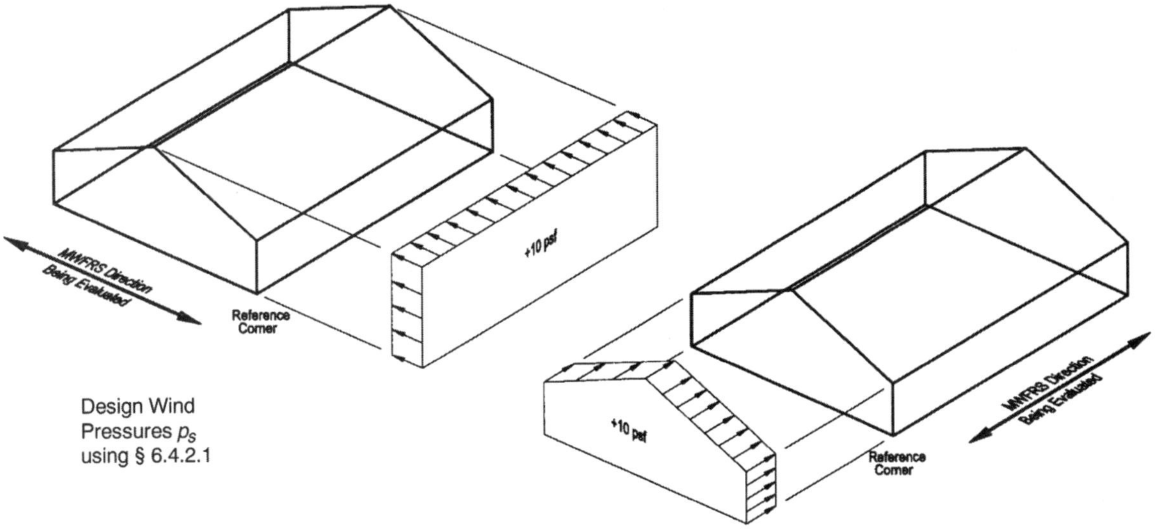

Design Wind Pressures p_S using § 6.4.2.1

Components and Cladding (C&C) (Everything except the MWFRS)

According to §6.1.1, "Buildings, structures, and parts thereof' must be designed for wind loads. Therefore, all parts of the exterior building envelope and any load paths, that are not part of the MWFRS, should be designed as C&C. For buildings such as this that qualify under §6.4.1.2, the C&C can be designed using §6.4.2.2, Equation 6-2.

Calculate the Edge Strip, *a*

Previously calculated in the MWFRS calculations, $a = 10$ ft

Calculate the design wind pressure on several components

Using Equation 6-2: $p_{net} = \lambda\, K_{zt}\, I\, p_{net30}$
Look up the base pressures directly from Figure 6-3, then modify for height, exposure, topography, and importance factors. With the mean roof height of 37.5 ft and exposure being B, the Height and Exposure Adjustment Factor from Figure 6-3, $\lambda = 1.0$. For a level building site, from §6.5.7, $K_{zt} = 1.0$. For a building Occupancy Category II, from Table 6-1, Importance Factor $I = 1.0$.

C&C – 90 mph, Exposure B, Height = 37.5 ft

Type	Zone	Item	Eff Wind Area	Direction	Interpolation	p_{net30} Base Press	λ Ht. & Exp. Factor	K_{zt} Topo-graphic Factor	I Impor-tance Factor	p_{net} Design Pressure
Roof 27° to 45°	Int (1)	Deck Screw	<10 sf	Positive	None Required	13.3	× 1.07	× 1.00	× 1.00	14.2
				Negative	None Required	–14.6	× 1.07	× 1.00	× 1.00	–15.6
		Roof Deck	12 sf	Positive	10 sf / 20 sf / 12 sf +13.3 / +13.0 / +13.2	13.2	× 1.07	× 1.00	× 1.00	14.1
				Negative	10 sf / 20 sf / 12 sf –14.6 / –13.8 / –14.4	–14.4	× 1.07	× 1.00	× 1.00	–15.4
		Joist	>100 sf	Positive	None Required	12.1	× 1.07	× 1.00	× 1.00	12.9
				Negative	None Required	–12.1	× 1.07	× 1.00	× 1.00	–12.9
	Edge (2)	Deck Screw	<10 sf	Positive	None Required	13.3	× 1.07	× 1.00	× 1.00	14.2
				Negative	None Required	–17.0	× 1.07	× 1.00	× 1.00	–18.2
		Roof Deck	12 sf	Positive	See Int (1)	13.2	× 1.07	× 1.00	× 1.00	14.1
				Negative	10 sf / 20 sf / 12 sf –17.0 / –16.3 / –16.9	–16.9	× 1.07	× 1.00	× 1.00	–18.1
		Joist	>100 sf	Positive	None Required	12.1	× 1.07	× 1.00	× 1.00	12.9
				Negative	None Required	–14.6	× 1.07	× 1.00	× 1.00	–15.6
	Corner (3)	Deck Screw	<10 sf	Positive	None Required	13.3	× 1.07	× 1.00	× 1.00	14.2
				Negative	None Required	–17.0	× 1.07	× 1.00	× 1.00	–18.2
		Roof Deck	12 sf	Positive	See Int (1)	13.3	× 1.07	× 1.00	× 1.00	14.2
				Negative	10 sf / 20 sf / 12 sf –17.0 / –16.3 / –16.9	–16.9	× 1.07	× 1.00	× 1.00	–18.1
		Joist	>100 sf	Positive	None Required	12.1	× 1.07	× 1.00	× 1.00	12.9
				Negative	None Required	–14.6	× 1.07	× 1.00	× 1.00	–15.6
Wall	Int (4)	Siding	<10 sf	Positive	None Required	14.6	× 1.07	× 1.00	× 1.00	15.6
				Negative	None Required	–15.8	× 1.07	× 1.00	× 1.00	–16.9
		Stud	17.3 sf	Positive	10 sf / 20 sf / 17.3 sf +14.6 / +13.9 / +14.1	14.1	× 1.07	× 1.00	× 1.00	15.1
				Negative	10 sf / 20 sf / 17.3 sf –15.8 / –15.1 / –15.3	–15.3	× 1.07	× 1.00	× 1.00	–16.4
	Edge (5)	Siding	<10 sf	Positive	None Required	14.6	× 1.07	× 1.00	× 1.00	15.6
				Negative	None Required	–19.5	× 1.07	× 1.00	× 1.00	–20.9
		Stud	17.3 sf	Positive	See Int (4)	14.1	× 1.07	× 1.00	× 1.00	15.1
				Negative	10 sf / 20 sf / 17.3 sf –19.5 / –18.2 / –18.6	–18.6	× 1.07	× 1.00	× 1.00	–19.9

The C&C pressures should be applied as described in Figure 6-3 and as shown in the following diagram.

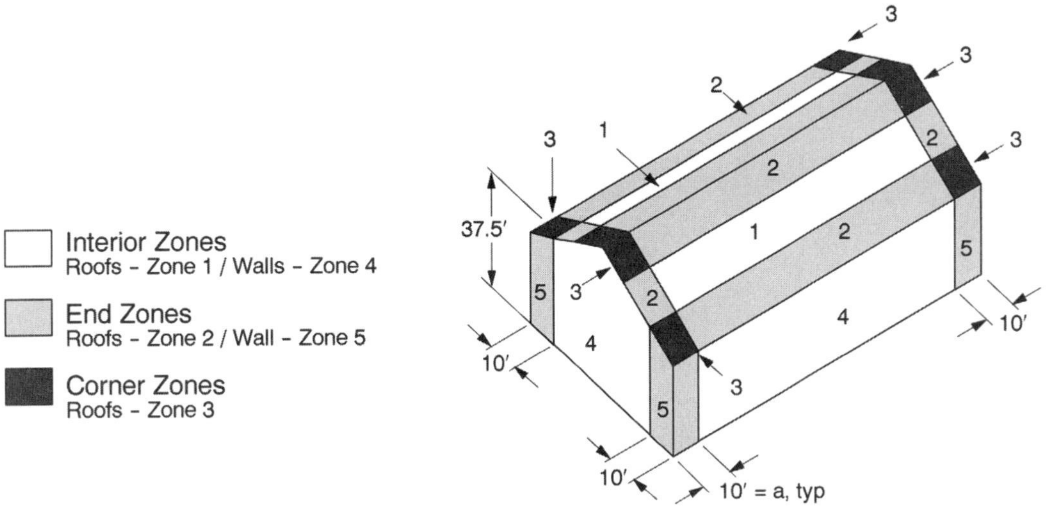
Interior Zones
Roofs - Zone 1 / Walls - Zone 4
End Zones
Roofs - Zone 2 / Wall - Zone 5
Corner Zones
Roofs - Zone 3
37.5′
10′
10′
10′
10′ = a, typ